An Introduction to Gröbner Bases

An Introduction to Gröbner Bases

Ralf Fröberg
Stockholm University, Sweden

JOHN WILEY & SONS
Chichester · New York · Weinheim · Brisbane · Singapore · Toronto

National 01243 779777
International (+44) 1243 779777

e-mail (for orders and customer service enquiries): cs-books@wiley.co.uk.
Visit our home page on http://www.wiley.co.uk
or http://www.wiley.com

Reprinted April 1998

Other Wiley Editorial Offices

John Wiley & Sons, Inc., 605 Third Avenue,
New York, NY 10158-0012, USA

WILEY-VCH Verlag GmbH, Pappelallee 3,
D-69469 Weinheim, Germany

Jacaranda Wiley Ltd, 33 Park Road, Milton,
Queensland 4064, Australia

John Wiley & Sons (Asia) Pte Ltd, 2 Clementi Loop #02-01,
Jin Xing Distripark, Singapore 0512

John Wiley & Sons (Canada) Ltd, 22 Worcester Road,
Rexdale, Ontario M9W 1L1, Canada

Library of Congress Cataloging in Publication Data

Fröberg, Ralf.
An introduction to Gröbner bases / Ralf Fröberg.
p. cm. — (Pure and applied mathematics)
Includes bibliographical references and index.
ISBN 0 471 97442 0 (alk. paper)
1. Gröbner bases. I. Title. II. Series: Pure and applied mathematics (John Wiley & Sons : Unnumbered)
QA251.3.F76 1997
512′.24—dc21 97-19668
CIP

British Library Cataloguing in Publication Data

A catalogue record for this book is available from the British Library

ISBN 0 471 97442 0

Produced from camera-ready copy supplied by the author
Printed and bound by Antony Rowe Ltd, Eastbourne
This book is printed on acid-free paper responsibly manufactured from sustainable forestation, for which at least two trees are planted for each one used for paper production.

Contents

Preface

When talking to students in mathematics at the beginner's level, one often gets the questions: How is it possible to do research in mathematics? Isn't all mathematics already explored? It is rather hard to be able to explain that the converse is true, that the research and knowledge increases at greater speed every decade. One reason for the difficulty to discuss this is of course that the level of abstraction of mathematics has constantly increased. This is true for most parts of mathematics, in particular for commutative algebra and (especially for) algebraic geometry. Since the 1960s there is however also a trend in the opposite direction in these fields. Interest in constructiveness and concrete calculations is also now increasing. One important reason for this is the growth in the capability of computers. It has become possible to carry out calculations that one could only dream of earlier. It is also possible to perform experiments on a large scale. But at least as important is the development of algorithms and of software. The work of Bruno Buchberger on Gröbner bases was a major breakthrough on the algorithmic side. The basic theory of Gröbner bases (but not all of its applications) is as unusual as any piece of elementary modern mathematics. The Gröbner basis program Macaulay by Dave Bayer and Michael Stillman has had a large impact. It has shown that calculations may be made efficiently, and the many scripts written by the authors and by David Eisenbud have shown the way to attack a wide range of problems. Following these pioneers has come a strong development on both the algorithmic side and the software side. All major computer algebra systems are now able to calculate Gröbner bases, and there are many good specialized programs.

This book is intended to be a concrete introduction to commutative algebra through Gröbner bases. The only prior knowledge we assume is a first course in algebra containing complex numbers and some elementary linear algebra. We have emphasized constructiveness above abstraction, and tried to give exemples for all new concepts and results. To make the theory accessible to a wider audience, we have intentionally confined ourselves to the more elementary parts. At certain points we suggest some further reading. And in the bibliography (which easily could be ten times as long), we have tried to limit ourselves to more accessible articles, although some do require further mathematical knowledge.

The audience we have in mind is, besides math students, students in technology and computer science, as well as people from applications interested in learning about this technique.

Much of the basic material is classical, and we have not found it easy to draw a time line for attributing results to their inventors. Thus we have taken the easy way out; the only times we have mentioned literature in the text are when we have suggested further reading.

Chapter 1

Rings, Fields, and Ideals

In this chapter we will introduce some basic concepts from commutative algebra that are important for the understanding of the following chapters. Our main object of study, will in the forthcoming chapters be *polynomials* in several variables with coefficients in an arbitrary *field.* Fields are patterned after, and are generalizations of, the complex numbers $\mathbb{C}$. The set of polynomials in n variables with coefficients from $\mathbb{C}$ is an example of a *ring* that plays an important role in both algebraic geometry and analysis. A ring is an algebraic system which is a generalization of the integers $\mathbb{Z}$.

1.1 Definition of a Ring

The set of integers $\mathbb{Z}$ is an algebraic system with two operations, addition and multiplication. These operations satisfy rules such as $a+b=b+a$, and $ab = ba$ for all integers a and b. Moreover, for every integer a there is an integer $-a$ such that $a+(-a)=0$. It is worth noting that the quotient a/b is usually *not* an integer. Roughly speaking, a ring is a set of objects that can be added, subtracted, and multiplied, more or less like integers. We will now define precisely what a ring is and check how well this definition agrees with our prototype $\mathbb{Z}$.

DEFINITION 1 (RING). A *ring* is a set R with two binary operations, denoted "+" and "·", such that for all elements a, b, c in R the following conditions are satisfied.

(a) (closed) $a+b \in R$, and $a \cdot b \in R$

(b) (commutative) $a+b=b+a$

(c) (associative) $(a + b) + c = a + (b + c)$, and $(a \cdot b) \cdot c = a \cdot (b \cdot c)$

(d) (additive identity) There is an element in R called *zero*, and written $\mathbf{0}$, such that $\mathbf{0} + a = a$.

(e) (additive inverse) Every element a has an additive inverse $-a$ such that $a + (-a) = \mathbf{0}$.

(f) (distributive) $a \cdot (b + c) = a \cdot b + a \cdot c$, and $(a + b) \cdot c = a \cdot c + b \cdot c$

It is easy to see that the set of integers forms a ring with the usual interpretation of the operations "+" and "·". What the definition does not guarantee is the existence of a multiplicative identity, i.e., an element $\mathbf{1}$ in R such that $a \cdot \mathbf{1} = \mathbf{1} \cdot a = a$. Neither can we be sure that $a \cdot b = b \cdot a$ for arbitrary rings. These observations lead us to the following two definitions.

DEFINITION 2 (RING WITH UNIT ELEMENT). If there is an element $\mathbf{1}$ in the ring R such that $a \cdot \mathbf{1} = \mathbf{1} \cdot a = a$ for every a in R, then R is called a *ring with unit element.*

DEFINITION 3 (COMMUTATIVE RING). The ring R is said to be *commutative* if the multiplication of R is such that $a \cdot b = b \cdot a$, for every $a, b \in R$.

We will now derive some immediate consequences of Definition 1 and check how well they agree with our prototype $\mathbb{Z}$. Our goal is to establish rules so that we can compute with elements in rings in much the same manner as we compute with integers, keeping in mind that in general $a \cdot b \neq b \cdot a$.

First we will show that *there cannot be more than one zero element.* Suppose that $\mathbf{0}' + a = a$ for every a in R. Then, in particular, $\mathbf{0}' + \mathbf{0} = \mathbf{0}$, but $\mathbf{0}' + \mathbf{0} = \mathbf{0} + \mathbf{0}'$ by (b), and $\mathbf{0} + \mathbf{0}' = \mathbf{0}'$. As $\mathbf{0}' + \mathbf{0}$ equals both $\mathbf{0}$ and $\mathbf{0}'$, we conclude that $\mathbf{0}' = \mathbf{0}$.

Similarly, assume that $\mathbf{1}$ and $\mathbf{1}'$ are both unit elements, then $\mathbf{1} = \mathbf{1} \cdot \mathbf{1}' = \mathbf{1}'$. Hence, *a ring has at most one unit element.*

We will establish another uniqueness result, namely *if a and b are given, the equation $a + x = b$ has a unique solution.* For by (e) every ring element a has an additive inverse, i.e., an element $-a$ such that $a + (-a) = \mathbf{0}$. Therefore, since $a + (-a + b) = (a + (-a)) + b = \mathbf{0} + b = b$, it follows that $-a + b$ is one solution. Again, supposing $b = a + x$, we have $-a + b = -a + (a + x) = (-a + a) + x = (a + (-a)) + x = \mathbf{0} + x = x$, which shows that $-a + b$ is the only solution.

A special case is the equation $a + x = \mathbf{0}$, with unique solution $x = -a$. If we apply this for the equation $\mathbf{0} + \mathbf{0} = \mathbf{0}$ we see that $\mathbf{0} = -\mathbf{0}$ (think of the equation as $\mathbf{0}_a + \mathbf{0}_x = \mathbf{0}$, with solution $\mathbf{0}_x = -\mathbf{0}_a$). Similarly, from

$-a + a = \mathbf{0}$ we see that $a = -(-a)$. Again, by (b), (c), and (e) we have $(a + b) + (-a) = b$, and if we add $-b$ to both sides of this equation we find that $(-a) + (-b)$ is the solution to $(a + b) + x = \mathbf{0}$, so we find that $(-a) + (-b) = -(a + b)$.

It is also true that $a \cdot \mathbf{0} = \mathbf{0} \cdot a = \mathbf{0}$ for arbitrary rings, i.e., the *additive* zero is also a *multiplicative* zero, or *annihilator*. To prove this, we write $\mathbf{0} \cdot a$ in two different ways and use the fact that the equation $a + x = b$ has a unique solution. To begin with $\mathbf{0} \cdot a = (\mathbf{0} + \mathbf{0}) \cdot a = \mathbf{0} \cdot a + \mathbf{0} \cdot a$, but it is also true that $\mathbf{0} \cdot a = \mathbf{0} \cdot a + \mathbf{0}$. It follows that $\mathbf{0} \cdot a = \mathbf{0}$. We can prove that $a \cdot \mathbf{0} = \mathbf{0}$ in a similar way.

There is a ring with only one element, called *the trivial ring*. By (d) a ring contains a zero element, and therefore the single element must be $\mathbf{0}$. In this ring $\mathbf{0} + \mathbf{0} = \mathbf{0} \cdot \mathbf{0} = \mathbf{0}$. Hence, $\mathbf{0}$ is also the unit element. In any other ring with unit element, R, we have $\mathbf{0} \neq \mathbf{1}$, since if $\mathbf{0} = \mathbf{1}$ we get $r = r \cdot \mathbf{1} = r \cdot \mathbf{0} = \mathbf{0}$ for any $r \in R$, so $\mathbf{0}$ is the only element in R.

From the distributive laws (f), the equalities $a \cdot b + (-a) \cdot b = (a + (-a)) \cdot b = \mathbf{0} \cdot b = \mathbf{0}$, and $a \cdot b + a \cdot (-b) = a \cdot (b + (-b)) = a \cdot \mathbf{0} = \mathbf{0}$, we see that $-(a \cdot b) = (-a) \cdot b = a \cdot (-b)$. If we now take the negatives of both sides of the equality $-(a \cdot b) = (-a) \cdot b$ we obtain $a \cdot b = -(-a) \cdot b = (-a)(-b)$.

Suppose that the ring has a unit element, then $a + (-\mathbf{1}) \cdot a = \mathbf{1} \cdot a + (-\mathbf{1}) \cdot a = (\mathbf{1} + (-\mathbf{1})) \cdot a = \mathbf{0} \cdot a = \mathbf{0}$. Thus, $(-\mathbf{1}) \cdot a = -a$. In particular if $a = (-\mathbf{1})$, $(-\mathbf{1}) \cdot (-\mathbf{1}) = -(-\mathbf{1}) = \mathbf{1}$.

The above observations permit us to compute with negatives, zero, and one as we have always done when working with the integers. For convenience $a + (-b)$ will henceforth be written $a - b$. Some immediate consequences of this notation are $-(a - b) = b - a$, $c \cdot (a - b) = c \cdot a - c \cdot b$, and $(a - b) \cdot c = a \cdot c - b \cdot c$. For neater notation, let us define a^i for any nonzero ring element a and any nonnegative integer i by $a^0 = \mathbf{1}$ (if the ring has a unit element), $a^1 = a$, $a^2 = a \cdot a$, $a^3 = a \cdot a^2$, and generally $a^n = a \cdot a^{n-1}$. We leave it to the reader to verify that the usual rules of exponents apply, i.e., for any positive integers m and n the product $a^m \cdot a^n = a^{m+n}$, and $(a^m)^n = a^{mn}$. Similarly, in any ring we shall abbreviate expressions like $a + a$ by $2a$ and more generally for any positive integer n put $na = a + a + \cdots + a$ (n terms). When $n < 0$, say $n = -m$, $m > 0$, we define na to mean $-ma$. It is easily verified that $ma + mb = m(a + b)$, and $m(na) = (mn)a$ for arbitrary integers m, n.

For ease of notation we will often drop the dot in $a \cdot b$ and simply write this product as ab. We have also seen that calculations with $\mathbf{0}$ and $\mathbf{1}$ do not give any surprises. We will in the sequel use the notation 0 and 1, respectively.

It is high time to pause and take a look at some concrete examples of rings.

EXAMPLE 4. The familiar number systems $\mathbb{Z}$, $\mathbb{Q}$ (the rational numbers), $\mathbb{R}$ (the real numbers), and $\mathbb{C}$ (the complex numbers) are all rings under the usual operations "+", "−", "·", and zero element 0. Each one of these rings is in fact a commutative ring with unit element 1. □

When we use 0 as we did in this example, it is understood that 0 is to be interpreted as one of $0_{\mathbb{Z}}$, $0_{\mathbb{Q}}$, $0_{\mathbb{R}}$, $0_{\mathbb{C}}$, the zero element of $\mathbb{Z}$, $\mathbb{Q}$, $\mathbb{R}$, and $\mathbb{C}$ respectively, depending on the context. If there is no danger of confusion, we will feel free to use this style. In situations where it is not clear which element or operation we are talking about, subscripts will be used, e.g., 1_R, $+_R$.

EXAMPLE 5. Let R be the set of even integers under the usual operations of addition and multiplication. Then R is a commutative ring but has no unit element. □

EXAMPLE 6. Let R be any ring and n any positive integer. The set of $n \times n$ matrices with entries $r_{ij} \in R$ forms a ring, R_n, under the usual operations, i.e., the sum of two matrices $C = A +_{R_n} B$ is a matrix with elements $c_{ij} = a_{ij} +_R b_{ij}$, and the product $C = A \cdot_{R_n} B$ is a matrix with elements $c_{ij} = \sum_{k=1}^{n} a_{ik} \cdot_R b_{kj}$. The zero element in R_n is the $n \times n$ matrix with every entry being equal to 0_R.

We note that if $n > 1$, and if R has more than one element, then R_n is noncommutative, even though the ring R itself may be commutative. If the R has a unit element, so has R_n. The unit element 1_{R_n} is the matrix (r_{ij}) with $r_{ij} = 0_R$ for $i \neq j$, and $r_{ij} = 1_R$ for $i = j$. □

EXAMPLE 7. We have seen that $\mathbb{Q}$, $\mathbb{R}$, and $\mathbb{C}$ are commutative rings with unit elements. These number systems have more structure than most rings since we can divide with nonzero elements. Such systems are called fields, see below. □

Exercises

1. Show that $\mathbb{Q}[\sqrt{2}] = \{ a + b\sqrt{2} \mid a, b \in \mathbb{Q} \}$ is a commutative ring with unit element.

2. Show that the set of matrices

$$\left\{ \begin{pmatrix} a & b \\ -b & a \end{pmatrix} \,\middle|\, a, b \in \mathbb{R} \right\}$$

is a commutative ring with unit element with the usual addition and multiplication of matrices.

3. Show that the set of pairs of real numbers $\{ (a, b) \mid a, b \in \mathbb{R} \}$ with addition $(a, b) + (c, d) = (a + c, b + d)$ and multiplication $(a, b) \cdot (c, d) = (ac, ad + bc)$ is a commutative ring with unit element.

1.2 Rings and Fields

In this section we will continue to study different examples of rings and introduce some special classes of rings.

1.2.1 Integral Domains and Fields

DEFINITION 8 (INTEGRAL DOMAIN). A commutative ring with unit element in which $ab = 0$ implies that $a = 0$ or $b = 0$ is called an *integral domain.*

EXAMPLE 9. The rings $\mathbb{Z}$, $\mathbb{Q}$, $\mathbb{R}$, and $\mathbb{C}$ are integral domains. □

DEFINITION 10 (FIELD). A *field* is a commutative ring with unit element in which every element $a \neq 0$ has a multiplicative inverse a^{-1}, so that $aa^{-1} = 1$. For practical reasons we exclude the trivial ring, $\{0\}$, from the set of fields.

The multiplicative inverse of a is unique, since if $ab = 1$, we get

$$\begin{aligned} a^{-1}(ab) &= a^{-1} \cdot 1 = a^{-1}, \text{ and} \\ a^{-1}(ab) &= (a^{-1}a)b = (aa^{-1})b = 1 \cdot b = b, \end{aligned}$$

hence $b = a^{-1}$. From $aa^{-1} = a^{-1}a = 1$ we also get $(a^{-1})^{-1} = a$.

PROPOSITION 11. *A field is an integral domain.*

Proof. If $ab = 0$ and $b \neq 0$ we get $a = a \cdot 1 = a(bb^{-1}) = (ab)b^{-1} = 0 \cdot b^{-1} = 0$. □

The most familiar examples of fields are the rational numbers $\mathbb{Q}$, the real numbers $\mathbb{R}$, and the complex numbers $\mathbb{C}$, but there are also fields with a finite number of elements that are of great importance. And now we prepare for the study of such fields.

PROPOSITION 12. *An integral domain with only a finite number of elements is a field.*

Proof. Let $R = \{x_1, x_2, \ldots, x_n\}$ be a finite integral domain and suppose $x_i \neq 0$. We claim that $\{ x_i x_k \mid k = 1, \ldots, n \}$ contains all elements of R. To show this, it is sufficient to show that $x_i x_k \neq x_i x_l$ if $k \neq l$. This is true, since if $x_i x_k = x_i x_l$ then $x_i(x_k - x_l) = 0$ which gives $x_k - x_l = 0$ since $x \neq 0$, so $x_k = x_l$. Since every element in R can be written as $x_i x_k$ for some k, we get that $x_i x_k = 1$ for some k so x_i has a multiplicative inverse. □

DEFINITION 13 (SUBFIELD). Suppose $k \subseteq K$ are fields, and the operations in k and K agree on the elements of k, i.e., that $a +_k b = a +_K b$ and $a \cdot_k b = a \cdot_K b$ if $a, b \in k$. T Then k is called a *subfield* of K, and K is called an *extension field* of k.

EXAMPLE 14. $\mathbb{Q}$ is a subfield of $\mathbb{R}$, and $\mathbb{R}$ is an extension field of $\mathbb{Q}$. □

We will use the word *subring* in a similar way as we do subfield. Thus, that R is a subring of S, does not only mean that $R \subseteq S$ as sets, but that we add and multiply elements in R in the same way regardless whether we think of them as elements in R or in S.

Exercises

1. Show that if a and b are nonzero elements in a field, then $(ab)^{-1} = a^{-1}b^{-1}$.
2. Show that the rings in Exercise 1 and Exercise 2 in Section 1.1 are fields.

1.2.2 A Convention

From now on, almost all the rings that we will consider will be commutative, and will all have a unit element. Therefore, we lay down the following convention:

> *Unless explicitly stated otherwise, the word* ring *will from now on mean a commutative ring with unit element.*

1.2.3 A Finite Ring, $\mathbb{Z}_n$

We will now give an important example of a finite ring. Let n be a positive integer. We will denote the ring by $\mathbb{Z}_n$ and its elements by $[0], [1], \ldots, [n-1]$, so there are n elements in $\mathbb{Z}_n$. The addition of elements is defined by $[i] + [j] = [k]$, where k is the remainder when $i + j$ is divided by n, and $0 \leq k \leq n - 1$. The multiplication is defined similarly, $[i] \cdot [j] = [m]$, where

Table 1.1: Addition and multiplication in $\mathbb{Z}_4$.

$+$	[0]	[1]	[2]	[3]
[0]	[0]	[1]	[2]	[3]
[1]	[1]	[2]	[3]	[0]
[2]	[2]	[3]	[0]	[1]
[3]	[3]	[0]	[1]	[2]

$\cdot$	[0]	[1]	[2]	[3]
[0]	[0]	[0]	[0]	[0]
[1]	[0]	[1]	[2]	[3]
[2]	[0]	[2]	[0]	[2]
[3]	[0]	[3]	[2]	[1]

m is the remainder when ij is divided by n, $0 \leq m \leq n-1$. It is clear that addition and multiplication are commutative, that [0] is the additive zero element, and that [1] is the multiplicative unit. It also follows that $[n-i]$ is the additive inverse to $[i]$ if $i > 0$. It remains to show that addition and multiplication are associative, and that the distributive law holds. Consider $([i]\cdot[j])\cdot[k]$. Here $[i]\cdot[j] = [r_1]$, where $ij = q_1n + r_1$. Hence $([i][j])[k] = [r_2]$, where $r_1k = q_2n + r_2$. Thus $ijk = kq_1n + q_2n + r_2 = (kq_1 + q_2)n + r_2$, so r_2 is the remainder when ijk is divided by n. In the same manner we see that $[i]([j][k]) = [r_2]$. Similarly one shows that both $([i] + [j]) + [k]$ and $[i] + ([j] + [k])$ equal $[r]$, where r is the remainder when $i + j + k$ is divided by n. Next consider $[i]([j] + [k])$. Here $[j] + [k] = [s_1]$, where $j + k = q_1n + s_1$. Hence $[i]([j] + [k]) = [s_2]$, where $is_1 = q_2n + s_2$. Thus $i(j + k) = iq_1n + q_2n + s_2 = (iq_1 + q_2)n + s_2$, so s_2 is the remainder when $i(j+k)$ is divided by n. In the same manner we see that $[i][j]+[i][k] = [s_2]$. The proof that $\mathbb{Z}_n$ is a ring is complete.

As a concrete example we give the tables for addition and multiplication in $\mathbb{Z}_4$, Table 1.1

We now show that the ring $\mathbb{Z}_n$ is a field if and only if n is a prime number. Suppose $n = n_1n_2$ is composite, i.e., not a prime. Then $[n_1][n_2] = [0]$ in $\mathbb{Z}_n$, so $\mathbb{Z}_n$ is not an integral domain and can therefore not be a field according to Proposition 11. Now suppose p is a prime. If $[a][b] = [0]$ in $\mathbb{Z}_p$, then by definition ab is divisible by p, which gives that either a or b is divisible by p. Hence either $[a]$ or $[b]$ equals [0] and hence $\mathbb{Z}_p$ is an integral domain. That $\mathbb{Z}_p$ is a field follows from Proposition 12. For the record we formulate this in a proposition.

THEOREM 15. *The ring $\mathbb{Z}_n$ is a field if and only if n is a prime number.*

Exercise

1. Construct tables for addition and multiplication for $\mathbb{Z}_6$ and $\mathbb{Z}_7$. Determine the multiplicative inverses to each element $\neq 0$ in $\mathbb{Z}_7$.

1.2.4 Polynomial Rings

The set of polynomials $a_0 + a_1x + a_2x^2 + \cdots + a_nx^n, a_i \in \mathbb{Q}$, n finite, is denoted $\mathbb{Q}[x]$. With natural addition and multiplication

$$\sum_i a_ix^i + \sum_i b_ix^i = \sum_i (a_i + b_i)x^i$$

and

$$\sum_i a_ix^i \cdot \sum_i b_ix^i = \sum_i (a_0 \cdot b_i + a_1 \cdot b_{i-1} + \cdots + a_i \cdot b_0)x^i$$

it is easily verified that $\mathbb{Q}[x]$ is a ring. In fact, if R is any ring, then the set $R[x]$ of polynomials $r_0 + r_1x + \cdots + r_nx^n$ with coefficients $r_i \in R$ is a ring with natural addition and multiplication. We can identify the ring R with the polynomials of degree zero in $R[x]$. Thus R is a subring of $R[x]$, in particular they have the same unit element.

Since $R[x]$ is a ring, it makes sense to consider the set $(R[x])[y]$, i.e., polynomials in y with coefficients in $R[x]$. We see that

$$\begin{aligned}(R[x])[y] &= \{\, f_0 + f_1y + \cdots + f_ny^n \mid f_i \in R[x] \,\} \\ &= \{r_0 + r_1x + r_2y + r_3x^2 + r_4xy + r_5y^2 + \cdots\} \\ &= \{\, g_0 + g_1x + g_2x^2 + \cdots + g_mx^m \mid g_i \in R[y] \,\} \\ &= (R[y])[x].\end{aligned}$$

So $(R[x])[y] = (R[y])[x]$, and can therefore be denoted by $R[x, y]$ without any risk for confusion. Likewise, we can define the polynomial ring over R in n variables $x_1, x_2, \ldots, x_n$. This polynomial ring is denoted by $R[x_1, x_2, \ldots, x_n]$.

DEFINITION 16 (DEGREE OF A POLYNOMIAL). The *degree* of an element $m = x_1^{i_1} \cdots x_n^{i_n}$ in a polynomial ring $R[x_1, \ldots, x_n]$ is $\deg(m) = i_1 + \cdots + i_n$. The degree of a nonzero polynomial $f(x_1, \ldots, x_n) = \sum r_{i_1,\ldots,i_n} x_1^{i_1} \cdots x_n^{i_n}$ equals $\deg(f) = \max\{\, \deg(x_1^{i_1} \cdots x_n^{i_n} \mid r_{i_1,\ldots,i_n} \neq 0 \,\}$. A polynomial of degree 0 is called a *constant*.

EXAMPLE 17. The degree of $f(x, y, z) = 2 + x^2 + xyz \in \mathbb{Z}[x, y, z]$ is 3. □

We can consider the ring R as a subring of $R[x_1, \ldots, x_n]$ consisting of constant polynomials.

REMARK. By far the most important classes of rings for us, in this book, will be *polynomial rings over a field*, i.e., rings of type $k[x_1, \ldots, x_n]$, where k is a field.

Exercises

1. Show that $R[x]$ is an integral domain if R is an integral domain.
2. Show that if $f, g \in R[x_1, \ldots, x_n] \setminus \{0\}$ and $f+g \neq 0$, then $\deg(f+g) \leq \max(\deg(f), \deg(g))$ and $\deg(fg) \leq \deg(f)+\deg(g)$. If R is an integral domain, show that $\deg(fg) = \deg(f) + \deg(g)$ but that there might be inequality if R is not an integral domain.
3. Show that the only invertible elements in $R[x]$, where R is an integral domain, are the constants which are invertible as elements in R. Show that there might be other invertible elements in $R[x]$ if R has zerodivisors.
4. Let k be a field and let R be the subset

$$\{ a_0 + a_1 x + \cdots + a_k x^k \mid a_1 = 0 \}$$

of $k[x]$. Show that R is a subring of $k[x]$.

1.2.5 Zerodivisors

Although a ring is a straightforward generalization of the integers, not every fact that we have become accustomed to in the ring of integers holds in general rings. As we have seen and as shown by the examples below, it is possible that $a \cdot b = 0$ with neither a nor b being zero.

DEFINITION 18 (ZERODIVISOR). An element $r \neq 0$ in a ring R is called a *zerodivisor* in R if $rs = 0$ for some element $s \in R, s \neq 0$. An element $r \neq 0$ which is not a zerodivisor is called a *nonzerodivisor*.

EXAMPLE 19. An integral domain is a ring in which every nonzero element is a nonzerodivisor. □

EXAMPLE 20. The element $[2]$ in $\mathbb{Z}_4$ is a zerodivisor since $[2][2] = [0]$. □

EXAMPLE 21. Let $R = \{ (a, b) \mid a, b \in \mathbb{Z} \}$ with addition, and multiplication defined by $(a,b)+(c,d) = (a+c, b+d)$ and $(a,b)(c,d) = (ac, bd)$, respectively. It is not difficult to check that R becomes a ring (often denoted $\mathbb{Z} \times \mathbb{Z}$) with zero element $(0,0)$ and unit element $(1,1)$. Here we have $(1,0)(0,1) = (0,0)$, so $(1,0)$ is a zerodivisor. □

DEFINITION 22 (CHARACTERISTIC). The *characteristic* of a ring R, here denoted $\operatorname{char}(R)$, is the smallest positive integer n such that $n \cdot 1 = 0$. If no such integer exists, the characteristic of the ring is 0.

EXAMPLE 23. The characteristic of $\mathbb{Z}, \mathbb{Q}[x]$ and $\mathbb{R}[x_1, \ldots, x_m]$ is 0. The characteristic of $\mathbb{Z}_n$ and $\mathbb{Z}_n[x_1, \ldots, x_m]$ is n. □

Exercises

1. If $\text{char}(R) = p \neq 0$ for an integral domain R, show that p is a prime number.
2. Show that the ring in Exercise 3 in Section 1.1 has zerodivisors.

1.3 Ideals and Rings

In this section we will introduce the important concept of ideal. An ideal is a special kind of subset of a ring, which behaves well with respect to the ring operations. Furthermore we will define and study certain important classes of rings. Finally we will introduce some operations, such as sum and product, on the set of ideals of a ring, and study the rules for these operations.

1.3.1 Definition of an Ideal

A subset $\mathfrak{a}$ of a ring R is called *closed under addition* if $a, a' \in \mathfrak{a}$ implies that $a + a' \in \mathfrak{a}$. It is *closed under multiplication with ring elements* if $a \in \mathfrak{a}$ and $r \in R$ implies that $a \cdot r \in \mathfrak{a}$. A nonempty subset of a ring which is closed under these two operations is called an *ideal.*

DEFINITION 24 (IDEAL). A nonempty subset $\mathfrak{a}$ of a ring R is called an *ideal of* R if

1. $a, a' \in \mathfrak{a} \Rightarrow a + a' \in \mathfrak{a}$,
2. $a \in \mathfrak{a}, r \in R \Rightarrow a \cdot r \in \mathfrak{a}$.

We note that every ideal is *closed under the operation of taking additive inverses*, i.e. if $a \in \mathfrak{a}$, then $-a \in \mathfrak{a}$, since $-a = a \cdot (-1)$. It also follows that every ideal contains the element 0, since $0 = a \cdot 0$. The set consisting of only 0 is an ideal, as is the whole ring R. An ideal $\mathfrak{a}$ coincides with R if and only if $1 \in \mathfrak{a}$, since if $1 \in \mathfrak{a}$, then $r = 1 \cdot r \in \mathfrak{a}$ for every $r \in R$.

Exercises

1. Show that $\{ 3n \mid n \in \mathbb{Z} \}$ is an ideal in $\mathbb{Z}$.
2. Let k be a field. Show that $\{ f \in k[x] \mid f(1) = 0 \}$ is an ideal in $k[x]$.
3. Let $(c_1, \ldots, c_n) \in k^n$, where k is a field. Show that
$$\{ f \in k[x_1, \ldots, x_n] \mid f(c_1, \ldots, c_n) = 0 \}$$
is an ideal in $k[x_1, \ldots x_n]$.
4. Determine all ideals in $\mathbb{Z}_6$.

1.3.2 The Ideal Generated by a Set

If S is an arbitrary nonempty subset of a ring R, then the set of finite linear combinations of elements in S, $\{ r_1 s_1 + \cdots + r_n s_n \mid r_i \in R, s_i \in S \}$, is an ideal denoted by $\langle S \rangle$. If for an ideal $\mathfrak{a}$ there is a finite set $S = \{s_1, \ldots, s_n\}$ such that $\mathfrak{a} = \langle S \rangle$, then $\mathfrak{a}$ is said to be *finitely generated* and denoted $\langle s_1, \ldots, s_n \rangle$.

Exercises

1. Show that the ideal in Exercise 1 in Section 1.3.1 is $\langle 3 \rangle$.
2. Show that the ideal in Exercise 2 in Section 1.3.1 is $\langle x - 1 \rangle$.
3. Show that the ideal in Exercise 3 in Section 1.3.1 is
$$\langle x_1 - c_1, \ldots, x_n - c_n \rangle .$$
4. Show that a ring R is a field if and only if the only ideals in R are $\{0\}$ and R.

1.3.3 Principal Ideals and Euclid's Algorithm

Let r be an element in a ring R. The set of all multiples of r, $\{ r \cdot r' \mid r' \in R \}$, constitutes an ideal. Such ideals are called ***principal ideals*** and r is called *a generator for the ideal.* The principal ideal generated by r is denoted rR or $\langle r \rangle$. The two ideals R and $\{0\}$ are both principal, $R = \langle 1 \rangle$ and $\{0\} = \langle 0 \rangle$.

We will now show that all ideals in $\mathbb{Z}$ are principal ideals. Suppose that $\mathfrak{a}$ is an ideal in $\mathbb{Z}$ different from $\langle 0 \rangle$. Then $\mathfrak{a}$ contains both positive and negative elements since $a \in \mathfrak{a}$ implies that $-a \in \mathfrak{a}$. Let n be the smallest positive number in $\mathfrak{a}$. We claim that $\mathfrak{a} = \langle n \rangle$, i.e., $\mathfrak{a}$ consists of all multiples of n. Let b be an element in $\mathfrak{a}$. Then, $b = qn + r$ for some integers q and r, $0 \leq r < n$, or stated otherwise $r = b - qn$. Since $n \in \mathfrak{a}$ it follows that both b and $-qn$ belong to $\mathfrak{a}$, so $r \in \mathfrak{a}$. But we picked n to be the smallest positive number in $\mathfrak{a}$ which forces r to be zero, so $b = qn + 0$ is a multiple of n.

Similar reasoning shows that every ideal in $\mathbb{Q}[x]$ is principal. Let $\mathfrak{a}$ be an ideal in $\mathbb{Q}[x]$ different from $\langle 0 \rangle$, and let $f \neq 0$ be a polynomial in $\mathfrak{a}$ of lowest degree. If g is another polynomial in $\mathfrak{a}$, we can find polynomials q and r such that $g = qf + r$ with $\deg(r) < \deg(f)$ or $r = 0$. As above, we see that $r \in \mathfrak{a}$ and conclude that r must be the zero polynomial since f was a polynomial in $\mathfrak{a}$ of lowest degree.

We will now show that the same conclusion can be drawn for $k[x]$, where k is an arbitrary field.

THEOREM 25 (THE DIVISION ALGORITHM). *Let k be any field and suppose $f, g \in k[x], f \neq 0$. Then there are uniquely defined polynomials $q, r \in k[x]$ such that $g = qf + r$ with $\deg(r) < \deg(f)$ or $r = 0$.*

Proof. We first show the existence of q and r. If $g = 0$ we can choose $q = r = 0$. Otherwise, let $f = a_n x^n + \cdots + a_0, a_n \neq 0$ and let $g = b_m x^m + \cdots + b_0, b_m \neq 0$. If $m < n$ we can choose $q = 0$ and $r = g$. If $m \geq n$ we see that $g = b_m a_n^{-1} x^{m-n} f + r_1$, where $\deg(r_1) < \deg(g)$ or $r_1 = 0$. If $r_1 \neq 0$ and $\deg(r_1) > \deg(f)$, say $r_1 = c_k x^k + \cdots + c_0$, we can continue and write $r_1 = c_k a_n^{-1} x^{k-n} f + r_2$ with $\deg(r_2) < \deg(r_1)$ or $r_2 = 0$, so $g = (b_m a_n^{-1} x^{m-n} + c_k a_n^{-1} x^{k-n}) f + r_2$. It is clear that in a finite number of steps we get a remainder which either is zero or has a smaller degree than $\deg(f)$. It remains to be shown that q and r are unique. Suppose $g = qf + r = q_1 f + r_1$. Then $(q - q_1)f = r_1 - r$. We have $\deg((q - q_1)f) \geq \deg(f)$ if $(q - q_1) \neq 0$, which is a contradiction since $\deg(r_1 - r) < \deg(f)$. Thus $q = q_1$. This gives $0 = 0 \cdot f = r_1 - r$, so $r_1 = r$. □

In the proof above it is essential that k is a field. There is no division algorithm in $\mathbb{Z}[x]$ as the example $g = x^2$ and $f = 2x$ shows. It is not easy to see a generalization of the division algorithm to $k[x_1, \ldots, x_n]$. One can say that the purpose of Gröbner bases is precisely to get a way to extend the division algorithm to polynomial rings in several variables.

We will now describe the Euclidean algorithm in $k[x]$. Let f and g be nonzero elements in $k[x]$, k a field. The division algorithm gives q_1 and r_1 such that $g = q_1 f + r_1$ with $\deg(r_1) < \deg(f)$ or $r_1 = 0$. If $r_1 \neq 0$ the division algorithm gives q_2 and r_2 so that $f = q_2 r_1 + r_2$ with $\deg(r_2) < \deg(r_1)$ or $r_2 = 0$. If $r_2 \neq 0$ we get $r_2 = q_3 r_2 + r_3$ a.s.o. If $r_3 \neq 0$ we have $\deg(r_3) < \deg(r_2) < \deg(r_1)$. Since the degrees decrease, this process is finite, so we must eventually get a remainder which is zero.

$$
\begin{aligned}
g &= q_1 f + r_1 \\
f &= q_2 r_1 + r_2 \\
r_1 &= q_3 r_2 + r_3 \\
&\vdots \\
r_{N-2} &= q_N r_{N-1} + r_N \\
r_{N-1} &= q_{N+1} r_N
\end{aligned}
$$

We claim that the last nonvanishing remainder r_N divides both f and g. The last equation gives that r_N divides r_{N-1}. The second equation from the bottom shows that r_N divides r_{N-2}, since r_N divides both terms on the right-hand side. Continuing like this upwards we eventually get the claim that r_N divides both f and g. Thus we have shown that r_N is a common divisor to f and g. Now suppose that h is another common divisor to f and g. The first equation shows that h divides $r_1 = g - q_1 f$, since h divides both g and f. The second equation then shows that h divides $r_2 = f - q_2 r_1$, since h divides both f and r_1. Continuing downwards like this we eventually get that h divides r_N. Thus we have shown that any common divisor to f and g must divide r_N. Since r_N is a common divisor to f and g, and since any other common divisor to f and g divides r_N, it is natural to say that r_N is the *greatest common divisor* to f and g, $r_N = \gcd(f, g)$.

DEFINITION 26 (GCD). Let f and g be nonzero polynomials in $k[x]$. Then h is a *greatest common divisor of f and g*, $\gcd(f, g)$, if h divides both f and g, and any other polynomial which divides both f and g also divides h.

THEOREM 27. *The last nonvanishing remainder in the Euclidean algorithm performed on f and g is a greatest common divisor to f and g. If h_1 and h_2 both are greatest common divisors to f and g, then $h_1 = ch_2$ for some $c \in k$.*

Proof. We only have to prove the last statement. Since h_1 is a common divisor to f and g and h_2 is a greatest common divisor to f and g we get that h_1 divides h_2, $h_2 = q_1 h_1$. Changing the roles of h_1 and h_2 we get $h_1 = q_2 h_2$. Thus $h_2 = q_1 q_2 h_2$, so $q_1 q_2 = 1$, hence $q_1 = q_2^{-1} \in k$. □

The Euclidean algorithm can be used to show that the greatest common divisor h to f and g is a linear combination of f and g, i.e. that there are polynomials h_1 and h_2 such that $h = h_1 f + h_2 g$. The second equation from the bottom in the Euclidean algorithm shows that $r_N = 1 \cdot r_{N-2} - q_N \cdot r_{N-1}$, hence r_N is a linear combination of r_{N-2} and r_{N-1}. The next equation from the bottom shows that $r_{N-1} = 1 \cdot r_{N-3} - q_{N-1} \cdot r_{N-2}$. Thus $r_N = -q_N \cdot r_{N-3} + (1 + q_N q_{N-1}) r_{N-2}$, hence r_N is a linear combination of r_{N-3} and r_{N-2}. Continuing upwards like this we eventually get the claim that r_N is a linear combination of f and g. This result is most easily described in terms of ideals.

THEOREM 28. *Let f and g be two nonzero polynomials in a polynomial ring $k[x]$ over a field k. Then $\langle f, g \rangle = \langle \gcd(f, g) \rangle$.*

Proof. Let $h = \gcd(f, g)$. We have shown that h is a linear combination of f and g which gives $h \in \langle f, g \rangle$. This gives that $\langle h \rangle \subseteq \langle f, g \rangle$, since $\langle h \rangle = \{ rh \mid r \in k[x] \}$, and if $h \in \langle f, g \rangle$, then $rh \in \langle f, g \rangle$. On the other hand, both f and g are multiples of h, thus $f, g \in \langle h \rangle$, which gives $\langle f, g \rangle = \{ r_1 f + r_2 g \mid r_1, r_2 \in k[x] \} \subseteq \langle h \rangle$. □

Let f be a polynomial of positive degree in $k[x]$ and let $c \in k \setminus \{0\}$. Since $f = c((1/c)f) = (1/c)(cf)$, we see that c and cf divide f. These divisors are called *trivial* divisors of f. If f is a polynomial of positive degree and the only divisors of f are the trivial ones, then f is called *irreducible.* All polynomials of degree one are irreducible, but there might be others, such as $x^2 + 1$ in $\mathbb{R}[x]$. Let $f \in k[x], \deg(f) > 0$. If f is not irreducible then $f = f_1 f_2$ with $\deg(f_i) < \deg(f), i = 1, 2$. If some f_i is not irreducible then this f_i can be factored further in nontrivial factors. Since the degrees of the factors decrease, it is clear that we eventually get $f = f_1 \cdots f_N$ with f_i irreducible. We will now show that this factorization is essentially unique.

LEMMA 29. *If f is irreducible and f divides $f_1 f_2$, then f divides f_1 or f divides f_2.*

Proof. Suppose that f does not divide f_1. Then $\gcd(f, f_1) = 1$ so $1 = h_1 f + h_2 f_1$ for some polynomials h_1, h_2, which gives $f_2 = h_1 f f_2 + h_2 f_1 f_2$. Since f divides both terms on the right-hand side, we get that f divides f_2. □

A polynomial $f = c_k x^k + c_{k-1} x^{k-1} + \cdots + c_0$ is called *monic* if $c_k = 1$.

THEOREM 30. *Suppose $\deg(f) > 0$. Then $f = c f_1 \cdots f_k$, where f_i are monic irreducible polynomials. This factorization is unique up to the ordering of the f_i's.*

Proof. We only need to show unicity. Let $f = c_1 f_1 \cdots f_k = c_2 g_1 \cdots g_m$. Being the coefficient of the largest power of x in f, it is clear that $c_1 = c_2$. Since f_1 divides $g_1 \cdots g_m$, we get that f_1 divides one of the g_i's. By reordering the g_i's we can suppose that f_1 divides g_1. Since both f_1 and g_1 are monic and irreducible, we get that $f_1 = g_1$. The equality $f_1 f_2 \cdots f_k = f_1 g_2 \cdots g_m$ gives $f_1(f_2 \cdots f_k - g_2 \cdots g_m) = 0$, so $f_2 \cdots f_k = g_2 \cdots g_m$ and we can continue in the same manner to get the desired conclusion. □

Let f and g be monic polynomials of positive degree and let $f_1, \ldots, f_N$ be all the monic irreducible factors that occur in the factorizations of f or g. Then we can write $f = f_1^{a_1} \cdots f_N^{a_N}$ and $g = f_1^{b_1} \cdots f_N^{b_N}$ with $a_i, b_i \geq 0$. It follows from the theorem that f divides g if and only if $a_i \leq b_i$ for all i. Thus $h = f_1^{\min(a_1, b_1)} \cdots f_N^{\min(a_N b_N)}$ divides both f and g, and any common divisor of f and g must divide h. Hence $h = \gcd(f, g)$.

EXAMPLE 31. If $f = (x-1)^3(x-2)(x^2+1)$ and $g = (x-1)^2(x-3)(x^2+1)(x^2+2)$, then $\gcd(f, g) = (x-1)^2(x^2+1)$. □

DEFINITION 32 (LCM). Let f and g be nonzero polynomials in $k[x]$. Then h is a *least common multiple* of f and g, $\mathrm{lcm}(f, g)$, if both f and g divide h and if any other polynomial which is a multiple of both f and g is a multiple of h.

We conclude this section with an application of the division algorithm. The following theorem is usually called the *Factor Theorem.* If $f = a_k x^k + \cdots + a_0$ and $c \in k$ then the *evaluation* of f at c is the element $f(c) = a_k c^k + \cdots + a_0 \in k$.

THEOREM 33 (THE FACTOR THEOREM). *Let k be a field and suppose that $f(x) \in k[x]$. Let $f(c) \in k$ denote the evaluation of f at c. Then $f(c) = 0$ if and only if $x - c$ divides $f(x)$.*

Proof. The division algorithm gives $q, r \in k[x]$ such that $f = q(x - c) + r$ and $\deg(r) < \deg(x - c) = 1$ or $r = 0$. Hence $r \in k$. Evaluation at c gives $f(c) = 0 + r$, so $f(c) = 0$ if and only if $r = 0$, hence if and only if $x - c$ divides f. □

COROLLARY 34. *A polynomial of degree n in $k[x]$ cannot have more than n zeros.*

Proof. If $c_1, \ldots, c_r$ are zeros to f then $(x - c_1) \cdots (x - c_r)$ is a factor of f which is impossible if $r > \deg(f)$. □

We also remind the reader of what is usually called the Fundamental Theorem of Algebra.

THEOREM 35 (THE FUNDAMENTAL THEOREM OF ALGEBRA). *Each polynomial $f \in \mathbb{C}[x]$ of degree at least one has a zero in $\mathbb{C}$.*

COROLLARY 36. *Any polynomial $f \in \mathbb{C}[x]$ with $\deg(f) = n > 0$ can be factored in linear factors, $f = c(x - c_1) \cdots (x - c_n)$.*

COROLLARY 37. *The polynomials $x - c$ are the only monic irreducible polynomials in $\mathbb{C}[x]$.*

If $(x - c)^m$ divides $f(x)$ but $(x - c)^{m+1}$ does not, we say that c is a root of multiplicity m to the equation $f(x) = 0$.

COROLLARY 38. *An equation $f(x) = 0, f \in \mathbb{C}[x], \deg(f) = n$, has exactly n roots, provided the number of roots are calculated with multiplicity.*

Fields with the property of the theorem above, and thus of the corollaries, are called *algebraically closed,* hence $\mathbb{C}$ is an algebraically closed field. It can be shown that any field has an extension field which is algebraically closed.

Exercises

1. Show the following division algorithm in $\mathbb{Z}$. Let $m, n \in \mathbb{Z}, n \neq 0$. Then there are $q, r \in \mathbb{Z}$ with $r < |n|$ such that $m = qn + r$.
2. Define the gcd of two elements in $\mathbb{Z}$ analogously to Definition 26.
3. Prove the analogue of the Euclidean algorithm for $\mathbb{Z}$.
4. Prove the following analogue of Theorem 27. The last nonvanishing remainder in the Euclidean algorithm performed on m and n is a greatest common divisor of m and n. If n_1 and n_2 are both greatest common divisors to m and n, then $n_1 = \pm n_2$. Show also that if $d = \gcd(m, n)$ there are $a, b \in \mathbb{Z}$ such that $d = am + bn$.
5. Determine $\gcd(x^4 + 3x^2 + 2, x^3 - 2x^2 + x - 2)$ in $\mathbb{Q}[x]$. Write $\gcd(x^4 + 3x^2 + 2, x^3 - 2x^2 + x - 2)$ as a linear combination of $x^4 + 3x^2 + 2$ and $x^3 - 2x^2 + x - 2$.
6. Let f and g be monic polynomials of positive degree and let $f_1, \ldots, f_N$ be all the monic irreducible factors that occur in the factorizations of f or g. Let $f = f_1^{a_1} \cdots f_N^{a_N}$ and $g = f_1^{b_1} \cdots f_N^{b_N}$ with $a_i, b_i \geq 0$. Show that $\operatorname{lcm}(f, g) = f_1^{\max(a_1, b_1)} \cdots f_N^{\max(a_N, b_N)}$.
7. If $f = (x-1)^3(x-2)(x^2+1)$ and $g = (x-1)^2(x-3)(x^2+1)(x^2+2)$, determine $\operatorname{lcm}(f, g)$.
8. If $r > 2$ we make the following recursive definition of $\gcd(f_1, \ldots, f_r)$. We define $\gcd(f_1, \ldots, f_r)$ to be $\gcd(\gcd(f_1, \ldots, f_{r-1}), f_r)$. Show that $\gcd(f_1 \ldots, f_r)$ divides f_i for $i = 1, \ldots, r$. Show that if h divides f_i for $i = 1, \ldots, r$ then h divides $\gcd(f_1, \ldots, f_r)$. Conclude that $\gcd(f_1, \ldots, f_r)$ does not depend on the order of the f_i's. Show that $\langle f_1, \ldots, f_r \rangle = \langle \gcd(f_1, \ldots, f_r) \rangle$.

1.3.4 Euclidean Rings

We can generalize the reasoning for $\mathbb{Z}$ and $k[x]$ in the following manner.

DEFINITION 39 (EUCLIDEAN RING). Suppose that R is an integral domain in which to each nonzero element a there is a nonnegative integer $d(a)$ so that

(1) $d(ab) \leq d(a)$ if $a, b \neq 0$
(2) For each pair a, b of elements, $a \neq 0$ there are elements q, r so that

$$b = qa + r$$

where either $r = 0$ or $d(r) < d(a)$.

Then R is called a *Euclidean ring*.

EXAMPLE 40. If $R = \mathbb{Z}$ we use $d(a) = |a|$. If $R = k[x]$ we use $d(f) = \deg(f)$. Thus $\mathbb{Z}$ and $k[x]$ are Euclidean rings. □

In the same way as for $\mathbb{Z}$ and $k[x]$ we can prove the following theorem.

THEOREM 41. *In a Euclidean ring every ideal is principal.*

Proof. Let $\mathfrak{a} \neq \langle 0 \rangle$ be an ideal and let $a \in \mathfrak{a}$ be an element with $d(a)$ minimal for elements in $\mathfrak{a}$. If $b \in \mathfrak{a}$ we have $b = qa + r$, and $r = b - qa \in \mathfrak{a}$. Since $d(a)$ is minimal we must have $r = 0$. Hence every element in $\mathfrak{a}$ is a multiple of a. □

Exercises

1. Consider the Gaussian integers $\mathbb{Z}[i] = \{ a + ib \mid a, b \in \mathbb{Z} \}$. Put $d(a + ib) = a^2 + b^2$. Show that for $\alpha, \beta \in \mathbb{Z}[i]$ we have $d(\alpha\beta) = d(\alpha)d(\beta)$. Show that $\mathbb{Z}[i]$ is a Euclidean ring. Hint: Suppose $\alpha = a + ib$ and $\beta = c + id, \beta \neq 0$, are Gaussian integers. Show that $\alpha/\beta = \gamma + q_1 + iq_2$ for some $\gamma \in \mathbb{Z}[i]$ and some $q_1, q_2 \in \mathbb{Q}$ with $|q_1| \leq |a|/2, |q_2| \leq |b|/2$.
2. Show that in a Euclidean domain $d(a)$ is minimal if and only if a is invertible.
3. Show that in a Euclidean domain one has $d(a) = d(b)$ if $a = bu$ for some invertible element u.

1.3.5 Ideals and their Calculus

We will now introduce some operations on the set of ideals in a ring. We will later show that, for ideals in a polynomial ring, these operations will have geometric meanings.

If $\mathfrak{a}$ and $\mathfrak{b}$ are ideals in R then

1. $\mathfrak{a} + \mathfrak{b} = \{ a + b \mid a \in \mathfrak{a}, b \in \mathfrak{b} \}$
2. $\mathfrak{a} \cap \mathfrak{b}$
3. $\mathfrak{a} \cdot \mathfrak{b} = \left\{ \sum_{i=1}^{k} a_i b_i \,\middle|\, a_i \in \mathfrak{a}, b_i \in \mathfrak{b} \right\}$
4. $\mathfrak{a} : \mathfrak{b} = \{ r \in R \mid rb \in \mathfrak{a} \quad \forall b \in \mathfrak{b} \}$

are all ideals. We prove this for $\mathfrak{a} : \mathfrak{b}$ and leave the other as exercises. If $r_1, r_2 \in \mathfrak{a} : \mathfrak{b}$ then $(r_1 + r_2)b = r_1 b + r_2 b \in \mathfrak{a}$ for every $b \in \mathfrak{b}$ since $r_1 b, r_2 b \in \mathfrak{b}$ and $\mathfrak{a}$ is closed under addition. Similarly, if $r_1 \in \mathfrak{a} : \mathfrak{b}$ and $r_2 \in R$, then $(r_2 r_1)b = r_2(r_1 b) \in \mathfrak{a}$ for every $b \in \mathfrak{b}$ since $r_1 b \in \mathfrak{a}$ and $\mathfrak{a}$ is closed under multiplication by elements in R.

DEFINITION 42 (THE RADICAL OF AN IDEAL). The *radical of an ideal* $\mathfrak{a}$ in a ring R is the set $\sqrt{\mathfrak{a}} = \{r \in R \mid r^n \in \mathfrak{a} \text{ for some } n\}$, where n is a positive integer.

DEFINITION 43 (RADICAL IDEAL). An ideal $\mathfrak{b}$ is a *radical ideal* if $\sqrt{\mathfrak{b}} = \mathfrak{b}$.

Note that $\mathfrak{a} \subseteq \sqrt{\mathfrak{a}}$ since $r \in \mathfrak{a}$ implies that $r^n \in \mathfrak{a}$ for $n = 1$, and so by definition, $r \in \sqrt{\mathfrak{a}}$.

THEOREM 44. *If $\mathfrak{a}$ is an ideal, then $\sqrt{\mathfrak{a}}$ is an ideal which contains $\mathfrak{a}$. Moreover, $\sqrt{\mathfrak{a}}$ is a radical ideal.*

Proof. We have already shown $\mathfrak{a} \subseteq \sqrt{\mathfrak{a}}$, so all that remains is to show $\sqrt{\mathfrak{a}}$ is an ideal which is radical.

First, we have to prove that $\sqrt{\mathfrak{a}}$ is an ideal. Suppose that $r_1, r_2 \in \sqrt{\mathfrak{a}}$. Then, by definition, there are two positive integers n_1 and n_2 such that $r_1^{n_1}, r_2^{n_2} \in \sqrt{\mathfrak{a}}$. If we expand the sum $(r_1 + r_2)^{n_1+n_2-1}$ with the binomial theorem, we see that every term is a multiple of some $r_1^{m_1} r_2^{m_2}$ with $m_1 + m_2 = n_1 + n_2 - 1$. Since either $m_1 \geq n_1$ or $m_2 \geq n_2$, either $r_1^{m_1}$ or $r_2^{m_2}$ is in $\mathfrak{a}$. Therefore, since all its terms are in $\mathfrak{a}$, $(r_1 + r_2)^{n_1+n_2-1} \in \mathfrak{a}$. We have proved that $r_1 + r_2 \in \sqrt{\mathfrak{a}}$, but we still have to show that if $r_1 \in \sqrt{\mathfrak{a}}$ and $r \in R$, then $rr_1 \in \sqrt{\mathfrak{a}}$. We have $r_1^n \in \mathfrak{a}$ for some n, hence $r^n r_1^n = (rr_1)^n \in \mathfrak{a}$, so $rr_1 \in \sqrt{\mathfrak{a}}$, hence $\sqrt{\mathfrak{a}}$ is an ideal. We have $\sqrt{\mathfrak{a}} \subseteq \sqrt{\sqrt{\mathfrak{a}}}$. Let $r \in \sqrt{\sqrt{\mathfrak{a}}}$. Then $r^n \in \sqrt{\mathfrak{a}}$ for some n, hence $r^{nm} = (r^n)^m \in \mathfrak{a}$ for some m, which gives $r \in \sqrt{\mathfrak{a}}$, so $\sqrt{\mathfrak{a}}$ is radical. □

There are lots of formulae connecting the different operations on ideals. We now prove some which will be needed later, and leave others as exercises.

PROPOSITION 45. *Let $\mathfrak{a}$, $\mathfrak{b}$, and $\mathfrak{c}$ be ideals in a ring R. Then*

1. $(\mathfrak{a} \cap \mathfrak{c}) + (\mathfrak{b} \cap \mathfrak{c}) \subseteq (\mathfrak{a} + \mathfrak{b}) \cap \mathfrak{c}$
2. $(\mathfrak{a} : \mathfrak{c}) + (\mathfrak{b} : \mathfrak{c}) \subseteq (\mathfrak{a} + \mathfrak{b}) : \mathfrak{c}$
3. $\mathfrak{a} : (\mathfrak{b} + \mathfrak{c}) = (\mathfrak{a} : \mathfrak{b}) \cap (\mathfrak{a} : \mathfrak{c})$
4. $\mathfrak{a} \subseteq \mathfrak{b} \Rightarrow \sqrt{\mathfrak{a}} \subseteq \sqrt{\mathfrak{b}}$
5. $\sqrt{\mathfrak{a}^k} = \sqrt{\mathfrak{a}}$
6. $\sqrt{\mathfrak{a}} \cap \sqrt{\mathfrak{b}} = \sqrt{\sqrt{\mathfrak{a}} \cap \sqrt{\mathfrak{b}}}$

Proof.
1. Let $r_1 \in \mathfrak{a} \cap \mathfrak{c}$ and $r_2 \in \mathfrak{b} \cap \mathfrak{c}$. Then $r_1 \in \mathfrak{a}$ and $r_2 \in \mathfrak{b}$ so $r_1 + r_2 \in \mathfrak{a} + \mathfrak{b}$. Since $r_1 \in \mathfrak{c}$ and $r_2 \in \mathfrak{c}$ we get $r_1 + r_2 \in \mathfrak{c}$.
2. Let $r \in \mathfrak{a} : \mathfrak{c} + \mathfrak{b} : \mathfrak{c}$. Then $r = r_1 + r_2$ where $r_1 c \in \mathfrak{a}$ and $r_2 c \in \mathfrak{b}$ for every $c \in \mathfrak{c}$. Thus $rc = r_1 c + r_2 c \in (\mathfrak{a} + \mathfrak{b})$ for every $c \in \mathfrak{c}$, so $r \in (\mathfrak{a} + \mathfrak{b}) : \mathfrak{c}$.

3. Let $r \in \mathfrak{a} : (\mathfrak{b} + \mathfrak{c})$. Then $r(b + c) \in \mathfrak{a}$ for every $b \in \mathfrak{b}$ and every $c \in \mathfrak{c}$. In particular $rb \in \mathfrak{a}$ for every $b \in \mathfrak{b}$ since $0 \in \mathfrak{c}$. In the same way we see that $rc \in \mathfrak{a}$ for every $c \in \mathfrak{c}$. Hence $r \in \mathfrak{a} : \mathfrak{b}$ and $r \in \mathfrak{a} : \mathfrak{c}$ so $r \in (\mathfrak{a} : \mathfrak{b}) \cap (\mathfrak{a} : \mathfrak{c})$ Now suppose $r \in (\mathfrak{a} : \mathfrak{b}) \cap (\mathfrak{a} : \mathfrak{c})$. Then $rb \in \mathfrak{a}$ for every $b \in \mathfrak{b}$ and $rc \in \mathfrak{a}$ for every $c \in \mathfrak{c}$. Hence $r(b + c) = rb + rc \in \mathfrak{a}$ for every $b \in \mathfrak{b}$ and every $c \in \mathfrak{c}$, so $r \in \mathfrak{a} : (\mathfrak{b} + \mathfrak{c})$.
4. If $r \in \sqrt{\mathfrak{a}}$, then $r^n \in \mathfrak{a} \subseteq \mathfrak{b}$ for some n, so $r \in \sqrt{\mathfrak{b}}$.
5. Since $\mathfrak{a}^k \subseteq \mathfrak{a}$ we have $\sqrt{\mathfrak{a}^k} \subseteq \sqrt{\mathfrak{a}}$. If $r \in \sqrt{\mathfrak{a}}$, then $r^n \in \mathfrak{a}$ for some n. Then $r^{nk} = (r^n)^k \in \mathfrak{a}^k$, so $r \in \sqrt{\mathfrak{a}^k}$.
6. Let $r \in \sqrt{\sqrt{\mathfrak{a}} \cap \sqrt{\mathfrak{b}}}$. Then $r^n \in \sqrt{\mathfrak{a}} \cap \sqrt{\mathfrak{b}}$ for some n. Then $r^{nk} \in \mathfrak{a}$ and $r^{nk} \in \mathfrak{b}$ for some k, hence $r \in \sqrt{\mathfrak{a}}$ and $r \in \sqrt{\mathfrak{b}}$, so $r \in \sqrt{\mathfrak{a}} \cap \sqrt{\mathfrak{b}}$. The other inclusion follows from (4).

□

Exercises

Let $\mathfrak{a}$, $\mathfrak{b}$, and $\mathfrak{c}$ be ideals in a ring R. Show that

1. $\mathfrak{a} + \mathfrak{b}, \mathfrak{a} \cap \mathfrak{b}$, and $\mathfrak{a} \cdot \mathfrak{b}$ are ideals. If $\mathfrak{a} = \langle S \rangle$ and $\mathfrak{b} = \langle T \rangle$, show that $\mathfrak{a} + \mathfrak{b} = \langle S \cup T \rangle$ and $\mathfrak{a} \cdot \mathfrak{b} = \langle st | s \in S, t \in T \rangle$.
2. $\mathfrak{a} + \mathfrak{b} = \mathfrak{b} + \mathfrak{a}, \ \mathfrak{a} + (\mathfrak{b} + \mathfrak{c}) = (\mathfrak{a} + \mathfrak{b}) + \mathfrak{c}$
3. $\mathfrak{a}\mathfrak{b} = \mathfrak{b}\mathfrak{a}, \ \mathfrak{a}(\mathfrak{b}\mathfrak{c}) = (\mathfrak{a}\mathfrak{b})\mathfrak{c}, \ \mathfrak{a}(\mathfrak{b} + \mathfrak{c}) = \mathfrak{a}\mathfrak{b} + \mathfrak{a}\mathfrak{c}$
4. $\mathfrak{a} \subseteq \mathfrak{a} + \mathfrak{b}, \ \mathfrak{a}\mathfrak{b} \subseteq \mathfrak{a} \cap \mathfrak{b}$
5. $(\mathfrak{a} : \mathfrak{b})\mathfrak{b} \subseteq \mathfrak{a}, \ \mathfrak{a} \subseteq \mathfrak{a} : \mathfrak{b}$
6. $(\mathfrak{a} : \mathfrak{b}) : \mathfrak{c} = \mathfrak{a} : (\mathfrak{b}\mathfrak{c})$
7. $\sqrt{\mathfrak{a} + \mathfrak{b}} = \sqrt{\sqrt{\mathfrak{a}} + \sqrt{\mathfrak{b}}}$
8. $\sqrt{\mathfrak{a}}\sqrt{\mathfrak{b}} = \sqrt{\mathfrak{a}\mathfrak{b}}$
9. $\sqrt{\mathfrak{a}\mathfrak{b}} = \sqrt{\sqrt{\mathfrak{a}}\sqrt{\mathfrak{b}}}$
10. $\sqrt{\mathfrak{a} : \mathfrak{b}} = \sqrt{\mathfrak{a}} : \sqrt{\mathfrak{b}}$
11. Determine $\sqrt{\langle 72 \rangle}$ in $\mathbb{Z}$.
12. Prove that if $\langle m \rangle$ and $\langle n \rangle$ are ideals in $\mathbb{Z}$, then
    (a) $\langle m \rangle + \langle n \rangle = \langle \gcd(n, m) \rangle$
    (b) $\langle m \rangle \cap \langle n \rangle = \langle \mathrm{lcm}(n, m) \rangle$
    (c) $\langle m \rangle \cdot \langle n \rangle = \langle mn \rangle$
    (d) $\langle m \rangle : \langle n \rangle = \langle m / \gcd(m, n) \rangle$

13. Prove that if $\langle f\rangle$ and $\langle g\rangle$ are ideals in $k[x]$, then
    (a) $\langle f\rangle + \langle g\rangle = \langle \gcd(f,g)\rangle$
    (b) $\langle f\rangle \cap \langle g\rangle = \langle \operatorname{lcm}(f,g)\rangle$
    (c) $\langle f\rangle \cdot \langle g\rangle = \langle fg\rangle$
    (d) $\langle f\rangle : \langle g\rangle = \langle f/\gcd(f,g)\rangle$

1.4 Equivalence Relations

Let M be a set and $M \times M$ the set of *ordered* pairs of elements in M. A subset S of $M \times M$ is called a *relation* on M and we write $a \sim_S b$ if the pair $(a,b) \in S$.

DEFINITION 46 (EQUIVALENCE RELATION). A relation S is called an *equivalence relation* if

1. $a \sim_S a$ for all a in M (reflexivity)
2. $a \sim_S b \Rightarrow b \sim_S a$ (symmetry)
3. $a \sim_S b$ and $b \sim_S c \Rightarrow a \sim_S c$ (transitivity)

EXAMPLE 47. The relation $a \sim b$ if a and b have the same parity, i.e., if a and b are both odd or both even, is an equivalence relation on $\mathbb{Z}$. □

PROPOSITION 48. *Let S be an equivalence relation and let $[a]$ be the subset of M consisting of all elements equivalent to a, i.e., $[a] = \{\, b \mid b \sim_S a \,\}$. Then $b \sim_S a$ implies $[a] = [b]$.*

Proof. If $c \in [a]$ then $c \sim_S a$ and since $a \sim_S b$ we get $c \sim_S b$, hence $c \in [b]$. Similarly we get the other inclusion. □

Thus M is partitioned in subsets called *equivalence classes*. Since $a \in [a]$, every element belongs to one and only one equivalence class. An element in a class $[a]$ is called a *representative* for $[a]$.

EXAMPLE 49. The equivalence relation on $\mathbb{Z}$, $a \sim b$ if a and b have the same parity yields the two classes, the odd numbers and the even numbers. Every even element is a representative for $[0]$ and every odd element is a representative for $[1]$. □

We now give a new description of $\mathbb{Z}_n$ in terms of an equivalence relation on $\mathbb{Z}$. Let n be a positive integer. Every element m can be written $m = qn + r$, for some quotient q and some remainder r, $0 \le r < n$. We get that the relation $m_1 \sim m_2$ if m_1 and m_2 have the same remainder when divided by n is an equivalence relation. Another way to define this equivalence relation is to say that $m_1 \sim m_2$ if $m_1 - m_2$ is divisible by n. The finite set of equivalence classes constitute $\mathbb{Z}_n$. The equivalence classes are $[0], [1], \ldots, [n-1]$, where the class $[i]$ consists of all integers with remainder i when divided by n. The addition in $\mathbb{Z}_n$ is then defined by $[a]+[b] = [a+b]$ and the multiplication by $[a][b] = [ab]$. These definitions seem to depend on the choice of representative for the classes, and we must show that they do not. If we choose other representatives a_1 and b_1, where $a_1 \sim a$ and $b_1 \sim b$, we must show that $[a_1 + b_1] = [a+b]$, i.e., that $a_1 + b_1 \sim a + b$. But $a_1 \sim a$ if and only if n divides $a_1 - a$ and $b_1 \sim b$ if and only if n divides $b_1 - b$. Then $(a_1 + b_1) - (a + b) = (a_1 - a) + (b_1 - b)$ is divisible by n, so $a_1 + b_1 \sim a + b$. Next we define a multiplication on $\mathbb{Z}_n$ by $[a] \cdot [b] = [ab]$. Again, we must show that $a_1 \sim a$ and $b_1 \sim b$ implies $a_1 b_1 \sim ab$. Now $a_1 b_1 - ab = a_1 b_1 - a_1 b + a_1 b - ab = a_1(b_1 - b) + b(a_1 - a)$. Hence, if both $b_1 - b$ and $a_1 - a$ are divisible by n we get that $a_1 b_1 - ab$ is divisible by n. It is now easy to check that all axioms for a ring are satisfied for $\mathbb{Z}_n$. For example, we see that $([a] \cdot [b]) \cdot [c] = [ab] \cdot [c] = [(ab)c] = [a(bc)] = [a] \cdot [bc] = [a] \cdot ([b] \cdot [c])$.

1.5 Field of Fractions of an Integral Domain

The field $\mathbb{Q}$ is the smallest field containing $\mathbb{Z}$. If we want to construct $\mathbb{Q}$ starting from $\mathbb{Z}$, we could introduce inverses n^{-1} to all nonzero elements in $\mathbb{Z}$ and let $\mathbb{Q} = \{ mn^{-1} \mid m \in \mathbb{Z}, n \in \mathbb{Z} \setminus \{0\} \}$. A problem is that this representation is not unique, e.g., $2 \cdot 3^{-1} = 4 \cdot 6^{-1}$. A solution to this problem is to consider pairs of integers and introduce an equivalence relation on the pairs so that $(2, 3) \sim (4, 6)$ say. Then $\mathbb{Q}$ is defined to be the set of equivalence classes of such pairs. We will perform such a construction for *any* integral domain.

Let R be an integral domain. We will construct a field, called *the fraction field of* R, that contains R and that does not have any subfield containing R. Consider the set of pairs $\{ (r_1, r_2) \mid r_1, r_2 \in R, r_2 \neq 0 \}$, and introduce a relation on this set of pairs by "$(r_1, r_2) \sim (r_1', r_2')$ if $r_1 r_2' = r_1' r_2$". We leave it to the reader to verify that this is an equivalence relation.

The set of equivalence classes associated with this equivalence relation can be made into a ring if we define the sum and product as follows. Let (r_1, r_2), and (r_1', r_2') be representatives for two equivalence classes, and define the sum of these classes to be the class containing $(r_1 r_2' + r_1' r_2, r_2 r_2')$, and define the product as the class containing $(r_1 r_1', r_2 r_2')$. To show that these operations are well defined one must check that the sum and product does not depend on the choice of representatives. Thereafter it must be verified that the axioms for a ring hold. All this is left to the reader. The zero element in this ring is the class containing $(0, 1)$ and the multiplicative unit element the class containing $(1, 1)$.

The class $[(r_1, r_2)]$ is different from the zero element if, and only if, $r_1 \neq 0$, and if so $[(r_1, r_2)] \cdot [(r_2, r_1)] = [(r_1 r_2, r_1 r_2)] = [(1, 1)]$. So every element different from the zero element has a multiplicative inverse, hence the ring is a field. We denote this field by $Q(R)$. The map $f : R \to Q(R)$, $f : r \mapsto [(r, 1)]$ is bijective, and shows that there is a copy of R inside $Q(R)$. It is easy to check that, if we identify R with its image in $Q(R)$, R is a subring of $Q(R)$. This means that addition and multiplication of elements are the same, regardless whether we do the operations in R or in $Q(R)$. For example is $[(r, 1)] + [(s, 1)] = [(r + s, 1)]$ in $Q(R)$. Since we have identified $r \in R$ with $[(r, 1)] \in Q(R)$, the addition is the same in R as in $Q(R)$. Similarly for multiplication. Any subfield of $Q(R)$ that contains R (with the identification we have made) will contain all equivalence classes $[(1, r_2)], r_2 \neq 0$, since these are inverses to elements in the image of the elements $[(r_2, 1)]$ in R. Hence, the subfield will contain $[(r_1, 1)] \cdot [(1, r_2)] = [(r_1, r_2)]$, and so is all of $Q(R)$.

EXAMPLE 50. If $R = k[x]$ then $Q(R)$ is denoted $k(x)$ and consists of all elements f/g, where $f, g \in k[x]$, and $g \neq 0$. Here we have identified $[(f, g)]$ with f/g. □

Exercises

1. Let R be an integral domain and consider the set of ordered pairs $\{ (r_1, r_2) \mid r_1, r_2 \in R, r_2 \neq 0 \}$. Show that the relation $(r_1, r_2) \sim (r_1', r_2')$ if $r_1 r_2' = r_1' r_2$ is an equivalence relation.
2. On the set of equivalence classes, define $[(a, b)] + [(c, d)] = [(ad + bc, bd)]$, and $[(a, b)] \cdot [(c, d)] = [(ab, cd)]$. Show that these operations are well defined, i.e., check that the sum and product do not depend on the choice of representatives.
3. Show that the set of equivalence classes associated with the equivalence relation in Exercise 1 with sum and product defined as in Exercise 2 form a ring with $[(0, 1)]$ as the zero element, and $[(1, 1)]$ as the multiplicative unit element.
4. Describe the fraction field of the ring $\mathbb{Z}[i]$ of Gaussian integers; see Exercise 1 in Section 1.3.4.

1.6 Unique Factorization Domains

We have seen that in a polynomial ring $k[x]$ over a field k, every polynomial of positive degree is a product of irreducible polynomials, and that this factorization is essentially unique. We will show that this is true also for polynomial rings in several variables over a field. It is not harder to be a bit more general; we will prove that if a ring R has unique factorization, then $R[x]$ has unique factorization. Note that if u is an invertible element in a ring R then $fg = (uf)(u^{-1}g)$, so we cannot expect perfect uniqueness if the ring has invertible elements other than 1.

DEFINITION 51. A nonzero element r in an integral domain R is said to be irreducible in R if $r = r_1r_2$ implies that r_1 or r_2 is invertible.

DEFINITION 52 (UNIQUE FACTORIZATION DOMAIN). An integral domain R is called a *unique factorization domain* if every element in R which is neither invertible nor 0 is a product of irreducible elements, and this factorization is unique in the following sense. If $r = r_1 \cdots r_k = s_1 \cdots s_m$ are two factorizations in irreducible elements then $k = m$, and after renumbering, for each i, $r_i = u_i s_i$ for some invertible elements $u_i \in R$.

To prove the main result we need some lemmas.

LEMMA 53. *Let p be an irreducible element in a unique factorization domain R. If ab is a multiple of p, then a or b is a multiple of p.*

Proof. We have $ab = pc$ for some c. Let $a = p_1 \cdots p_k, b = q_1 \cdots q_l, c = r_1 \cdots r_m$ be factorizations in irreducible elements. Then $p_1 \cdots p_k q_1 \cdots q_l = pr_1 \cdots r_m$, and the uniqueness gives that $up = p_i$ or $up = q_i$ for some i. Hence either a or b is a multiple of p. □

LEMMA 54. *If R is a unique factorization domain, then two elements a and b in R have a greatest common divisor $\gcd(a,b)$ which is unique up to a multiplication with an invertible element.*

Proof. If c divides both a and b, then each irreducible factor of c divides both a and b. Hence the product of all irreducible factors which occur in the factorizations of both a and b is $\gcd(a,b)$. □

DEFINITION 55 (CONTENT OF A POLYNOMIAL). Let R be a unique factorization domain, and let $f(x) = a_0 + \cdots + a_k x^k \in R[x]$. Then the *content* of f is $c(f) = \gcd(a_0, \ldots, a_k)$.

LEMMA 56. *If R is a unique factorization domain and $f, g \in R[x]$, then $c(fg) = c(f)c(g)$.*

Proof. Let $f = a_0 + \cdots + a_k x^k$ and $g = b_0 + \cdots + b_m x^m$. Then $c(f)$ divides all a_i and $c(g)$ divides all b_j, hence $c(f)c(g)$ divides all $a_i b_j$ and thus also all coefficients of fg. We have proved that $c(f)c(g)$ divides $c(fg)$. Suppose that there is an irreducible element p which divides $c(fg)$ but does not divide $c(f)c(g)$, i.e., does not divide either $c(f)$ or $c(g)$. Suppose that p divides $a_0, \ldots, a_{i-1}$ but not a_i, and that p divides $b_0, \ldots, b_{j-1}$ but not b_j. Consider the coefficient $a_0 b_{i+j} + \cdots + a_i b_j + \cdots + a_{i+j} b_0$ of x^{i+j} in fg. We see that p divides all terms in this sum except $a_i b_j$, which gives that p does not divide the sum, a contradiction. □

THEOREM 57. *If R is a unique factorization domain, then $R[x]$ is a unique factorization domain.*

Proof. Let $f(x) \in R[x]$ be a polynomial which is neither invertible nor 0. If $\deg(f) = 0$ then $f \in R$, so f is a unique product of irreducibles, since R has unique factorization. It is clear that this is a unique factorization into irreducibles also in $R[x]$. If $\deg(f) > 0$ we write $f = c(f) f_1$. Then $c(f_1) = 1$. Let $k = Q(R)$ be the fraction field of R. Every polynomial in $R[x]$ can be considered as a polynomial in $k[x]$. We know that $k[x]$ is a unique factorization domain by Theorem 30 in Section 1.3.3. Thus $f_1(x)$ has a unique factorization into irreducible factors in $k[x]$. It is clear that all we have to prove is the following. If $f(x) \in R[x]$ is a polynomial with content 1, and $f(x)$ is irreducible in $R[x]$, then $f(x)$ is irreducible in $k[x]$. So suppose that $f(x) = g(x)h(x)$ in $k[x]$ and $\deg(g) < \deg(f), \deg(h) < \deg(f)$. Then $g(x) = (c_1/c_2)G(x)$ and $h(x) = (c_3/c_4)H(x)$ for some $c_i \in R$ and some $G, H \in R[x]$ with contents 1 (factor out the least common denominators of the coefficients of g and h, respectively, to get polynomials in $R[x]$, and then factor out the content). Then $(c_2 c_4) f(x) = (c_1 c_3) G(x) H(x)$ and $c(GH) = 1$ by Lemma 56. Thus $f(x) = G(x)H(x)$ is a factorization in $R[x]$. □

COROLLARY 58. *If k is a field then $k[x_1, \ldots, x_n]$ is a unique factorization domain.*

Proof. This is immediate by induction on n. □

REMARK. From the proof of the theorem it follows that if $f(x)$ is irreducible in $R[x]$ of positive degree, then $f(x)$ is irreducible in $k[x]$. So for example if $f(x) \in \mathbb{Z}[x]$ can be factored with factors of lower degree in $\mathbb{Q}[x]$, then $f(x)$ can be factored with factors of lower degree in $\mathbb{Z}[x]$.

Exercises

1. Show that $x^3 + x + 1$ is irreducible in $\mathbb{Q}[x]$.
2. Prove Eisenstein's irreducibility criterion:
   Let $f(x) = a_0 + a_1 x + \cdots + a_n x^n \in \mathbb{Z}[x]$. If for some prime p we have that p divides a_i for $i < n$, p does not divide a_n, and p^2 does not divide a_0, then $f(x)$ is irreducible in $\mathbb{Z}[x]$.
3. Show that $x^m - p$ is irreducible in $\mathbb{Q}[x]$ for any m if p is a prime.
4. Show that $q(x) = x^{p-1} + x^{p-2} + \cdots + x + 1$ is irreducible in $\mathbb{Q}[x]$ if p is a prime.
5. Determine all monic irreducible polynomials of degree at most 3 in $\mathbb{Z}_2[x]$. Determine all monic irreducible polynomials of degree at most 2 in $\mathbb{Z}_3[x]$. Show that $x^5 + x + 1$ is irreducible in $\mathbb{Z}_2[x]$. Show that $x^5 + 2x + 2$ is irreducible in $\mathbb{Z}_3[x]$.

1.7 Factor Rings and Homomorphisms

We will now imitate the construction of $\mathbb{Z}_n$ from $\mathbb{Z}$ to get, from any ideal $\mathfrak{a}$ in any ring R, a new ring, the *factor ring* of R with respect to $\mathfrak{a}$. Let $\mathfrak{a}$ be an ideal in a ring R. Then the relation $a_1 \sim a_2$ if $a_1 - a_2 \in \mathfrak{a}$ is an equivalence relation on R. An equivalence class $[a]$ consists of the set $\{ a + a' \mid a' \in \mathfrak{a} \}$ and is denoted $a + \mathfrak{a}$. These equivalence classes are often called *cosets* of $\mathfrak{a}$. If $a + \mathfrak{a} = b + \mathfrak{a}$, i.e., if $a - b \in \mathfrak{a}$ we say that a is equivalent to b *modulo* $\mathfrak{a}$. The set of equivalence classes is denoted $R/\mathfrak{a}$. We make $R/\mathfrak{a}$ into a ring by defining $(a_1 + \mathfrak{a}) + (a_2 + \mathfrak{a}) = (a_1 + a_2) + \mathfrak{a}$, and $(a_1 + \mathfrak{a}) \cdot (a_2 + \mathfrak{a}) = a_1 \cdot a_2 + \mathfrak{a}$. As for $\mathbb{Z}_n = \mathbb{Z}/\langle n \rangle$, one easily checks that the addition and multiplication are well defined. With these operations $R/\mathfrak{a}$ becomes a ring, *the factor ring of R modulo* $\mathfrak{a}$.

EXAMPLE 59. Let $f(x) \in \mathbb{Q}[x]$ be a polynomial of degree d. Any element $g(x) \in \mathbb{Q}[x]$ can be written $g(x) = f(x)q(x) + r(x)$ with $\deg(r(x)) < d$ or $r(x) = 0$. Hence, any element in $\mathbb{Q}[x]$ is equivalent to either a polynomial of degree less than d or to 0 modulo $\langle f(x) \rangle$. On the other hand, two different polynomials $f_1(x)$ and $f_2(x)$ of degree less than d cannot be equivalent modulo $f(x)$ since $f_1(x) - f_2(x)$ cannot be divisible by $f(x)$ for degree reasons. Hence, the equivalence classes different from $\langle f(x) \rangle$ are in 1-1 correspondence to the set of polynomials of degree less than d. □

EXAMPLE 60. Let $f(x) = x^2 + 1 \in \mathbb{Q}[x]$. The equivalence classes are $\{ ax + b + \langle x^2 + 1 \rangle \mid a, b \in \mathbb{Q} \}$. Addition and multiplication are given by $(a_1x + b_1 + \langle x^2 + 1 \rangle) + (a_2x + b_2 + \langle x^2 + 1 \rangle) = (a_1 + a_2)x + (b_1 + b_2) + \langle x^2 + 1 \rangle$ and $(a_1x + b_1 + \langle x^2 + 1 \rangle) \cdot (a_2x + b_2 + \langle x^2 + 1 \rangle) = (a_1b_2 + b_1a_2)x + (b_1b_2 - a_1a_2) + \langle x^2 + 1 \rangle$ since $(a_1x + b_1)(a_2x + b_2) + \langle x^2 + 1 \rangle = a_1a_2x^2 + (a_1b_2 + a_2b_1)x + b_1b_2 + \langle x^2 + 1 \rangle = a_1a_2(x^2 + 1) + (a_1b_2 + a_2b_1)x + b_1b_2 - a_1a_2 + \langle x^2 + 1 \rangle = (a_1b_2 + b_1a_2)x + (b_1b_2 - a_1a_2) + \langle x^2 + 1 \rangle$. □

We now introduce a special kind of map between rings. A map $f : R \longrightarrow S$ is called a *homomorphism* (or *ring homomorphism*) if it respects the ring structures, i.e., if $f(r + s) = f(r) + f(s)$ and $f(rs) = f(r)f(s)$. We recall the concepts of injection, surjection, and bijection. A map $f : X \longrightarrow Y$ is *injective* if different elements have different images, i.e., if $x_1 \neq x_2$ then $f(x_1) \neq f(x_2)$. It is *surjective* if every element in Y is the image of some element in X. It is *bijective* if it is both injective and surjective. A bijective homomorphism is called an *isomorphism*. If there is an isomorphism between two rings R and S, they are called *isomorphic* and this is denoted $R \simeq S$. Sometimes there is no need to separate between isomorphic rings; they can be viewed as different representations of the same ring, and hence can be identified. It follows from the definition that if $f : R \longrightarrow S$ is a homomorphism, then the image of the zero element in R is the zero element in S, i.e., $f(0_R) = 0_S$, because for any r we have $f(r) = f(r + 0_R) = f(r) + f(0_R)$ and hence $f(0_R) = 0_S$. It is also true that $f(-r) = -f(r)$ for each r in R since $0 = f(0) = f(r + (-r)) = f(r) + f(-r)$. The *image* of a homomorphism $f : R \longrightarrow S$ is $\text{im}(f) = \{ s \in S \mid s = f(r) \}$ for some $r \in R$. It is easy to see that $\text{im}(f)$ is itself a ring. We now investigate surjective homomorphisms. First an example. Let R be a ring and $\mathfrak{a}$ an ideal in R. The map $p : R \longrightarrow R/\mathfrak{a}$ defined by $p(r) = r + \mathfrak{a}$ is a homomorphism since $p(r + s) = r + s + \mathfrak{a} = (r + \mathfrak{a}) + (s + \mathfrak{a}) = p(r) + p(s)$ and $p(rs) = rs + \mathfrak{a} = (r + \mathfrak{a})(s + \mathfrak{a}) = p(r)p(s)$. We will show that this is a typical example of a surjective homomorphism. If $f : R \longrightarrow S$ is a homomorphism we define the *kernel* of f as $\ker(f) = \{ r \in R \mid f(r) = 0 \}$.

THEOREM 61. *Let $f : R \rightarrow S$ be a homomorphism. Then $\ker(f)$ is an ideal in R. If f is surjective, then S is isomorphic to $R/\ker(f)$.*

Proof. Let $\mathfrak{a} = \ker(f)$. That $\mathfrak{a}$ is an ideal follows from the fact that if $r_1, r_2 \in \mathfrak{a}$ and $r \in R$, then $f(r_1 + r_2) = f(r_1) + f(r_2) = 0 + 0 = 0$ and $f(rr_1) = f(r)f(r_1) = f(r) \cdot 0 = 0$. We now define a map $g : R/\mathfrak{a} \longrightarrow S$ by $g(r + \mathfrak{a}) = f(r)$. The first thing we have to check is whether g is well defined. This means, if we happen to choose another representative for $r + \mathfrak{a}$, say $s + \mathfrak{a} = r + \mathfrak{a}$, is it then true that $f(r) = f(s)$? That this *is* true follows from the fact that $s+\mathfrak{a} = r+\mathfrak{a}$ implies $r-s \in \mathfrak{a}$ so $f(r-s) = 0$, which gives $f(r) - f(s) = 0$, hence $f(r) = f(s)$. We have shown that g is well defined. That g is a homomorphism follows from the following calculations:

$$\begin{aligned} g((r_1 + \mathfrak{a}) + (r_2 + \mathfrak{a})) &= g(r_1 + r_2 + \mathfrak{a}) \\ &= f(r_1 + r_2) \\ &= f(r_1) + f(r_2) \\ &= g(r_1 + \mathfrak{a}) + g(r_2 + \mathfrak{a}) \end{aligned}$$

$$\begin{aligned} g((r_1 + \mathfrak{a})(r_2 + \mathfrak{a})) &= g(r_1 r_2 + \mathfrak{a}) \\ &= f(r_1 r_2) \\ &= f(r_1) f(r_2) \\ &= g(r_1 + \mathfrak{a}) g(r_2 + \mathfrak{a}) \end{aligned}$$

It remains to show that g is bijective. That g is surjective follows from the asumption that f is surjective. Suppose $g(r_1 + \mathfrak{a}) = g(r_2 + \mathfrak{a})$. Then $f(r_1) = f(r_2)$ so $f(r_1 - r_2) = 0$, hence $r_1 - r_2 \in \ker(f)$. But then $r_1 + \mathfrak{a} = r_2 + \mathfrak{a}$, so g is also injective. □

We have defined polynomials in $k[x]$ as formal expressions $a_n x^n + \cdots + a_0, a_i \in k$. The addition and multiplication of these formal expressions have been defined so they satisfy the ring axioms. Some readers are perhaps more used to considering polynomials as functions, e.g., from $\mathbb{R}$ to $\mathbb{R}$. If the field has an infinite number of elements, there is no great risk of confusion, since $f = g$ in $k[x]$ if and only if $f = g$ considered as functions on k. For finite fields one has to be a bit more cautious. A function $\phi_f : k \longrightarrow k$ is called a *polynomial function* if there is a polynomial $f = a_n x^n + \cdots + a_0$ so that $\phi_f(c) = a_n c^n + \cdots + a_0$ for $c \in k$. The sum and product of polynomial functions are polynomial functions, and it is easy to see that the set of polynomial functions constitute a ring $\mathbf{Pol}(k)$.

PROPOSITION 62. *The map $\phi : k[x] \longrightarrow \mathbf{Pol}(k), f \mapsto \phi_f$ is a homomorphism. If the number of elements in k is infinite then ϕ is an isomorphism; if k is finite then ϕ is not an isomorphism.*

Proof. That ϕ is a homomorphism is immediate from the definition. We have $\ker(\phi) = \{ f \in k[x] \mid f(c) = 0 \text{ for all } c \in k \}$. If k is infinite it follows from Theorem 33 (the Factor Theorem) that only the zero polynomial has all elements in k as zeros. But if k is finite, $k = \{c_1, \ldots, c_n\}$, then $\ker(\phi) = \langle (x - c_1) \cdots (x - c_n) \rangle$. □

Exercises

1. Let k be a field and let $c \in k$. Show that $\phi : k[x] \longrightarrow k, \phi(f) = f(c)$ is a homomorphism. Determine $\ker(\phi)$.

2. Let k be a field and let $c_1, \dots, c_n \in k$. Show that $\phi : k[x_1, \dots, x_n] \longrightarrow k, \phi(f) = f(c_1, \dots, c_n)$ is a homomorphism. Determine $\ker(\phi)$.

3. Let $f : Q(\mathbb{Z}) \longrightarrow \mathbb{Q}, [(m, n)] \mapsto m/n$. Show that f is an isomorphism.

4. Let R be an integral domain and k a field, and suppose that R is a subring of k. Show that k has a subfield isomorphic to $Q(R)$.

5. Show that a field of characteristic 0 has a smallest subfield which is isomorphic to $\mathbb{Q}$. Let p be a prime number. Show that a field of characteristic p has a smallest subfield which is isomorphic to $\mathbb{Z}_p$.

6. Show that $\mathbb{R}[x]/\langle x^2 + 1\rangle \simeq \mathbb{C}$.

7. Show that $R = \mathbb{Z}_2[x, y]/\langle x^2, xy, y^2\rangle$ is a ring with 8 elements. Determine a multiplication table for R.

8. Show that $\mathbb{Q}[x]/\langle x^2 - 2\rangle \simeq \mathbb{Q}[\sqrt{2}]$.

9. Show that $R = \mathbb{Z}_3[x]/\langle x^2 + 1\rangle$ is a field with 9 elements. Determine a multiplication table for R.

10. Show that the homomorphism f is injective if and only if $\ker(f) = \langle 0\rangle$.

11. Show that if f is a surjective homomorphism, then $f(1)$ is the multiplicative unit element in $\mathrm{im}(f)$.

12. Let R be the ring in Exercise 2 in Section 1.1. Let $f : R \longrightarrow \mathbb{C}$ be defined by $f(\begin{pmatrix} a & b \\ -b & a \end{pmatrix}) = a + ib$. Show that f is an isomorphism.

13. Let R be the ring in Exercise 3 in Section 1.1. Show that the map $f : R \longrightarrow \mathbb{R}$, where f is defined by $f((a, b)) = a$, is a homomorphism. Determine $\ker(f)$. Show that $R \simeq \mathbb{R}[x]/\langle x^2\rangle$.

14. Let $\phi : k[x_1, x_2] \longrightarrow k[x]$ be defined by $\phi(f(x_1, x_2)) = f(x^2, x^3)$. Show that ϕ is a homomorphism. Show that the image of ϕ is the ring R in Exercise 4 in Section 1.2.4. Show that $R \simeq k[x_1, x_2]/\langle x_1^3 - x_2^2\rangle$.

15. Suppose that $\phi : R \to S$ is a surjective ring homomorphism. Show that there is a 1-1 correspondence between ideals in R which contain $\ker\phi$ and all ideals in S. Show that if $\mathfrak{a}$ is an ideal in R containing $\ker\phi$, then $R/\mathfrak{a} \simeq S/\phi(\mathfrak{a})$.

16. Let $\phi : \mathbb{Z}[x] \longrightarrow \mathbb{Z}_p[x]$, where p is a prime, be defined by $\phi(a_0 + a_1x + \cdots + a_nx^n) = [a_0] + [a_1]x + \cdots + [a_n]x^n$.

    (a) Show that ϕ is a surjective homomorphism.

(b) Show that if $\phi(f)$ is irreducible, then f is irreducible.

(c) Show that $x^5 - x^2 + 1$ is irreducible in $\mathbb{Q}[x]$.

(d) Show that $x^4 - 3x^3 + 3x^2 + 7$ is irreducible in $\mathbb{Q}[x]$.

17. This exercise shows that factorization in $\mathbb{Q}[x]$ is algorithmic. Let $f(x) \in \mathbb{Z}[x]$ be a polynomial of degree n. If f is reducible it has a factor g of degree at most $[n/2]$ (the largest integer $\leq n/2$). Take any integers $a_1, \ldots, a_{[n/2]+1}$. We have that $g(a_i)$ divides $f(a_i)$, so there are only a finite number of possibilities for $g(a_i)$. A polynomial of degree k is determined by its values in $k+1$ points. Thus there are only finitely many possible factors g to test. Use this to factor $x^4 - x^3 + 2x^2 - x + 1$ in $\mathbb{Q}[x]$.

1.8 Prime Ideals and Maximal Ideals

Two particular types of ideals will be very important in the sequel, prime ideals and maximal ideals. In polynomial rings they will correspond to building blocks for geometric sets.

DEFINITION 63 (PRIME IDEAL). An ideal $\mathfrak{p} \neq R$ in a ring R is a *prime ideal* if $rs \in \mathfrak{p}$ implies that $r \in \mathfrak{p}$ or $s \in \mathfrak{p}$.

LEMMA 64. *We have that* $\mathfrak{p}$ *is a prime ideal if and only if* $\mathfrak{a}_1 \cdots \mathfrak{a}_k \subseteq \mathfrak{p}$ *implies that some* $\mathfrak{a}_i \subseteq \mathfrak{p}$.

Proof. Suppose $\mathfrak{p}$ is a prime ideal. By induction on k it is clear that it suffices to consider the case when $k = 2$. Let $\mathfrak{a}_1\mathfrak{a}_2 \subseteq \mathfrak{p}$ and suppose $\mathfrak{a}_1 \not\subseteq \mathfrak{p}$. Take an $x \in \mathfrak{a}_1 \setminus \mathfrak{p}$. For each $a \in \mathfrak{a}_2$ we have $xa \in \mathfrak{p}$ which gives $a \in \mathfrak{p}$ so $\mathfrak{a}_2 \subseteq \mathfrak{p}$. For the converse we note that $xy \in \mathfrak{p}$ is equivalent to $\langle x \rangle \langle y \rangle \subseteq \mathfrak{p}$. Hence if $xy \in \mathfrak{p}$ then $\langle x \rangle \langle y \rangle \subseteq \mathfrak{p}$ which gives $\langle x \rangle \subseteq \mathfrak{p}$ or $\langle y \rangle \subseteq \mathfrak{p}$, i.e. $x \in \mathfrak{p}$ or $y \in \mathfrak{p}$. □

The following property of prime ideals is often used.

LEMMA 65. *Let* $\mathfrak{a}$ *be an ideal and let* $\mathfrak{p}_i$ *be prime ideals for* $i = 1, \ldots, s$. *If* $\mathfrak{a} \subseteq \cup_{i=1}^{s} \mathfrak{p}_i$, *then* $\mathfrak{a} \subseteq \mathfrak{p}_i$ *for some* i.

Proof. We use induction over s. If $s = 2$ and $\mathfrak{a} \subseteq \mathfrak{p}_1 \cup \mathfrak{p}_2, \mathfrak{a} \not\subseteq \mathfrak{p}_1, \mathfrak{a} \not\subseteq \mathfrak{p}_2$ there exists $a_1 \in \mathfrak{a} \setminus \mathfrak{p}_1, a_2 \in \mathfrak{a} \setminus \mathfrak{p}_2$. Then $a_1 + a_2 \in \mathfrak{a}$ but $a_1 + a_2 \notin \mathfrak{p}_1$ because $a_1 \notin \mathfrak{p}_1$ and $a_2 \in \mathfrak{p}_1$. In the same way we see that $a_1 + a_2 \notin \mathfrak{p}_2$, which gives a contradiction. (Note how in this part we haven't used that $\mathfrak{p}_i$ are prime ideals, only that they are ideals.) Now suppose that $s > 2$ and that $\mathfrak{a} \cap \mathfrak{p}_i \not\subseteq \cup_{j \neq i} \mathfrak{p}_j$ for all i (so all $\mathfrak{p}_i$'s are really needed). Take for each

i an $a_i \in (\mathfrak{a} \cap \mathfrak{p}_i) \setminus \cup_{j \neq i} \mathfrak{p}_j$. Then we get that $a = a_1 + a_2 \cdots a_s \in \mathfrak{a}$. We have $a_2 \cdots a_s \notin \mathfrak{p}_1$ (here we use that $\mathfrak{p}_1$ is a prime ideal) so $a \notin \mathfrak{p}_1$ since $a_1 \in \mathfrak{p}_1$. If $i > 1$ then $a_1 \notin \mathfrak{p}_i$ and $a_2 \cdots a_s \in \mathfrak{p}_i$ so $a \notin \mathfrak{p}_i$, which gives us a contradiction. □

DEFINITION 66 (MAXIMAL IDEAL). Let R be a ring. An ideal $\mathfrak{m} \neq R$ in R is a *maximal ideal* of R if the only ideal of R which strictly contains $\mathfrak{m}$ is the whole ring R.

Prime ideals and maximal ideals can be characterized by means of their factor rings.

THEOREM 67. *An ideal $\mathfrak{p} \neq R$ in a ring R is a prime ideal if and only if $R/\mathfrak{p}$ is an integral domain. An ideal $\mathfrak{m} \neq R$ is a maximal ideal if and only if $R/\mathfrak{m}$ is a field. Maximal ideals are prime.*

Proof. If $\mathfrak{a}$ is an ideal in R, we denote the element $r + \mathfrak{a}$ in the factor ring $R/\mathfrak{a}$ by $\bar{r}$. In particular the coset $\mathfrak{a}$ in $R/\mathfrak{a}$, which is the zero element in $R/\mathfrak{a}$, is denoted $\bar{0}$. Suppose $\mathfrak{p}$ is an ideal and that $rs \in \mathfrak{p}$. In $R/\mathfrak{p}$ this means that $\bar{r}\bar{s} = \bar{0}$. The equation $\bar{r}\bar{s} = \bar{0}$ is equivalent to $\bar{r} = \bar{0}$ or $\bar{s} = \bar{0}$ if and only if $R/\mathfrak{p}$ is an integral domain, which means that $rs \in \mathfrak{p}$ is equivalent to $r \in \mathfrak{p}$ or $s \in \mathfrak{p}$ if and only if $R/\mathfrak{p}$ is an integral domain. Suppose $\mathfrak{m}$ is a maximal ideal. If $r \notin \mathfrak{m}$ we get $\mathfrak{m} + \langle r \rangle = R$ since $\mathfrak{m} + \langle r \rangle$ strictly contains $\mathfrak{m}$. Thus $1 = m + rs$ for some $m \in \mathfrak{m}$ and $s \in R$. In $R/\mathfrak{m}$ this gives $\bar{1} = \bar{r}\bar{s}$ so any nonzero element $\bar{r}$ in $R/\mathfrak{m}$ has a multiplicative inverse, so $R/\mathfrak{m}$ is a field. Conversely, if $R/\mathfrak{m}$ is a field and $\bar{r} \neq \bar{0}$ in $R/\mathfrak{m}$, then $\bar{r}\bar{s} = \bar{1}$ for some $\bar{s} \in R/\mathfrak{m}$, which gives $rs + m = 1$ for some $m \in \mathfrak{m}$, hence $\mathfrak{m} + \langle r \rangle = \langle 1 \rangle = R$ for any $r \notin \mathfrak{m}$. Thus an ideal which strictly contains $\mathfrak{m}$ equals R and $\mathfrak{m}$ is a maximal ideal. Since fields are integral domains we get that maximal ideals are prime. □

EXAMPLE 68. The prime ideals in $\mathbb{Z}$ are $\langle p \rangle$, where p is a prime number and $\langle 0 \rangle$. All these but $\langle 0 \rangle$ are maximal. The prime ideals in $k[x]$, where k is a field, are $\langle f(x) \rangle$ where f is irreducible and $\langle 0 \rangle$. All these but $\langle 0 \rangle$ are maximal. □

Exercises

1. Let $(c_1, \ldots, c_n) \in k^n$ where k is a field and k^n is the set of n-tuples of elements in k. Show that the ideal $\langle x_1 - c_1, \ldots, x_n - c_n \rangle$ is a maximal ideal in $k[x_1, \ldots, x_n]$.
2. Show that $\langle x_1^2 + 1, x_2 \rangle$ is a maximal ideal in the ring $\mathbb{R}[x_1, x_2]$.

3. Show that $\langle x_1^3 - x_2^2 \rangle$ is a prime ideal in $k[x_1, x_2]$. Hint: Exercise Exercise 14 in Section 1.7.

4. Show the following improvement of Lemma 65. Let $\mathfrak{a} \subseteq \cup_{i=1}^{s} \mathfrak{a}_i$. If all but at most two of the $\mathfrak{a}_i$'s are prime ideals, then $\mathfrak{a} \subseteq \mathfrak{a}_i$ for some i.

5. If $\mathfrak{p}$ is a prime ideal, then $\sqrt{\mathfrak{p}} = \mathfrak{p}$.

6. Let R be the ring in Exercise 3 in Section 1.1. Show that $\{(0, b)\}$ is a maximal ideal.

† 7. Let $\phi : R \to S$ be a surjective ring homomorphism. Show that, in the correspondence of ideals in R and S (see Exercise 15 in Section 1.7), we have that prime ideals correpond to prime ideals and maximal ideals to maximal ideals.

1.9 Vector Spaces

We suppose that the reader is familiar with the concept of real vector spaces. It is possible to define vector spaces over *any* field, and the elementary theory gives no surprises. We therefore refrain from giving proofs of the most elementary facts.

DEFINITION 69 (VECTOR SPACE). A *vector space* over a field k is a set V with an operation $+$ and admitting a multiplication with elements in k. The elements in V are called vectors. The following rules are satisfied:

1. $v_1 + v_2 \in V$ if $v_i \in V$
2. $cv \in V$ if $c \in k$ and $v \in V$
3. $v_1 + v_2 = v_2 + v_1$ and $v_1 + (v_2 + v_3) = (v_1 + v_2) + v_3$ if $v_i \in V$
4. There exists a zero vector $\mathbf{0}$ such that $v + \mathbf{0} = v$ if $v \in V$
5. Each vector v has an additive inverse $-v$ such that $v + (-v) = \mathbf{0}$
6. $1 \cdot v = v$ if $v \in V$ and 1 is the multiplicative unit element in k
7. $(c_1 c_2)v = c_1(c_2 v)$ if $c_i \in k$ and $v \in V$
8. $c_1(v_1 + v_2) = c_1 v_1 + c_1 v_2$ and $(c_1 + c_2)v_1 = c_1 v_1 + c_2 v_1$ if $c_i \in k$ and $v_i \in V$

A set of vectors $\{v_i\}_{i \in I}$ is said to be linearly independent if any linear combination $c_1 v_1 + \cdots + c_n v_n = \mathbf{0}$ of elements in $\{v_i\}$ implies that all coefficients $c_i = 0$. A set of vectors $\{v_i\}_{i \in I}$ is called a generating set for V if every element in V is a finite linear combination of elements in $\{v_i\}$. A set which is both linearly independent and a generating set is called a *basis* for V. If V has a finite basis then V is called finite dimensional, and in that case every basis has the same number of elements. This number is called the dimension of V. Any linearly independent set can be extended to a basis and any generating set has a subset which is a basis. The set k^n of n-tuples $(c_1, \ldots, c_n)$ of elements $c_i \in k$ is a vector space with addition $(c_1, \ldots, c_n) + (c'_1, \ldots, c'_n) = (c_1 + c'_1, \ldots, c_n + c'_n)$ and multiplication $c(c_1, \ldots, c_n) = (cc_1, \ldots, cc_n)$. k^n has dimension n since $\{(1, 0, \ldots, 0), (0, 1, \ldots, 0), \ldots, (0, 0, \ldots, 1)\}$ is a basis.

EXAMPLE 70. If $k \subseteq K$ are fields then K is a vector space over k. As an example, $\mathbb{C}$ is a vector space over $\mathbb{R}$ of dimension 2 with basis $\{1, i\}$. □

EXAMPLE 71. The ring $k[x]$ is a vector space over k with basis $\{1, x, x^2, \ldots\}$. More generally, $k[x_1, \ldots, x_n]$ is a vector space over k with a basis consisting of all monomials $\{x_1^{i_1} \cdots x_n^{i_n}\}$. □

EXAMPLE 72. Let $f \in k[x]$ be of degree d. Then $k[x]/\langle f\rangle$ is a vector space over k of dimension d with basis $B = \{1 + \langle f\rangle, x + \langle f\rangle, \ldots, x^{d-1} + \langle f\rangle\}$. This follows from the fact that every element in $k[x]/\langle f\rangle$ can be written as $g(x) + \langle f\rangle$ with $\deg(g) < d$, hence B is a generating set. Furthermore the elements in B are linearly independent, since a nontrivial linear combination which is the zero element $\langle f\rangle$ in $k[x]/\langle f\rangle$ would yield an element of degree $< d$ which is a multiple of f. □

DEFINITION 73 (SUBSPACE). A nonempty subset of a vector space V, that is closed under addition and multiplication with field elements is called a *subspace* of V.

If V is a subspace of W, we define an equivalence relation on W by $w_1 \sim w_2$ if $w_1 - w_2 \in V$. The set of equivalence classes $\{w + V \mid w \in W\}$ constitute the *factor space* W/V with addition $(w_1 + V) + (w_2 + V) = (w_1 + w_2) + V$ and multiplication $c(w + V) = cw + V$. A map $f : V \longrightarrow W$ between two vector spaces is called *linear* or a *homomorphism* if $f(v_1 + v_2) = f(v_1) + f(v_2)$ and $f(cv) = cf(v)$ for all $v_1, v_2, v \in V, c \in k$. If $f : V \longrightarrow W$ is a homomorphism then $\ker(f) = \{v \in V \mid f(v) = \mathbf{0}\}$, and $\ker(f)$ is a subspace of V. A surjective homomorphism is called an *isomorphism* if it is bijective. Two vector spaces V_1 and V_2 are called *isomorphic*, $V_1 \simeq V_2$ if there exists an isomorphism $f : V_1 \longrightarrow V_2$. If $f : V \longrightarrow W$ is a surjective homomorphism then $W \simeq V/\ker(f)$, and f is an isomorphism if and only if $\ker(f) = \{0\}$.

THEOREM 74. *Suppose V is finite dimensional and that $f : V \longrightarrow W$ is surjective. Then* $\dim(\ker(f)) + \dim(W) = \dim(V)$.

Proof. Take a basis $\{e_1, \dots, e_k\}$ for $\ker(f)$ and extend it to a basis for V, $\{e_1, \dots, e_k, \dots, e_n\}$ for V. We claim that $B = \{f(e_{k+1}), \dots, f(e_n)\}$ is a basis for W. Since f is surjective any element in W has the form $f(c_1e_1 + \cdots + c_ne_n) = c_1f(e_1) + \cdots + c_nf(e_n) = c_{k+1}f(e_{k+1}) + \cdots + c_nf(e_n)$, thus B is a generating set for W. Suppose $c_{k+1}f(e_{k+1}) + \cdots + c_nf(e_n) = \mathbf{0}$. Then $f(c_{k+1}e_{k+1} + \cdots + c_ne_n) = \mathbf{0}$, so $c_{k+1}e_{k+1} + \cdots + c_ne_n \in \ker(f)$, thus $c_{k+1}e_{k+1} + \cdots + c_ne_n = c_1e_1 + \cdots + c_ke_k$, so $c_1e_1 + \cdots + c_ke_k - c_{k+1}e_{k+1} - \cdots - c_ne_n = \mathbf{0}$. Since $\{e_1, \dots, e_n\}$ is a basis for V we get that $c_i = 0$ for all i, hence $\{f(e_{k+1}), \dots, f(e_n)\}$ are linearly independent. □

DEFINITION 75 (DIRECT PRODUCT). Let $V_1, V_2, \dots$ be a sequence (finite or infinite) of vector spaces over k. Let $\prod V_i = \{ (v_1, v_2, \dots) \mid v_i \in V_i \}$. If we define addition in $\prod V_i$ and multiplication of an element in $\prod V_i$ with an element in k componentwise, $(v_1, v_2, \dots) + (w_1, w_2, \dots) = (v_1 + w_1, v_2 + w_2, \dots)$ and $c(v_1, v_2, \dots) = (cv_1, cv_2, \dots)$, then $\prod V_i$ is a vector space, the *direct product* of the V_i's.

If we have a finite sequence $V_1, \dots, V_n$, the direct product is often denoted $V_1 \times \cdots \times V_n$. If furthermore $V_i = V$ for all i then $\prod V_i$ is denoted V^n.

DEFINITION 76 (DIRECT SUM). Let $V_1, V_2, \dots$ be a sequence (finite or infinite) of vector spaces over k. Let $\oplus V_i = \{ (v_1, v_2, \dots) \mid v_i \in V_i \}$, where $v_i = \mathbf{0}$ for all but finitely many i. If we define addition in $\oplus V_i$ and multiplication of an element in $\oplus V_i$ with an element in k componentwise, $(v_1, v_2, \dots) + (w_1, w_2, \dots) = (v_1 + w_1, v_2 + w_2, \dots)$ and $c(v_1, v_2, \dots) = (cv_1, cv_2, \dots)$, then $\oplus V_i$ is a vector space, the *direct sum* of the V_i's.

In case the sequence is finite, the condition that $v_i = \mathbf{0}$ for all but finitely many i is an empty condition. Hence, in this case the direct sum coincides with the direct product.

THEOREM 77. *If V and W are finite dimensional, then* $\dim(V \times W) = \dim(V) + \dim(W)$.

Proof. If $\{e_i\}$ is a basis for V and $\{e'_i\}$ is a basis for W, we claim that $\{(e_i, \mathbf{0})\} \cup \{(\mathbf{0}, e'_i)\}$ is a basis for $V \times W$. Suppose $\sum c_i(e_i, \mathbf{0}) + \sum c'_i(\mathbf{0}, e'_i) = (\mathbf{0}, \mathbf{0})$. The left-hand side equals $(\sum c_ie_i, c'_ie'_i)$ so $\sum c_ie_i = \mathbf{0}$ in V and $\sum c'_ie'_i = \mathbf{0}$ in W. Since $\{e_i\}$ are linearly independent in V, we get $c_i = 0$ for all i. Similarly $c'_i = 0$ for all i, so $\{(e_i, \mathbf{0})\} \cup \{(\mathbf{0}, e'_i)\}$ is linearly independent. Finally $\{(e_i, \mathbf{0})\} \cup \{(\mathbf{0}, e'_i)\}$ is a generating set for $V \times W$ since an element in $V \times W$ is of the form (v, w) with $v \in V$ and $w \in W$. Since $\{e_i\}$ generates V, we have $v = \sum c_ie_i$ for some c_i and similarly we have $w = \sum c'_ie'_i$ for some c'_i. Hence $(v, w) = (\sum c_ie_i, c'_ie'_i) = \sum c_i(e_i, \mathbf{0}) + \sum c'_i(\mathbf{0}, e'_i)$. □

The most important property of $\oplus V_i$ is that the coordinates are unique, if $(v_1, v_2, \ldots) = (w_1, w_2, \ldots)$ then $v_i = w_i$ for all i. Our construction, from a (finite or infinite) set of vector spaces V_i define a new vector space $\oplus V_i$, is sometimes called the *external* direct sum of the V_i's. Now suppose that we have subspaces $V_1, V_2, \ldots$ of a vector space V. The *sum* $\sum V_i$ of the vector spaces V_i consists of all sums $\sum c_i v_i$, where $v_i \in V_i, c_i \in k$ and $c_i = 0$ for all but finitely many i. It is easy to see that $\sum V_i$ is the smallest subspace of V that contains all V_i. We now suppose that $V = \sum V_i$ and that for all i, $V_i \cap \sum_{j \neq i} V_j = \{\mathbf{0}\}$. Then we claim that $V \simeq \oplus V_i$, and V is called the internal direct sum of the V_i's. To prove this claim, we consider the map $\phi : \oplus V_i \longrightarrow V$ defined by $\phi((v_1, v_2, \ldots)) = v_1 + v_2 + \cdots$. (Even if there is an infinite number of V_i's, this sum is a finite sum.) It is clear that ϕ is a homomorphism and that ϕ is surjective. Suppose $\phi((v_1, v_2, \ldots)) = \mathbf{0}$. Then $v_1 + v_2 + \cdots = \mathbf{0}$ so $v_i = -v_1 - v_2 - \cdots - v_{i-1} - v_{i+1} - \cdots$, so $v_i \in V_i \cap \sum_{j \neq i} V_j$ which gives $v_i = \mathbf{0}$. This means that ϕ is also injective. Note that if V is the internal direct sum of $\{V_i\}$, then every element $v \in V$ can be written $v = v_1 + v_2 + \cdots, v_i \in V_i$ in a unique way.

EXAMPLE 78. Let $V = k[x]$. Then $V_i = \{ cX^i \mid c \in k \}$ is a one-dimensional vector space with basis X^i and V is the (internal) direct sum of the V_i's. □

For future reference we formulate a lemma.

LEMMA 79. *Let V_i be subspaces of a vector space V. Then V is the internal direct sum of the V_i's if and only if each element in $v \in V$ can in a unique way be written as a finite sum $v = \sum v_i$, where $v_i \in V_i$.*

Proof. We have already shown "only if", so suppose each element in V has a unique expression $\sum v_i, v_i \in V_i$. Then $V_i \cap \sum_{j \neq i} V_j = \{\mathbf{0}\}$ since otherwise, if $v_i \in V_i \cap \sum_{j \neq i} V_j, v_i \neq \mathbf{0}$, we would have a contradiction to unicity. □

Exercises

1. Show that the zero vector and the inverse are unique and that $0v = \mathbf{0}$ if $v \in V$.

2. Prove that for a vector space V over k, we have $\dim(V) = n$ if and only if $V \simeq k^n$.

3. Determine a basis for, and the dimension of, $\mathbb{Q}[\sqrt{2}]$ (cf. Exercise 1 in Section 1.1) as a vector space over $\mathbb{Q}$.

4. Determine a basis for, and the dimension of, the ring in Exercise 7 in Section 1.7 as a vector space over the field k.

5. Determine a basis for, and the dimension of, the ring $k[x, y]/\langle x^3, y^4\rangle$ as a vector space over the field k.
6. Show that if $V_1, V_2, \ldots$ are vector spaces over k, then $\prod V_i$ and $\oplus V_i$ are vector spaces over k.
7. Show that if $V_1, V_2, \ldots$ are subspaces of a vector space V, then $\sum V_i$ is the smallest subspace of V that contains all V_i.

Chapter 2

Monomial Ideals

In this chapter we will consider a particularly simple type of ideal in a polynomial ring, called *monomial ideals*. A monomial ideal is an ideal generated by a set of *monomials*, i.e., elements in $k[x_1, \ldots, x_n]$ of the form $x_1^{e_1} \ldots x_n^{e_n}$. In all examples we give, the monomial ideals will be finitely generated. This is no coincidence. We will show in the next chapter that *all* monomial ideals (in fact *all* ideals in $k[x_1, \ldots, x_n]$) are finitely generated.

Monomial ideals will be important in the sequel, and we will now study them in order to be familiar with them. Another good reason to study monomial ideals is that they are much easier to understand and to compute with than arbitrary polynomial ideals, so they constitute a good class of ideals to start with. For example, if $\langle f_1, \ldots, f_r \rangle$ is an ideal and f_i divides f_j for some $i \neq j$, then f_j is not needed as a generator. For monomial ideals this also gives a sufficient condition to get a minimal generating set. Thus, since it is trivial to see if a monomial divides another, it is easy to reduce a generating set to a minimal generating set, minimal in the sense that no generator can be cancelled without changing the ideal. The essential feature of monomial ideals, which explains why they are easier to handle, is formulated in the following theorem.

THEOREM 1. *Let $\mathfrak{a}$ be a monomial ideal in $k[x_1, \ldots, x_n]$. Let $f = \sum c_i m_i$, where $c_i \in k \setminus \{0\}$ and m_i are different monomials. If $f \in \mathfrak{a}$ then $m_i \in \mathfrak{a}$ for each i.*

Proof. We introduce the concept of fine grading or *multigrading* of the polynomial ring $k[x_1, \ldots, x_n]$. If $cm = cx_1^{i_1} \cdots x_n^{i_n}, c \in k \setminus \{0\}$, we set $\mathrm{mdeg}(cm) = (i_1, \ldots, i_n)$. Let $\mathfrak{a} = \langle n_1, \ldots n_s \rangle$ be a monomial ideal, and suppose $f = \sum c_i m_i \in \mathfrak{a}$. Then $f = g_1 n_1 + \cdots + g_s n_s$ for some $g_i = \sum c_{ij} m_{ij} \in k[x_1, \ldots, x_n]$. Let $c_i m_i$ be a nonzero term in f. Then $c_i m_i$

equals the sum of all elements $c_{ij}m_{ij}n_i$ which are of the same multidegree as $c_i m_i$. Hence $c_i m_i$ is a linear combination of the n_i's, so $c_i m_i \in \mathfrak{a}$. □

2.1 Sums and Products of Monomial Ideals

Remember that for any two ideals, $\mathfrak{a} = \langle a_1, \ldots, a_r \rangle$ and $\mathfrak{b} = \langle b_1, \ldots, b_s \rangle$, the sum is

$$\mathfrak{a} + \mathfrak{b} = \langle a_1, \ldots, a_r, b_1, \ldots, b_s \rangle ,$$

and the product is

$$\mathfrak{a}\mathfrak{b} = \langle a_1 b_1, \ldots, a_1 b_s, \ldots, a_r b_1, \ldots, a_r b_s \rangle ,$$

see Exercise 1 in Section 1.3.5.

Throughout the chapter we will look at some concrete computations involving the two monomial ideals $\mathfrak{a} = \langle x^3, xy, y^4 \rangle$ and $\mathfrak{b} = \langle x^2, xy^2 \rangle$. Thus, for our particular monomial ideals, we get

$$\mathfrak{a} + \mathfrak{b} = \langle x^3, xy, y^4 \rangle + \langle x^2, xy^2 \rangle = \langle x^3, xy, y^4, x^2, xy^2 \rangle .$$

But x^2 divides x^3 and xy divides xy^2, so we can remove the generators x^3, and xy^2 without changing the ideal. The final answer thus becomes

$$\mathfrak{a} + \mathfrak{b} = \langle x^3, xy, y^4 \rangle + \langle x^2, xy^2 \rangle = \langle x^3, xy, y^4, x^2, xy^2 \rangle = \langle xy, x^2, y^4 \rangle .$$

The computation of the product $\mathfrak{a}\mathfrak{b}$ is also straightforward.

$$\begin{aligned} \mathfrak{a}\mathfrak{b} &= \langle x^3, xy, y^4 \rangle \langle x^2, xy^2 \rangle \\ &= \langle x^5, x^4y^2, x^3y, x^2y^3, x^2y^4, xy^6 \rangle \\ &= \langle x^5, x^3y, x^2y^3, xy^6 \rangle. \end{aligned}$$

2.2 Intersections of Monomial Ideals

It is a bit harder to determine the intersection and quotient of monomial ideals. We note that if the ideals $\mathfrak{a} = \langle m \rangle$ and $\mathfrak{b} = \langle n \rangle$ are both principal, i.e., generated by a single element, then their intersection $\mathfrak{a} \cap \mathfrak{b} = \langle \operatorname{lcm}(m, n) \rangle$. Thus, $\langle x^2 y \rangle \cap \langle xy^3 \rangle = \langle \operatorname{lcm}(x^2y, xy^3) \rangle = \langle x^2y^3 \rangle$.

For any three ideals, $\mathfrak{a}$, $\mathfrak{b}$, and $\mathfrak{c}$, we have seen that

$$(\mathfrak{a} + \mathfrak{b}) \cap \mathfrak{c} \supseteq (\mathfrak{a} \cap \mathfrak{c}) + (\mathfrak{b} \cap \mathfrak{c}),$$

see Proposition 45 in Section 1.3.5. But if $\mathfrak{a}$, $\mathfrak{b}$, and $\mathfrak{c}$ are monomial ideals, the relation becomes an equality,

$$(\mathfrak{a} + \mathfrak{b}) \cap \mathfrak{c} = (\mathfrak{a} \cap \mathfrak{c}) + (\mathfrak{b} \cap \mathfrak{c}).$$

To prove this, let $f = f_1 + f_2 \in \mathfrak{c}$ with $f_1 \in \mathfrak{a}$ and $f_2 \in \mathfrak{b}$. Since $\mathfrak{a}$ and $\mathfrak{b}$ are monomial ideals, each monomial occuring in f_1 belongs to $\mathfrak{a}$ and each monomial occuring in f_2 belongs to $\mathfrak{b}$. If the same monomial occurs in both f_1 and f_2, say $c_1 m$ is a term of f_1 and $c_2 m$ is a term of f_2, we can write $f_1 + f_2 = (f_1 - c_1 m) + (f_2 + c_1 m)$ where $f_1 - c_1 m \in \mathfrak{a}$ and $f_2 + c_1 m \in \mathfrak{b}$ by Theorem 1. Hence we may assume that the monomials occuring in f_1 are different from those occuring in f_2. Since $\mathfrak{c}$ is a monomial ideal, we get that each monomial occuring in f_1 or f_2 belongs to $\mathfrak{c}$. Hence $f_1 \in \mathfrak{c}$ and $f_2 \in \mathfrak{c}$. Thus $f_1 + f_2 \in (\mathfrak{a} \cap \mathfrak{c}) + (\mathfrak{b} \cap \mathfrak{c})$.

Consequently, for monomial ideals we get

$$\langle m_1, \ldots, m_r \rangle \cap \langle n_1, \ldots, n_s \rangle = \sum_{i=1}^{r} \sum_{j=1}^{s} \langle m_i \rangle \cap \langle n_j \rangle = \sum_{i=1}^{r} \sum_{j=1}^{s} \langle \operatorname{lcm}(m_i, n_j) \rangle .$$

In particular, for our monomial ideals $\mathfrak{a} = \langle x^3, xy, y^4 \rangle$, and $\mathfrak{b} = \langle x^2, xy^2 \rangle$, we find that

$$\begin{aligned} \langle x^3, xy, y^4 \rangle \cap \langle x^2, xy^2 \rangle &= \langle \operatorname{lcm}(x^3, x^2), \operatorname{lcm}(x^3, xy^2), \ldots, \operatorname{lcm}(y^4, xy^2) \rangle \\ &= \langle x^3, x^3y^2, x^2y, xy^2, x^2y^4, xy^4 \rangle \\ &= \langle x^3, x^2y, xy^2 \rangle . \end{aligned}$$

2.3 Quotients of Monomial Ideals

We now turn to the quotient $\mathfrak{a} : \mathfrak{b}$ of monomial ideals. For arbitrary ideals $\mathfrak{a}$, $\mathfrak{b}$, and $\mathfrak{c}$, we have seen that the following relation holds:

$$(\mathfrak{a} + \mathfrak{b}) : \mathfrak{c} \supseteq (\mathfrak{a} : \mathfrak{c}) + (\mathfrak{b} : \mathfrak{c}),$$

see Proposition 45 in Section 1.3.5. But for the monomial ideal $\mathfrak{a} = \langle m_1, \ldots, m_r \rangle$ and a monomial n

$$\langle m_1, \ldots, m_r \rangle : \langle n \rangle = \langle m_1 \rangle : \langle n \rangle + \cdots + \langle m_r \rangle : \langle n \rangle .$$

This can be seen as follows. It suffices to check $f \in \langle m_1, \ldots, m_r \rangle : \langle n \rangle$ for monomials, since $fn \in \langle m_1, \ldots, m_r \rangle$ if and only if $mn \in \langle m_1, \ldots, m_r \rangle$ for every monomial m that occurs in f. If $mn \in \langle m_1, \ldots, m_r \rangle$ then mn is a multiple of some m_i, i.e., $mn \in \langle m_i \rangle$, hence $m \in \langle m_i \rangle : \langle n \rangle$.

For arbitrary ideals $\mathfrak{a}, \mathfrak{b}_1, \mathfrak{b}_2$, we have seen that

$$\mathfrak{a} : (\mathfrak{b}_1 + \mathfrak{b}_2) = (\mathfrak{a} : \mathfrak{b}_1) \cap (\mathfrak{a} : \mathfrak{b}_2),$$

see Proposition 45 in Section 1.3.5, so by induction we get

$$\mathfrak{a} : (\mathfrak{b}_1 + \cdots + \mathfrak{b}_s) = \cap_{i=1}^{s} (\mathfrak{a} : \mathfrak{b}_i).$$

Let $\mathfrak{a} = \langle m_1, \ldots, m_r \rangle$, $\mathfrak{b} = \langle n_1, \ldots, n_s \rangle$, and $\mathfrak{b}_i = \langle n_i \rangle$. Then $\mathfrak{b} = \mathfrak{b}_1 + \cdots + \mathfrak{b}_s$. It is easily seen that for any two monomials, m and n, the quotient ideal $\langle m \rangle : \langle n \rangle = \langle m / \gcd(m, n) \rangle$. We conclude that

$$\begin{aligned}
\mathfrak{a} : \mathfrak{b} &= \mathfrak{a} : (\mathfrak{b}_1 + \cdots + \mathfrak{b}_s) \\
&= \mathfrak{a} : (\langle n_1 \rangle + \cdots + \langle n_s \rangle) \\
&= (\mathfrak{a} : \langle n_1 \rangle) \cap \cdots \cap (\mathfrak{a} : \langle n_s \rangle) \\
&= (\langle m_1, \ldots, m_r \rangle : \langle n_1 \rangle) \cap \cdots \cap (\langle m_1, \ldots, m_r \rangle : \langle n_s \rangle) \\
&= (\langle m_1 \rangle : \langle n_1 \rangle + \cdots + \langle m_r \rangle : \langle n_1 \rangle) \cap \cdots \\
&\qquad \cap (\langle m_1 \rangle : \langle n_s \rangle + \cdots + \langle m_r \rangle : \langle n_s \rangle) \\
&= (\langle m_1 / \gcd(m_1, n_1) \rangle + \cdots + \langle m_r / \gcd(m_r, n_1) \rangle) \cap \cdots \\
&\qquad \cap (\langle m_1 / \gcd(m_1, n_s) \rangle + \cdots + \langle m_r / \gcd(m_r, n_s) \rangle) \\
&= \langle m_1 / \gcd(m_1, n_1), \ldots, m_r / \gcd(m_r, n_1) \rangle \cap \cdots \\
&\qquad \cap \langle m_1 / \gcd(m_1, n_s), \ldots, m_r / \gcd(m_r, n_s) \rangle .
\end{aligned}$$

Hence, for our monomial ideals, $\mathfrak{a} = \langle x^3, xy, y^4 \rangle$ and $\mathfrak{b} = \langle x^2, xy^2 \rangle$, the quotient ideal becomes

$$\begin{aligned}
\mathfrak{a} : \mathfrak{b} &= \langle x^3, xy, y^4 \rangle : \langle x^2, xy^2 \rangle \\
&= \langle x^3 / \gcd(x^3, x^2), xy / \gcd(xy, x^2), y^4 / \gcd(y^4, x^2) \rangle \cap \\
&\qquad \langle x^3 / \gcd(x^3, xy^2), xy / \gcd(xy, xy^2), y^4 / \gcd(y^4, xy^2) \rangle \\
&= \langle x^3 / x^2, xy / x, y^4 / 1 \rangle \cap \langle x^3 / x, xy / xy, y^4 / y^2 \rangle \\
&= \langle x, y, y^4 \rangle \cap \langle x^2, 1, y^2 \rangle \\
&= \langle x, y \rangle \cap \langle 1 \rangle \\
&= \langle x, y \rangle .
\end{aligned}$$

2.4 Prime Ideals

If some minimal generator m of a monomial ideal $\mathfrak{a}$ is of degree $\deg(m) > 1$ then $\mathfrak{a}$ is not a prime ideal, since if $m = x_i m'$ then $x_i \notin \mathfrak{a}$ and $m' \notin \mathfrak{a}$ but $x_i m' \in \mathfrak{a}$. On the other hand, if all generators are of degree one, so that $\mathfrak{a}$ is generated by a set of variables, then $\mathfrak{a}$ is prime since $k[x_1, \ldots, x_n] / \langle x_1 \ldots, x_l \rangle$ is a polynomial ring $k[x_{l+1}, \ldots, x_n]$ which is an integral domain. The only monomial ideal which is maximal in $k[x_1, \ldots, x_n]$ is $\langle x_1, \ldots, x_n \rangle$.

2.5 The Radical of a Monomial Ideal

Finally we consider the radical of a monomial ideal. But first we need to do some preparation.

DEFINITION 2 (SQUAREFREE MONOMIAL). A monomial m is called *squarefree* if $m = x_{i_1} \cdots x_{i_k}, i_1 < \cdots < i_k$. A monomial ideal generated by squarefree monomials is called a *squarefree monomial ideal.*

LEMMA 3. *A squarefree monomial ideal is radical.*

Proof. Let $\mathfrak{a} = \langle m_1, \ldots, m_t \rangle, m_1 = x_{i_1} \cdots x_{i_k}$. Then

$$\mathfrak{a} = \langle x_{i_1}, m_2, \ldots, m_t \rangle \cap \cdots \cap \langle x_{i_k}, m_2, \ldots, m_t \rangle,$$

see Section 2.2. If $m_2 = x_{j_1} \cdots x_{j_l}$, then

$$\langle x_{i_j}, m_2, \ldots, m_t \rangle = \cap_{m=1}^{l} \langle x_{i_j}, x_{j_m}, m_3, \ldots, m_t \rangle .$$

Continuing like this we see that $\mathfrak{a}$ is an intersection of ideals of type

$$\langle x_{m_1}, \ldots, x_{m_s} \rangle = \mathfrak{p}_m,$$

i.e., ideals generated by a set of variables. These ideals are prime according to Section 2.4. Thus

$$\sqrt{\mathfrak{a}} = \sqrt{\mathfrak{p}_1 \cap \cdots \cap \mathfrak{p}_N} = \sqrt{\mathfrak{p}_1} \cap \cdots \cap \sqrt{\mathfrak{p}_N} = \mathfrak{p}_1 \cap \cdots \cap \mathfrak{p}_N = \mathfrak{a}.$$

□

Now we turn to the general case.

PROPOSITION 4. *Let $\mathfrak{a} = \langle m_1, \ldots, m_s \rangle$ be a monomial ideal with $m_i = x_{i1}^{c_{i1}} \cdots x_{ik_i}^{c_{ik_i}}, i = 1, \ldots, s$ and all $c_{ik} \neq 0$. Then $\sqrt{\mathfrak{a}} = \langle n_1, \ldots, n_s \rangle$, where $n_i = x_{i1} \cdots x_{ik_i}, i = 1, \ldots, s$.*

Proof. Let $c_i = \max \{ c_{ik} \mid k = 1, \ldots, k_i \}$. Then $n_i^{c_i} \in \mathfrak{a}$, and hence we have $\langle n_1, \ldots, n_s \rangle^N \subseteq \mathfrak{a}$ with $N = c_1 + \cdots + c_s - s + 1$ since $(f_1 n_1 + \cdots + f_s n_s)^N$ is a sum of terms $g_i n_1^{d_1} \cdots n_s^{d_s}$ with $d_1 + \cdots + d_s = N$, so $d_i \geq c_i$ for at least one i, therefore every term belongs to $\mathfrak{a}$. Since

$$\langle n_1, \ldots, n_s \rangle^N \subseteq \mathfrak{a} \subseteq \langle n_1, \ldots, n_s \rangle$$

we get

$$\sqrt{\langle n_1, \ldots, n_s \rangle^N} \subseteq \sqrt{\mathfrak{a}} \subseteq \sqrt{\langle n_1, \ldots, n_s \rangle},$$

but

$$\sqrt{\langle n_1, \ldots, n_s \rangle^N} = \sqrt{\langle n_1, \ldots, n_s \rangle} = \langle n_1, \ldots, n_s \rangle$$

since $\langle n_1, \ldots, n_s \rangle$ is squarefree. □

EXAMPLE 5.

$$\sqrt{\langle x_1^3, x_2^2x_3^3, x_2^3x_3^2, x_2^2x_4, x_4^3\rangle} = \langle x_1, x_2x_3, x_2x_3, x_2x_4, x_4\rangle = \langle x_1, x_2x_3, x_4\rangle .$$

□

Exercises

1. Let $\mathfrak{a} = \langle x_1^3, x_2x_4, x_3^2\rangle$ and $\mathfrak{b} = \langle x_2^2, x_3x_4\rangle$ be ideals in $k[x_1, x_2, x_3, x_4]$. Determine $\mathfrak{a}+\mathfrak{b}, \mathfrak{a}\mathfrak{b}, \mathfrak{a}\cap\mathfrak{b}, \mathfrak{a} : \mathfrak{b}$, and $\sqrt{\mathfrak{a}}$.
2. Let $\mathfrak{a} = \langle x_1^2, x_1x_2^2\rangle$ and $\mathfrak{b} = \langle x_1x_2, x_2x_3\rangle$ be ideals in $k[x_1, x_2, x_3]$. Determine $\mathfrak{a}+\mathfrak{b}, \mathfrak{a}\mathfrak{b}, \mathfrak{a}\cap\mathfrak{b}, \mathfrak{a} : \mathfrak{b}$, and $\sqrt{\mathfrak{a}}$.
3. Write $\langle x_1x_4, x_2x_4\rangle$ as an intersection of prime ideals.
4. Write $\langle x_1^3, x_2^2x_3^3, x_2^3x_3^2, x_2^2x_4, x_4^3\rangle$ as an intersection of ideals of type $\langle x_{i_1}^{n_1}, \ldots, x_{i_k}^{n_k}\rangle$.
5. Let $\mathfrak{a}$ be an ideal generated by monomials in $k[x_1, \ldots, x_n]$. Show that $k[x_1, \ldots, x_n]/\mathfrak{a}$ is a finite dimensional vector space over k if and only if for each i there is a $d_i > 0$, such that $x_i^{d_i} \in \mathfrak{a}$.
6. Let $\mathfrak{a}$ be a monomial ideal such that $\left\{x_1^{d_1}, \ldots, x_n^{d_n}\right\}$ is a part of a minimal system of generators for $\mathfrak{a}$, where d_i are positive. Show that

$$d_1 + \cdots + d_n - n + 1 \leq \dim_k k[x_1, \ldots, x_n]/\mathfrak{a} \leq d_1d_2\cdots d_n$$

Chapter 3

Gröbner Bases

Our primary objects of study are factor rings $k[x_1, \ldots, x_n]/\mathfrak{a}$ over a field k. We want to be able to perform calculations in such rings. Elements in $k[x_1, \ldots, x_n]/\mathfrak{a}$ are cosets $f + \mathfrak{a}$ where $f \in k[x_1, \ldots, x_n]$. The first problem to overcome is that we have no unique way to represent elements in $R = k[x_1, \ldots, x_n]/\mathfrak{a}$; we have that $f + \mathfrak{a} = g + \mathfrak{a}$ in R if and only if $f - g \in \mathfrak{a}$. It is clear that a unique way to represent elements in R, a *normal form*, would be desirable. When $\mathfrak{a}$ is generated by monomials we have an ideal situation. If $f + \mathfrak{a}$ is a coset then f is a linear combination of monomials. We can delete all monomials which belong to $\mathfrak{a}$, i.e., those which are multiples of generators for $\mathfrak{a}$. Thus each element in $k[x_1, \ldots, x_n]/\mathfrak{a}$ can be represented as $f + \mathfrak{a}$, where f is a linear combination of monomials outside $\mathfrak{a}$. This representation is unique, cf. the introduction to Chapter 2. It is this situation we want to imitate in the general case. It is also easy to achieve a normal form in the case of a single variable. An ideal in $k[x]$ is generated by a single polynomial, $\mathfrak{a} = \langle g(x) \rangle$. The division algorithm shows that any polynomial $f(x) \in k[x]$ can be written $f(x) = q(x)g(x) + r(x)$, with $\deg(r(x)) < \deg(g(x))$ (or $r(x) = 0$) and that $r(x)$ is uniquely defined by $f(x)$. Thus each element in $k[x]/\langle g(x) \rangle$ has a representation $f_1(x) + \langle g(x) \rangle$ with $\deg(f_1) < \deg(g)$, and this representation is unique. We will point out some problems we have to overcome when extending the notion of normal form to the case of several variables, the multivariate case. But before doing that, we will look at an example of how to get normal forms in the one-variable case slightly differently. Suppose $R = k[x]/\langle x^2 - x + 1 \rangle$ and that we want to calculate the normal form of $x^4 + 2x + 3$. We can view the generator $x^2 - x + 1$ for the ideal as a rule for rewriting polynomials, namely since $x^2 - x + 1 + \langle x^2 - x + 1 \rangle = 0 + \langle x^2 - x + 1 \rangle$, we have $x^2 + \langle x^2 - x + 1 \rangle = x - 1 + \langle x^2 - x + 1 \rangle$. Thus, when x^2 occurs in a polynomial, we can replace it with $x-1$. In the same way we can replace any polynomial $x^2h(x)$ which is divisible by x^2 with $(x-1)h(x)$ since $(x^2-x+1)h(x) + \langle x^2-x+1\rangle = 0 + \langle x^2-x+1\rangle$, which gives $x^2h(x) + \langle x^2-x+1\rangle = (x-1)h(x) + \langle x^2-x+1\rangle$. We write this as a rule $x^2 \to x-1$. Hence we have $x^4+2x+3+\langle x^2-x+1\rangle = x^2x^2+2x+3+\langle x^2-x+1\rangle = (x-1)x^2+2x+3+\langle x^2-x+1\rangle = x^3-x^2+2x+3+\langle x^2-x+1\rangle = x^2(x-1)+2x+3+\langle x^2-x+1\rangle = (x-1)(x-1)+2x+3+\langle x^2-x+1\rangle = x^2+4+\langle x^2-x+1\rangle = x-1+4+\langle x^2-x+1\rangle = x+3+\langle x^2-x+1\rangle$ or $x^4+2x+3 \to (x-1)x^2+2x+3 = x^3-x^2+2x+3 = x^2(x-1)+2x+3 \to (x-1)^2+2x+3 = x^2+4 \to x-1+4 = x+3$. We now give an example which shows that already in one variable we must be a little cautious when translating a generator to a rewrite rule. Let $R = k[x]/\langle x^3-x^2\rangle$. It could seem that we have two choices to make a rule, namely $x^3 \to x^2$ or $x^2 \to x^3$. But the second choice would give never-terminating sequences of replacements, e.g., $x^2 \to x^3 \to x^4 \to \cdots$. In one variable we should always replace a monomial x^n with a polynomial of lower degree. In several variables we have to order *all* monomials and make sure that we replace a monomial with "smaller" terms. This will be carried through in the next section. Next we consider the example $R = k[x,y]/\langle x^2-xy\rangle$. Here we have two choices to make a rewrite rule, either $x^2 \to xy$ or $xy \to x^2$. It will turn out that both choices will work, we just have to decide which one to use and then stick to that choice. Which one to use will depend on which kind of questions we want to answer. The first choice will give normal forms of the type $g_1(y) + xg_2(y) + \langle x^2-xy\rangle$, the second will give normal forms of type $g(y) + xf(x) + \langle x^2-xy\rangle$. Note that in the first choice the normal forms will be linear combinations of 1, y^d and xy^{d-1} for $d \geq 1$; in the second we get linear combinations of $1, y^d, x^d, d \geq 1$; so the number of monomials used in each degree is the same for both choices. We will show that this is no coincidence. The last example will pinpoint a more serious difficulty we have to overcome. Let $R = k[x,y,z]/\langle x^2-xz, xy-z^2\rangle$. Suppose that we decide to use the rules $x^2 \to xz$ and $xy \to z^2$. If we try to get a normal form of x^2y we could use the first rule to get $x^2y \to xyz$ and then the second rule to get $xyz \to z^3$. On the other hand we could start with the second rule to get $x^2y \to xz^2$. Neither z^3 nor xz^2 can be rewritten by means of the rules. This will be resolved by adding a new rule. As soon as we have an ambiguity like this, in a systematic way we will add a new rule which gives a common answer. In the example we add the rule $xz^2 \to z^3$. This is okay since $xz^2 - z^3 = y(x^2-xz) + (z-x)(xy-z^2) \in \langle x^2-xz, xy-z^2\rangle$, so $xz^2 + \langle x^2-xz, xy-z^2\rangle = z^3 + \langle x^2-xz, xy-z^2\rangle$. One way of looking at Gröbner bases is as a sufficiently large set of rules to get unique normal forms of elements. This will be systematically developed later on.

3.1 Monomial Orderings

Let k be a field and $A = k[x_1, \ldots, x_n]$. An element $x_1^{i_1} \cdots x_n^{i_n}$ in A is called a *monomial* and an element $cx_1^{i_1} \cdots x_n^{i_n}$ with $c \in k \setminus \{0\}$ is called a *term*. (This terminology is not universally agreed on.) Let $\mathfrak{M}$ denote the set of all monomials in A. We will introduce an order relation $\prec$ on $\mathfrak{M}$. We define $\prec$ to be an *admissible* order relation on $\mathfrak{M}$ if it is a *total order* which is *compatible* with the multiplication of monomials, i.e.,

- For any pair of monomials m, n we have $m \prec n$ or $n \prec m$ or $m = n$
- If $m_1 \prec m_2$ and $m_2 \prec m_3$ then $m_1 \prec m_3$
- $1 \prec m$ for any monomial $m \neq 1$
- If $m_1 \prec m_2$ then $mm_1 \prec mm_2$ for any monomial m

We next give some examples of admissible orderings.

- The lexicographical ordering, Lex. Here $x_1^{i_1} \cdots x_n^{i_n} \prec x_1^{j_1} \cdots x_n^{j_n}$ if $i_1 = j_1, \ldots, i_k = j_k, i_{k+1} < j_{k+1}$ for some k. In other words $m \prec n$ if the first variable with different exponents in m and n has lower exponent in m.
- The degree lexicographical ordering, Deglex. Here $m = x_1^{i_1} \cdots x_n^{i_n} \prec x_1^{j_1} \cdots x_n^{j_n} = n$ if $\deg(m) = i_1 + \cdots + i_n < j_1 + \cdots + j_n = \deg(n)$ or if $\deg(m) = \deg(n)$ and $m \prec n$ in Lex.
- The degree reverse lexicographical ordering, Degrevlex. Here $m = x_1^{i_1} \cdots x_n^{i_n} \prec x_1^{j_1} \cdots x_n^{j_n} = n$ if $\deg(m) < \deg(n)$ or if $\deg(m) = \deg(n)$ and $i_n = j_n, i_{n-1} = j_{n-1}, \ldots, i_k = j_k, i_{k-1} > j_{k-1}$ for some k. In other words $m \prec n$ if $\deg(m) < \deg(n)$ or if $\deg(m) = \deg(n)$ and the last variable with different exponents in m and n has higher exponent in m.

There are different ways to vary these orderings to get new ones. All examples we have given imply the ordering $x_n \prec x_{n-1} \prec \cdots \prec x_1$ of the variables. The ordering of the variables can be varied in $n!$ ways which all induce new orderings. Another way to get a new ordering is to divide the variables into two or more blocks and compare each block in order. These orderings are called *block orderings* or *product orderings*. As an example, let $B_1 = \{x_1, \ldots, x_k\}$ and $B_2 = \{x_{k+1} \ldots, x_n\}$. Then every monomial can be written $m = m_1 m_2$ with all variables in m_i in block $B_i, i = 1, 2$. To compare two monomials m and n we make the partition of them in blocks, $m = m_1 m_2$ and $n = n_1 n_2$. We let $m \prec n$ if $m_1 \prec_1 n_1$ in some admissible ordering $\prec_1$ of the monomials of B_1. If $m_1 = n_1$ we let $m \prec n$ if $m_2 \prec_2 n_2$ in some admissible ordering $\prec_2$ of the monomials of B_2. This ordering can also be preceded by a degree check, $m \prec n$ if $\deg(m) < \deg(n)$.

DEFINITION 1. We call an ordering a *degree ordering* if the most important criterion for comparison is the degree of the monomials.

For future use we give the following lemma.

LEMMA 2. *Let $\prec$ be an arbitrary ordering. We define $m_1 \prec_g m_2$ if m_i are monomials with* $\deg m_1 < \deg m_2$ *or* $\deg m_1 = \deg m_2$ *and $m_1 \prec m_2$. Then $\prec_g$ is a monomial ordering. (We call $\prec_g$ the grading of $\prec$.)*

Proof. The four conditions in the definition of a monomial ordering are easily checked. □

Exercises

1. Prove Lemma 2.
2. Show that Deglex and Degrevlex agree on the monomials in $k[x_1, x_2]$, but not on the monomials in $k[x_1, x_2, x_3]$.
3. Show that $x_1^{i_1} \cdots x_n^{i_n} \prec x_1^{j_1} \cdots x_n^{j_n}$ if $i_n = j_n, i_{n-1} = j_{n-1}, \ldots, i_k = j_k, i_{k-1} < j_{k-1}$ for some k, does *not* define an admissible ordering.

3.1.1 A Classification of Orderings

It is possible to classify the set of *all* admissible orderings on the monomials of $k[x_1, \ldots, x_n]$. We quote only the result and refer the reader to articles for further reading. For each monomial $m = x_1^{k_1} \cdots x_n^{k_n}$ we recall that the multidegree of m is defined as $\text{mdeg}(m) = (k_1, \ldots, k_n)$. Then any admissible ordering is defined by an ordered set of orthogonal vectors $\mathbf{a}_1, \ldots, \mathbf{a}_r \in \mathbb{R}^n$, where $r \leq n$. Now $m \prec n$ if $\mathbf{a}_1 \cdot \text{mdeg}(m) < \mathbf{a}_1 \cdot \text{mdeg}(n)$ or if $\mathbf{a}_1 \cdot \text{mdeg}(m) = \mathbf{a}_1 \cdot \text{mdeg}(n)$ and $\mathbf{a}_2 \cdot \text{mdeg}(m) < \mathbf{a}_2 \cdot \text{mdeg}(n)$ or if $\mathbf{a}_1 \cdot \text{mdeg}(m) = \mathbf{a}_1 \cdot \text{mdeg}(n)$ and $\mathbf{a}_2 \cdot \text{mdeg}(m) = \mathbf{a}_2 \cdot \text{mdeg}(n)$ and $\mathbf{a}_3 \cdot \text{mdeg}(m) < \mathbf{a}_3 \cdot \text{mdeg}(n)$ and so on. Here $\mathbf{a}_1 \cdot \text{mdeg}(m)$ denotes the usual scalar product in $\mathbb{R}^n$. The vectors $\mathbf{a}_1, \ldots, \mathbf{a}_r \in \mathbb{R}^n$ must satisfy the condition which gives $1 \prec x_i$ for all i, so in particular, all coordinates of $\mathbf{a}_1$ must be nonnegative. We give some examples. The Lex ordering is determined by the n coordinate vectors $\mathbf{a}_1 = (1, 0, \ldots, 0), \ldots, \mathbf{a}_n = (0, 0, \ldots, 1)$. As a concrete example we let $n = 2$. Then $x_1^a x_2^b \prec x_1^c x_2^d$ if $(1, 0) \cdot (a, b) = a < c = (1, 0) \cdot (c, d)$ or if $a = c$ and $(0, 1) \cdot (a, b) = b < d = (0, 1) \cdot (c, d)$. In every degree ordering, we have $\mathbf{a}_1 = (1, 1, \ldots, 1)$, since $\deg(m) < \deg(n)$ if and only if $(1, 1, \ldots, 1) \cdot \text{mdeg}(m) < (1, 1, \ldots, 1) \cdot \text{mdeg}(n)$. For example, Deglex is determined by $\mathbf{a}_1 = (1, 1, \ldots, 1), \mathbf{a}_2 = (n-1, -1, \ldots, -1), \mathbf{a}_3 = (0, n-2, -1, \ldots, -1), \mathbf{a}_3 = (0, 0, n-3, \ldots, -1) \ldots, \mathbf{a}_n = (0, 0, \ldots, 1, -1)$.

Further Reading

L. Robbiano, *Term orderings on the polynomial ring*, Proc. Eurocal'85, Lect. Notes Comp. Science **204**, pp. 513–517, Springer 1985.

Exercises

1. Show that Deglex is defined as claimed above.
2. Determine vectors $\mathbf{a}_1, \ldots, \mathbf{a}_n$ which determine Degrevlex.
3. Describe the ordering on the monomials of $k[x_1, x_2, x_3]$, determined by $\mathbf{a}_1 = (1, 1, 0), \mathbf{a}_2 = (1, -1, 0), \mathbf{a}_3 = (0, 0, 1)$.
4. In all examples of orderings we have shown so far, it was necessary to use as many vectors $\mathbf{a}_i$ as the number of variables. Here is an example where this is not true. Let $n = 2$ and let $x_1^a x_2^b \prec x_1^c x_2^d$ if $a + b\sqrt{5} < c + d\sqrt{5}$, so here we use the single vector $\mathbf{a}_1 = (1, \sqrt{5})$. Show that this determines an admissible ordering. Determine the five smallest monomials in this ordering.

3.2 Dickson's Lemma and Some Applications

Let S be a set of monomials in $k[x_1, \ldots, x_n]$ and let $\mathfrak{a} = \langle S \rangle$ be the monomial ideal generated by S. The result in this section will justify our choice of S to be a finite set, in all our examples. We will show that, for any S, there is a finite subset S' such that $\langle S' \rangle = \langle S \rangle$.

Dickson was an American mathematician. He proved the following lemma in somewhat different form around 1910.

3.2.1 Dickson's Lemma

LEMMA 3 (DICKSON'S LEMMA). *Every monomial ideal in $k[x_1, \ldots, x_n]$, a polynomial ring over a field k, is finitely generated.*

Proof. We proceed by induction on n. Since every ideal in $k[x]$ is principal, the lemma is true for $n = 1$. Suppose it is true for $n - 1$ variables, and let $\mathfrak{a}$ be a monomial ideal in $k[x_1, \ldots, x_n]$. Let $\mathfrak{b}_j = (\mathfrak{a} : \langle x_n^j \rangle) \cap k[x_1, \ldots, x_{n-1}] = \langle S_j \rangle$. Since $\mathfrak{b}_j$ is an ideal in $k[x_1, \ldots, x_{n-1}]$, we can choose S_j to be finite. We have $\mathfrak{b}_0 \subseteq \mathfrak{b}_1 \subseteq \cdots$. It follows easily that $\cup \mathfrak{b}_j$ is an ideal $\mathfrak{b}$

in $k[x_1, \ldots, x_{n-1}]$ and hence finitely generated, $\mathfrak{b} = \langle S \rangle$. If $m \in \mathfrak{a}$ is a monomial then $m = m'x_n^k$ for some monomial $m' \in k[x_1, \ldots, x_{n-1}]$ and some k. Since $m'x_n^k \in \mathfrak{a}$ we get $m' \in \mathfrak{a} : \langle x_n^k \rangle$, so $m \in \langle x_n^k S_k \rangle$. Thus $S' = S_0 \cup x_n S_1 \cup x_n^2 S_2 \cup \cdots$ is a generating set for $\mathfrak{a}$. But for some r we have that $S_k = S_r$ if $k \geq r$, so $S_0 \cup x_n S_1 \cup \cdots \cup x_n^r S_r$ is a (finite) generating set for $\mathfrak{a}$. □

Exercise

1. Let $\mathfrak{a}_0 \subseteq \mathfrak{a}_1 \subseteq \cdots$ be an increasing sequence of ideals in a ring. Show that $\cup \mathfrak{a}_i$ is an ideal.

3.2.2 Applications of Dickson's Lemma

Let $f \in A = k[x_1, \ldots, x_n], f \neq 0$, and suppose $\prec$ is an admissible ordering of the monomials in A. Then f can be uniquely written $f = c_1 m_1 + \cdots + c_N m_N$ with monomials $m_1 \succ m_2 \succ \cdots \succ m_N$ and $c_i \neq 0, i = 1, \ldots, N$. We define the *support* of f to be $\operatorname{supp}(f) = \{m_i \mid i = 1, \ldots, N\}$, the *leading monomial* of f to be $\operatorname{lm}(f) = m_1$, the *leading term* of f to be $\operatorname{lt}(f) = c_1 m_1$, and the *leading coefficient* of f to be $\operatorname{lc}(f) = c_1$. As an example, let $f = 3x_1^2 x_2 + 2x_1 - x_2^2$ and let $\prec$ be $\prec_{Lex}$. Then $\operatorname{supp}(f) = \{x_1^2 x_2, x_1, x_2^2\}, \operatorname{lm}(f) = x_1^2 x_2, \operatorname{lt}(f) = 3x_1^2 x_2$, and $\operatorname{lc}(f) = 3$. If $\mathfrak{a}$ is a nonzero ideal in A, we define the *ideal of leading monomials* of $\mathfrak{a}$ to be $l(\mathfrak{a}) = \langle \operatorname{lm}(f) : f \in \mathfrak{a} \rangle = \langle \operatorname{lt}(f) : f \in \mathfrak{a} \rangle$. If $\mathfrak{a} = \langle f_1, \ldots, f_r \rangle$ then clearly $\operatorname{lm}(f_i) \in l(\mathfrak{a})$ for $i = 1, \ldots, r$. Since $\operatorname{lm}(mf) = m \operatorname{lm}(f)$ for a monomial m and a polynomial f, we get that $\langle \operatorname{lm}(f_1), \ldots, \operatorname{lm}(f_r) \rangle \subseteq l(\mathfrak{a})$. There might well be strict inclusion as the following example shows. Let $\mathfrak{a} = \langle x_1^2, x_1 x_2 - x_2^2 \rangle$ and let $\prec$ be $\prec_{Lex}$. Then $\langle \operatorname{lm}(x_1^2), \operatorname{lm}(x_1 x_2 - x_2^2) \rangle = \langle x_1^2, x_1 x_2 \rangle$. But $x_2^3 = x_2 \cdot x_1^2 - (x_1 + x_2) \cdot (x_1 x_2 - x_2^2) \in \mathfrak{a}$, so $\operatorname{lm}(x_2^3) = x_2^3 \in l(\mathfrak{a})$ and $x_2^3 \notin \langle x_1^2, x_1 x_2 \rangle$.

LEMMA 4. *Let $\prec$ be an admissible ordering on the monomials of A. Then any nonempty set S of monomials has a smallest element, i.e., there is an $m_0 \in S$ such that $m_0 \prec n$ for any $n \in S, n \neq m_0$.*

Proof. Let $\mathfrak{a} = \langle S \rangle$ be the ideal generated by S. By Dickson's Lemma, $\mathfrak{a}$ is finitely generated, $\mathfrak{a} = \langle m_1, \ldots, m_r \rangle$. Since any monomial m in S is a multiple of some m_i, $m = m_i m'$ for some i and some m', we can choose m_0 to be the smallest monomial in $\{m_1, \ldots, m_r\}$ since any multiple of m_i is larger than m_i. □

The following lemma is merely a reformulation.

LEMMA 5. *Any strictly decreasing sequence of monomials in an admissible ordering $\succ$ is finite.*

Proof. Let $m_1 \succ m_2 \succ \cdots$ be a strictly decreasing sequence of monomials and let $S = \{m_i \mid i = 1, 2, \dots\}$. Since S has a smallest element, the sequence must be finite. □

REMARK. If $\prec$ is a degree ordering, then for any monomial m there are only finitely many monomials smaller than m. This is *not* true in general. In Lex we have $x_2^i \prec x_1$ for all i if $x_2 \prec x_1$.

3.3 The Reduction Process

The division algorithm for polynomials in one variable states that if f and g are polynomials, $g \neq 0$, there exist polynomials q and r so that $f = qg + r$ where either $r = 0$ or $\deg(r) < \deg(g)$. This means that we can represent each polynomial $f \in k[x]$ with a polynomial r with $\deg(r) < \deg(g)$ or with 0, modulo the ideal $\langle g \rangle$. This representation is unique, since if $f = q_1 g + r_1 = q_2 g + r_2$, we have that $r_1 - r_2$ is a multiple of g, and this is only possible if $r_1 - r_2 = 0$ for degree reasons. Hence we have a unique representation of polynomials modulo the ideal $\langle g \rangle$. It is this unique representation of elements which makes it possible to perform calculations modulo $\langle g \rangle$, i.e., in the factor ring $k[x]/\langle g \rangle$. If we want to extend this to factor rings of polynomial rings in several variables, we can see several obstacles. The first thing we can note is that there is only one ordering of the monomials in $k[x]$ which is compatible with the multiplication of monomials, namely $1 \prec x \prec x^2 \prec \cdots$, but there are many possibilities for polynomial rings in several variables. This obstacle we have already overcome by introducing the concept of admissible orderings, and we will see that different choices of orderings are good for different purposes. A second obstacle is that ideals in polynomial rings in several variables are no longer principal. This means it is necessary to make a good choice of the generating set for the ideal, and will lead to the concept of Gröbner basis for an ideal. We are now ready to generalize the division algorithm to the case where f is a polynomial in several variables and $g = (g_1, \dots, g_s)$ is an ordered sequence of polynomials in several variables. We fix an admissible ordering of the monomials. In order to introduce the difficulties one at a time, we first consider the case of several variables but where g consists of a single polynomial g_1, and we will define the remainder of f with respect to g_1. If $f = 0$ we let the remainder of f be 0. Otherwise, let $m = \mathrm{lm}(g_1)$ and let $S_f = \{n \in \mathrm{supp}(f) \mid m \text{ divides } n\}$. Let p_0 be the largest element in S_f and let the coefficient of p_0 in f be c_{p_0}. Let $f_1 = f - ((c_{p_0} p_0)/(\mathrm{lc}(g_1) m)) g_1$.

EXAMPLE 6. If $f = 2x_1^3 + x_1^2 x_2 + x_2^3$, $g_1 = x_1^2 - x_1 x_2$ and $\prec$ is Lex, then $S_f = \{x_1^3, x_1^2 x_2\}$, $p_0 = x_1^3$ and $c_{p_0} = 2$. We get $f_1 = 2x_1^3 + x_1^2 x_2 + x_2^3 -$

$(2x_1^3/x_1^2)(x_1^2 - x_1x_2) = x_1^2x_2 + x_2^3 + 2x_1^2x_2 = 3x_1^2x_2 + x_2^3$. □

Let $S_{f_1} = \{ n \in \text{supp}(f_1) \mid m \text{ divides } n \}$. We claim that if p_1 is the largest element in S_{f_1} then $p_1 \prec p_0$. This follows from the fact that $\text{supp}(f_1) \subseteq (\text{supp}(f) \cup \text{supp}((p_0/m)g_1)) \setminus \{p_0\}$. If $p_1 \in \text{supp}(f) \setminus \{p_0\}$ then obviously $p_1 \prec p_0$. We have $\text{lm}((p_0/m)g_1) = (p_0/m)\,\text{lm}(g_1) = (p_0/m)m = p_0$. Hence also if $p_1 \in \text{supp}((p_0/m)g_1)) \setminus \{p_0\}$ we get $p_1 \prec p_0$. We continue this process and define $f_2 = f_1 - ((c_{p_1}p_1)/(\text{lc}(g_1)m))g_1$ and $p_2 = \max(S_{f_2})$. The sequence $p_0 \succ p_1 \succ p_2 \succ \cdots$ is finite according to Lemma 5, hence after a finite number of steps no element in $\text{supp}(f_N)$ is divisible by m. We have $f_N = f - hg_1$ for some polynomial h, and we define f_N to be the *remainder* of f with respect to g_1. In the case of one variable we have the usual remainder in the division algorithm.

EXAMPLE 7. Let again $f = x_1^3 + x_1^2x_2 + x_2^3$, $g_1 = x_1^2 - x_1x_2$ and let $\prec$ be Lex. Then $f_1 = 3x_1^2x_2 + x_2^3$ by the example above, so $f_2 = f_1 - 3x_2(x_1^2 - x_1x_2) = 3x_1x_2^2 + x_2^3$, which is the remainder of f with respect to g_1 since no monomial in f_2 is divisible with x_1^2. □

We will now formulate the process we have just described in a way which makes it easy to generalize to the case of remainder with respect to an ordered sequence of polynomials $(g_1, \ldots, g_s)$. The remainder can recursively be described as follows. If $\text{lm}(g_1)$ divides $\text{lm}(f)$, let $\text{rem}(f, (g_1)) = \text{rem}(f - ng_1, (g_1))$, where n is a term chosen such that $\text{lt}(f) = \text{lt}(ng_1)$. If $\text{lm}(g_1)$ does not divide $\text{lm}(f)$, let $\text{rem}(f, (g_1)) = \text{lt}(f) + \text{rem}(f - \text{lt}(f), (g_1))$. This gives a recursive definition of $\text{rem}(f)$ with respect to g_1. Now let $(g_1, \ldots, g_s)$ be an ordered sequence of nonzero polynomials and f a polynomial. We recursively define $\text{rem}(f, (g_1, \ldots, g_s)) = \text{rem}(f - ng_k, (g_1, \ldots, g_s))$, where k is the smallest index such that $\text{lm}(g_k)$ divides $\text{lm}(f)$ and n is a term chosen so that $\text{lt}(f) = \text{lt}(ng_k)$. If no $\text{lm}(g_i)$ divides $\text{lm}(f)$ we define $\text{rem}(f, (g_1, \ldots, g_s)) = (\text{lt}(f) + \text{rem}(f - \text{lt}(f), (g_1, \ldots, g_s))$. The process is finite since in both cases the leading monomial drops.

Recursive Formula.

3.4 Definition of Gröbner Bases

For a general sequence $(g_1, \ldots, g_s)$ the remainder is not very well behaved. For instance the result depends on the order of the elements in the sequence. This is seen by the following example.

EXAMPLE 8. Let $f = x_1^2x_2, (g_1, g_2) = (x_1^2, x_1x_2 - x_2^2)$ and let $\prec$ be Lex. Then $\text{rem}(f, (g_1, g_2)) = 0$ but

$$\text{rem}(f, (g_2, g_1)) = \text{rem}(x_1^2x_2 - x_1(x_1x_2 - x_2^2), (g_2, g_1))$$

$$
\begin{aligned}
&= \operatorname{rem}(x_1x_2^2, (g_2, g_1)) \\
&= \operatorname{rem}(x_1x_2^2 - x_2(x_1x_2 - x_2^2), (g_2, g_1)) \\
&= \operatorname{rem}(x_2^3, (g_2, g_1)) = x_2^3.
\end{aligned}
$$

□

We have already noted the following fact, but since it will be used several times we formulate it as a lemma.

LEMMA 9. *Let $f, g_1, \ldots, g_s \in k[x_1, \ldots, x_n]$. Then $f - \operatorname{rem}(f, (g_1, \ldots g_s)) \in \langle g_1, \ldots g_s \rangle$. In particular, if $\operatorname{rem}(f, (g_1, \ldots g_s)) = 0$ then $f \in \langle g_1, \ldots g_s \rangle$.*

Our next aim is now to choose a generating set $g_1', \ldots, g_N'$ for $\mathfrak{a} = \langle g_1, \ldots g_s \rangle$ so that the remainder of a polynomial with respect to $(g_1', \ldots, g_N')$ is unique in the sense that it only depends on the ideal, so we want to achieve that $\operatorname{rem}(f_1, (g_1', \ldots, g_N')) = \operatorname{rem}(f_2, (g_1', \ldots, g_N'))$ if and only if $f_1 - f_2 \in \mathfrak{a}$.

DEFINITION 10. Let $\mathfrak{a}$ be an ideal in $k[x_1, \ldots, x_n]$. A set $\{g_1, \ldots g_s\}$ of elements in $\mathfrak{a}$ such that $\langle \operatorname{lm}(g_1), \ldots, \operatorname{lm}(g_s) \rangle = l(\mathfrak{a})$ is called a *Gröbner basis* for $\mathfrak{a}$.

← ideal of leading monomials of a

First we note that a Gröbner basis for an ideal is a generating set for the ideal.

LEMMA 11. *If $\{g_1, \ldots g_s\}$ is a Gröbner basis for $\mathfrak{a}$, then $\langle g_1, \ldots g_s \rangle = \mathfrak{a}$.*

Proof. Clearly $\langle g_1, \ldots g_s \rangle \subseteq \mathfrak{a}$ since $g_i \in \mathfrak{a}$ for all i. Let $f \in \mathfrak{a}$. Then $\operatorname{lm}(f) \in \langle \operatorname{lm}(g_1), \ldots, \operatorname{lm}(g_s) \rangle$, hence $\operatorname{lm}(f - ng_k) \prec \operatorname{lm}(f)$ for some g_k and some term n. Since $f - ng_k \in \mathfrak{a}$, we get by recursiveness that $f \in \langle g_1, \ldots g_s \rangle$. □

3.5 Hilbert's Basis Theorem and Noetherian Rings

We get as a corollary of Lemma 11 that *every* ideal $\mathfrak{a}$ in a polynomial ring is finitely generated. This is a classical and important result by the German mathematician David Hilbert from 1891.

COROLLARY 12 (HILBERT'S BASIS THEOREM). *Every ideal in a polynomial ring over a field is finitely generated.*

Proof. Being a monomial ideal, we know that $l(\mathfrak{a})$ is finitely generated by Dickson's Lemma. Let $l(\mathfrak{a}) = \langle m_1 \ldots, m_n \rangle$, and choose $f_i \in \mathfrak{a}$ such that $\operatorname{lm}(f_i) = m_i, i = 1, \ldots, N$. Then $\mathfrak{a} = \langle f_1, \ldots, f_N \rangle$ by Lemma 11. □

A ring in which every ideal is finitely generated is called *Noetherian* after the German mathematician Emmy Noether. There are different ways to state the condition of being Noetherian.

Noetherian Ring

PROPOSITION 13. *Let A be a ring. The following conditions are equivalent.*

- *Every ideal in A is finitely generated.*
- *Every strictly ascending sequence of ideals $\mathfrak{a}_1 \subset \mathfrak{a}_2 \subset \cdots$ is finite.*
- *Every ascending sequence of ideals $\mathfrak{a}_1 \subseteq \mathfrak{a}_2 \subseteq \cdots$ is eventually constant, i.e., there is an n such that $\mathfrak{a}_n = \mathfrak{a}_{n+1} = \cdots$.*
- *Every nonempty set of ideals $M = \{\mathfrak{a}_\alpha\}$ has a maximal element, i.e., there is an $\mathfrak{a}_{\alpha_0} \in M$ such that no other ideal in M strictly contains $\mathfrak{a}_{\alpha_0}$.*

REMARK. The second (or third) condition is called the *ascending chain condition* for ideals, the ACC. The last condition is called the *maximum condition* for ideals.

Proof of Proposition 13. Suppose that every ideal in A is finitely generated and that $\mathfrak{a}_1 \subset \mathfrak{a}_2 \subset \cdots$ is a strictly ascending sequence of ideals in A. Then $\mathfrak{a} = \cup \mathfrak{a}_i$ is an ideal. Since $\mathfrak{a}$ is finitely generated the chain must stop at $\mathfrak{a}_n$ as soon as all generators of $\mathfrak{a}$ are contained in $\mathfrak{a}_n$. That the second and third conditions are equivalent is trivial. If some nonempty set M of ideals doesn't have a maximal element, we can construct an infinite strictly ascending chain of ideals in M. Thus the second condition implies the fourth. Assume that the maximum condition holds and that $\mathfrak{a}_1 \subset \mathfrak{a}_2 \subset \cdots$ is a strictly ascending sequence of ideals in A. Let $\mathfrak{a}_n$ be a maximal element in $M = \{\mathfrak{a}_i\}$. Then the sequence stops with $\mathfrak{a}_n$, so the maximum condition implies ACC. If some ideal $\mathfrak{a}$ in A can't be generated by finitely many elements, we could construct an infinite strictly increasing sequence of ideals in A by adding new generators one at a time, so ACC implies that every ideal is finitely generated. □

3.6 Gröbner Bases and Normal Forms

The importance of a Gröbner basis for an ideal comes from the fact that for every polynomial in $k[x_1, \ldots, x_n]$ one can calculate a unique *normal form* with respect to the ideal with the aid of the Gröbner basis.

PROPOSITION 14. *Let $\{g_1, \ldots g_s\}$ be a Gröbner basis for the ideal $\mathfrak{a} = \langle g_1, \ldots g_s \rangle$. Then $\operatorname{rem}(f_1, (g_1, \ldots g_s)) = \operatorname{rem}(f_2, (g_1, \ldots g_s))$ if and only if $f_1 - f_2 \in \mathfrak{a}$. In particular $\operatorname{rem}(f, (g_1, \ldots, g_s)) = 0$ if and only if $f \in \mathfrak{a}$.*

Proof. Suppose $\text{rem}(f_1, G) = \text{rem}(f_2, G)$, where $G = (g_1, \dots g_s)$. Since $f_i - \text{rem}(f_i, G) \in \mathfrak{a}, i = 1, 2$, cf. Lemma 9 in Section 3.4, we get $f_1 - \text{rem}(f_1, G) - (f_2 - \text{rem}(f_2, G)) = f_1 - f_2 \in \mathfrak{a}$. Now suppose $f_1 - f_2 \in \mathfrak{a}$. Then $\text{rem}(f_1, G) - \text{rem}(f_2, G) = (f_2 - \text{rem}(f_2, G)) - (f_1 - \text{rem}(f_1, G)) + (f_1 - f_2) \in \mathfrak{a}$ since $f_i - \text{rem}(f_i, G) \in \mathfrak{a}, i = 1, 2$. But $\text{rem}(f_i, G), i = 1, 2$, is a linear combination of monomials outside $l(\mathfrak{a})$. If $\text{rem}(f_1, G) \neq \text{rem}(f_2, G)$ we would have $\text{lm}(\text{rem}(f_1, G) - \text{rem}(f_2, G)) \notin \mathfrak{a}$. But $\text{rem}(f_1, G) - \text{rem}(f_2, G) \in \mathfrak{a}$ gives $\text{lm}(\text{rem}(f_1, G) - \text{rem}(f_2, G)) \in \mathfrak{a}$, a contradiction. □

We will formulate what we have shown in a somewhat different way. Since $f - \text{rem}(f, G) \in \mathfrak{a}$ we could represent each element $\bar{f}$ in $k[x_1, \dots, x_n]/\mathfrak{a}$ with a polynomial $\text{rem}(f, G)$ which is a linear combination of monomials outside $l(\mathfrak{a})$, so these monomials generate $k[x_1, \dots, x_n]/\mathfrak{a}$ as a vector space over k. The last argument in the proof shows that these monomials are linearly independent modulo $\mathfrak{a}$, hence the monomials in $k[x_1, \dots, x_n] \setminus l(\mathfrak{a})$ constitute a vector space basis for $k[x_1, \dots, x_n]/\mathfrak{a}$. We have derived a way to obtain a unique representation of the elements in $k[x_1, \dots, x_n]/\mathfrak{a}$ (if we have a Gröbner basis). For reference we formulate a proposition.

PROPOSITION 15. *Let $\mathfrak{a}$ be an ideal in $k[x_1, \dots, x_n]$. The monomials outside $l(\mathfrak{a})$ constitute a k-basis of $k[x_1, \dots, x_n]/\mathfrak{a}$.*

It follows from the definition of Gröbner basis that any permutation of a Gröbner basis is a Gröbner basis. Since for any $f \in k[x_1, \dots, x_n]$ we have that $\text{rem}(f, G)$ is a linear combination of monomials outside $l(\mathfrak{a})$, and these monomials are linearly independent modulo $\mathfrak{a}$, we get that $\text{rem}(f, G) = \text{rem}(f, G')$ if G and G' are two Gröbner bases of $\mathfrak{a}$ (for the same ordering). In particular the remainder does not depend on the order of the elements in a Gröbner basis. Thus, given a fixed ordering, $\text{rem}\,(f, G)$ will be uniquely defined and does not depend on the Gröbner basis G of $\mathfrak{a}$. We will call this unique element the *normal form* of f and denote it by $\text{NF}(f)$. For practical reasons it is convenient to have a Gröbner basis as small as possible. We will see that there is a natural choice. We will give conditions on the Gröbner basis which will force the Gröbner basis to be unique.

3.7 Reduced Gröbner Bases

Gröbner bases are not unique. For any Gröbner basis G of $\mathfrak{a}$ and any element $f \in \mathfrak{a}$ we can extend G with f and get a new Gröbner basis for $\mathfrak{a}$. We will now introduce the concept of *reduced Gröbner basis* and prove that reduced Gröbner bases are unique. If $\mathfrak{a} = \langle g_1, \dots, g_s \rangle$ and $G = \{g_1, \dots, g_s\}$ is a Gröbner basis, we define G to be reduced if

- $\{\mathrm{lm}(g_1), \ldots, \mathrm{lm}(g_s)\}$ constitutes a minimal set of generators for $l(\mathfrak{a})$
- g_i are monic, i.e., $\mathrm{lc}(g_i) = 1, i = 1, \ldots, s$
- no $\mathrm{lm}(g_i)$ divides any monomial in $\mathrm{supp}(g_j), i \neq j$

Given a Gröbner basis G it is easy to construct a reduced Gröbner basis. First we pick a subset $G' = \{g_{i_1}, \ldots, g_{i_k}\}$ of G so that the first condition in the definition is fulfilled. Then we multiply each g_{i_t} with $\mathrm{lc}(g_{i_t})^{-1}$ so that we get monic elements. To finish, we take the remainder of each g_{i_t} with respect to $G' \setminus \{g_{i_t}\}$. Thus reduced Gröbner bases exist. They are also unique since if $\{g_1, \ldots, g_s\}$ and $\{h_1, \ldots, h_s\}$ are two reduced Gröbner bases for $\mathfrak{a}$ with $\mathrm{lm}(g_i) = \mathrm{lm}(h_i)$, we get that $g_i - h_i \in \mathfrak{a}$. If $g_i - h_i \neq 0$ we would have $\mathrm{lm}(g_i - h_i) \in l(\mathfrak{a})$, but the leading terms in g_i and h_i are equal, hence cancel in $g_i - h_i$. But $g_i - h_i$ is a linear combination of monomials outside $l(\mathfrak{a})$, which gives a contradiction.

EXAMPLE 16. If G is the reduced Gröbner basis for $\mathfrak{a}$ in any ordering, then $\mathfrak{a} = k[x_1, \ldots, x_n]$ if and only if $G = \{1\}$. □

EXAMPLE 17. Different orderings of course give different Gröbner bases. Let $\mathfrak{a} = \langle x_1^2, x_1x_3 - x_2^3 \rangle$. Then $\{x_1^2, x_2^3 - x_1x_3\}$ is the reduced Gröbner basis for $\mathfrak{a}$ in Degrevlex. In Lex or Deglex, however, the reduced Gröbner basis is $\{x_1^2, x_1x_2^3, x_1x_3 - x_2^3, x_2^6\}$. □

Exercise

1.

$$\{x_1^2 + x_2 + x_1x_2^3 + x_1x_2, x_1x_2^2 + x_1, x_2^3 + x_2, x_1^2x_2^2 - x_2, x_1x_2^3 + x_1x_2\}$$

is a Gröbner basis for the ideal $\mathfrak{a} = \langle x_1^2 + x_2, x_1x_2^2 + x_1 \rangle$ in Lex. Construct the reduced Gröbner basis for $\mathfrak{a}$.

3.8 Construction of Gröbner Bases

A Gröbner basis for an ideal $\mathfrak{a}$ gives us normal forms for elements modulo $\mathfrak{a}$. This will help us to solve lots of problems. We have seen that Gröbner bases do exist, but this does not help us much if we don't know how to calculate them. The remaining parts of this chapter will be dedicated to how we construct a Gröbner basis for an ideal $\mathfrak{a} = \langle f_1, \ldots, f_k \rangle$ given by generators. However, we require some new concepts and some theory, so a little patience may be needed.

3.9 Free Modules and Syzygies

Let A be a ring and let $A^n = \{ (a_1, \ldots, a_n) \mid a_i \in A \}$. We define an addition of elements in A^n by $(a_1, \ldots, a_n) + (b_1, \ldots, b_n) = (a_1 + b_1, \ldots, a_n + b_n)$ and a multiplication of elements in A^n with elements in A by $a(a_1, \ldots, a_n) = (aa_1, \ldots, aa_n)$. With these operations A^n is called a *free module*. A^n has a *basis* $e_1 = (1, 0, \ldots, 0), e_2 = (0, 1, \ldots, 0) \ldots, e_n = (0, 0, \ldots, 1)$ in the sense that every element in A^n is a linear combination, with coefficients in A, of $e_1, \ldots, e_n$ and furthermore that $e_1, \ldots, e_n$ are linearly independent, i.e., if $a_1e_1 + \cdots + a_ne_n = (0, 0, \ldots, 0)$ then $a_i = 0$ for $i = 1, \ldots, n$. All rules for addition and multiplication which are true for vector spaces are also true for free modules. In fact, if A is a field then A^n is nothing but an n-dimensional vector space. On the other hand, if $n = 1$ we can identify A^1 with A. Thus the concept of free module is a common generalization of the concepts of vector space and ring. A nonempty subset M of A^n is called a *submodule* of A^n if M is closed under addition and multiplication with elements in A. Thus $M \subseteq A^n$ is a submodule of A^n if $M \neq \emptyset$ and the following two conditions apply:

- $(a_1, \ldots, a_n), (b_1, \ldots, b_n) \in M$ implies that $(a_1 + b_1, \ldots, a_n + b_n) \in M$
- $(a_1, \ldots, a_n) \in M, a \in A$ implies that $(aa_1, \ldots, aa_n) \in M$

In case A is a field, a submodule is just a subspace of A^n. In case $n = 1$, a submodule is just an ideal in A.

Let $F = (f_1 \ldots, f_n) \in A^n$. An element $s = (s_1 \ldots, s_n)$ in A^n such that $s_1f_1 + \cdots + s_nf_n = 0$ is called a *syzygy* of F. The set of all syzygies of F constitute a submodule of A^n. From now on let $A = k[x_1, \ldots, x_n]$. Let $F = (f_1, \ldots, f_k)$ be a fixed element of A^k. We say that an element $H = (h_1, \ldots, h_k) \in A^k$ *represents* $f \in A$ with respect to F, if $h_1f_1 + \cdots + f_kh_k = f$. An element $S \in A^k$ representing 0 with respect to F is just a syzygy of F. If H and H' both represent f with respect to F, then their difference $H - H'$ is a syzygy of F. If H represents f and S is a syzygy of F, then $H + S$ represents f with respect to F.

syzygy

Exercises

1. Let $F \in A^n$. Prove that the set of syzygies of F constitute a submodule of A^n.
2. Let $A = k[x_1, x_2]$ and let $F = (x_1, x_2)$. Determine all syzygies of F.
3. Let $F = (f_1, \ldots, f_k)$ be a fixed element of A^k. Show that if H and H' both represent f with respect to F, then their difference $H - H'$ is a

syzygy of F. Show that if H represents f with respect to F and S is a syzygy of F, then $H + S$ represents f with respect to F.

3.10 Syzygies of Sequences of Monomials

In general it is not a trivial task to determine the syzygies of a sequence of elements $F = (f_1, \ldots, f_k) \in A^k$. If the f_i's are monomials the task is much easier. We begin with an example. Let $F = (m_1, m_2) = (x_1^2 x_2 x_3, x_2^2 x_3^2)$ and suppose (a, b) is a syzygy of F. We have $am_1 + bm_2 = 0$, so $am_1 = -bm_2 \in \langle m_2 \rangle$, hence $a \in \langle m_2 \rangle : \langle m_1 \rangle = \langle m_2 / \gcd(m_1, m_2) \rangle = \langle x_1^2 x_2 x_3 / x_2 x_3 \rangle = \langle x_1^2 \rangle$, so $a = x_1^2 a_1$. In the same way we see that $b \in \langle m_1 \rangle : \langle m_2 \rangle = \langle m_1 / \gcd(m_1, m_2) \rangle = \langle x_2^2 x_3^2 / x_2 x_3 \rangle = \langle x_2 x_3 \rangle$, so $b = x_2 x_3 b_1$. We get $am_1 + bm_2 = x_1^2 a_1 m_1 + x_2 x_3 b_1 m_2 = x_1^2 x_2^2 x_3^2 a_1 + x_1^2 x_2^2 x_3^2 b_1 = x_1^2 x_2^2 x_3^2 (a_1 + b_1) = 0$, so $a_1 + b_1 = 0$. Hence every syzygy is a multiple of

$$(m_2 / \gcd(m_1, m_2), -m_1 / \gcd(m_1, m_2)) = (x_1^2, -x_2 x_3).$$

If $F = (m_1, m_2, m_3)$ is a sequence of three monomials it is easy to find three syzygies, namely

$$\begin{aligned}
s_{12} &= (m_2/\gcd(m_1,m_2), -m_1/\gcd(m_1,m_2), 0) \\
&= (m_2/\gcd(m_1,m_2))e_1 - (m_1/\gcd(m_1,m_2))e_2, \\
s_{13} &= (m_3/\gcd(m_1,m_3), 0, -m_1/\gcd(m_1,m_3)) \\
&= (m_3/\gcd(m_1,m_3))e_1 - (m_1/\gcd(m_1,m_3))e_3, \text{ and} \\
s_{23} &= (0, m_3/\gcd(m_2,m_3), -m_2/\gcd(m_2,m_3)) \\
&= (m_3/\gcd(m_2,m_3))e_2 - (m_2/\gcd(m_2,m_3))e_3.
\end{aligned}$$

If for example $F = (x_1^2 x_2 x_3, x_2^2 x_3, x_1^2 x_3^2)$ then $(x_2, -x_1^2, 0), (x_3, 0, -x_2)$ and $(0, x_1^2 x_3, -x_2^2)$ are syzygies. The next proposition shows that for sequences with three or more monomials it suffices to calculate these pairwise syzygies, since every syzygy is a linear combination of them. Note that $n/\gcd(m, n) = \operatorname{lcm}(m, n)/m$.

PROPOSITION 18. *Let $m_1, \ldots, m_k$ be monomials. If $S = (s_1, \ldots, s_k)$ is a syzygy of $(m_1, \ldots, m_k)$ then S is a linear combination of the syzygies*

$$s_{ij} = \frac{\operatorname{lcm}(m_i, m_j)}{m_i} e_i - \frac{\operatorname{lcm}(m_i, m_j)}{m_j} e_j,$$

where $1 \le i < j \le k$.

Proof. If $s_1m_1 + \cdots + s_km_k = 0$ then each mdeg is 0. Hence we can assume that the s_i's are terms $t_i = c_in_i, c_i \in k, n_i$ a monomial, and that $\mathrm{mdeg}(n_im_i)$ is the same, say equal to $\mathrm{mdeg}(m)$, for all i. Hence we have $c_1 + \cdots + c_k = 0$. The solutions to $c_1 + \cdots + c_k = 0$ are linear combinations of $e_i - e_j$, $1 \leq i < j \leq k$. This means that $(t_1, \ldots, t_k)$ are linear combinations of $(m/m_i)e_i - (m/m_j)e_j$, which is a multiple of s_{ij}. □

Exercise

1. Determine all syzygies of (x_1x_2, x_1x_3, x_2x_3).

3.11 S-polynomials

The following easy construction will be a main ingredient in the determination of a Gröbner basis from a set of generators of an ideal.

DEFINITION 19. Let (f_1, f_2) be a fixed pair of monic elements from A with an ordering $\prec$ given. The *S-polynomial* of (f_1, f_2) in $\prec$ is

$$S_{1,2} = S(f_1, f_2) = \frac{\mathrm{lcm}(\mathrm{lm}(f_1), \mathrm{lm}(f_2))}{\mathrm{lm}(f_1)} f_1 - \frac{\mathrm{lcm}(\mathrm{lm}(f_1), \mathrm{lm}(f_2))}{\mathrm{lm}(f_2)} f_2.$$

Note that s_{12} in Proposition 18 represents $S_{1,2}$ with respect to (f_1, f_2).

EXAMPLE 20. If $(f_1, f_2) = (x_1^2, x_1x_2 - x_2^2)$ and the ordering is Lex, then $S(f_1, f_2) = x_2 \cdot x_1^2 - x_1 \cdot (x_1x_2 - x_2^2) = x_1x_2^2$. □

3.12 A Criterion for Gröbner Basis

We are now going to give a test criterion for the question: Is $\{f_1, \ldots, f_k\}$ a Gröbner basis for the ideal generated by the f_i's? Once this criterion is given, it is easy to construct an algorithm for determining a Gröbner basis for an ideal given by generators. It is clear that $\{f_1, \ldots, f_k\}$ is a Gröbner basis if and only if $\{\mathrm{lc}(f_1)^{-1}f_1, \ldots, \mathrm{lc}(f_k)^{-1}f_k\}$ is a Gröbner basis, so we may assume that the f_i's are monic. The main point with a Gröbner basis for an ideal is that for every element $f \in A$, we can determine a unique normal form of f with respect to the ideal. In particular the reduction process with respect to the sequence $(f_1, \ldots, f_k)$ should not depend on the order of the elements in the sequence. If we have an element f which can be reduced both by means of f_1 and f_2, i.e., if both $\mathrm{lm}(f_1)$ and $\mathrm{lm}(f_2)$ divide $\mathrm{lm}(f)$, then we have a potential ambiguity, when f reduces to $f - (\mathrm{lt}(f)/\mathrm{lm}(f_1))f_1$ and to $f - (\mathrm{lt}(f)/\mathrm{lm}(f_2))f_2$, respectively, in the first step of the reduction process. The difference of these reductions is $(\mathrm{lt}(f)/\mathrm{lm}(f_2))f_2 - (\mathrm{lt}(f)/\mathrm{lm}(f_1))f_1$, which is a multiple of the S-polynomial $S(f_1, f_2)$. If $S(f_1, f_2)$ reduces to 0 this ambiguity disappears, since there is a common second reduction.

THEOREM 21. *Let $(f_1, \ldots, f_k)$ be a sequence of monic elements in A and let $S_{i,j} = S(f_i, f_j)$ be the S-polynomials. Then $\{f_1, \ldots, f_k\}$ is a Gröbner basis of $\langle f_1, \ldots, f_k\rangle$ if and only if $\mathrm{rem}(S_{i,j}, (f_1, \ldots, f_k)) = 0$ for all i, j.*

Proof. First assume that $(f_1, \ldots, f_k)$ is a Gröbner basis. Since $S_{i,j} \in \langle f_1, \ldots, f_k\rangle$, they all reduce to 0; see Proposition 14. Now assume that $\mathrm{rem}(S_{i,j}, (f_1, \ldots, f_k)) = 0$ for all i, j and let $f \in \langle f_1, \ldots, f_k\rangle$. We have to show that $\mathrm{lm}(f) \in \langle \mathrm{lm}(f_1), \ldots, \mathrm{lm}(f_k)\rangle$. Since $f \in \langle f_1, \ldots, f_k\rangle$, we can write $f = q_1 f_1 + \cdots + q_k f_k$ for some $q_i \in A$. It is clear that $\max\{\mathrm{lm}(q_i f_i)\} \succ \mathrm{lm}(f)$ or $\max\{\mathrm{lm}(q_i f_i)\} = \mathrm{lm}(f)$. If $\max\{\mathrm{lm}(q_i f_i)\} = \mathrm{lm}(f)$ we are ready since then $\mathrm{lm}(f) = \mathrm{lm}(q_i)\,\mathrm{lm}(f_i)$ for some i, so $\mathrm{lm}(f) \in \langle \mathrm{lm}(f_1), \ldots, \mathrm{lm}(f_k)\rangle$. Now assume that $m = \max\{\mathrm{lm}(q_i f_i)\} \succ \mathrm{lm}(f)$ and then we denote the set $\{\, i \mid \mathrm{lm}(q_i f_i) = \max\{\mathrm{lm}(q_i f_i)\}\,\} = \{i_1, \ldots, i_t\} = I$. The idea is now to add a syzygy Z to $Q = (q_1, \ldots, q_k)$ to get another representation $Q' = Q + Z$ for f with $\max\{\mathrm{lm}(q_i' f_i)\} \prec \max\{\mathrm{lm}(q_i f_i)\}$. Since there is no infinite decreasing sequence of monomials, see Lemma 5, we will after a finite number of steps come to the first case $\max\{\mathrm{lm}(q_i f_i)\} = \mathrm{lm}(f)$, and we can conclude the result. If $\max\{\mathrm{lm}(q_i f_i)\} \succ \mathrm{lm}(f)$ we must have that the leading terms in $\sum_{i\in I} q_i f_i$ cancel. But then $S_Q = (s_1, \ldots, s_k)$, where $s_i = \mathrm{lt}(q_i)$ if $i \in I$ and $s_i = 0$ otherwise, is a syzygy for $(\mathrm{lm}(f_1), \ldots, \mathrm{lm}(f_k))$. Hence S_Q is a linear combination of s_{ij}'s according to Proposition 18, say $S_Q = \sum_{1\le i<j\le k} t_{ij} s_{ij}$, where t_{ij} are terms. We see that s_{ij} represents $S_{i,j}$ with respect to $F = (f_1, \ldots, f_k)$, so we get $s_1 f_1 + \cdots + s_k f_k = \sum t_{ij} S_{i,j}$. We have assumed that $S_{i,j}$ reduces to 0 with respect to $(f_1, \ldots, f_k)$, so we get that $S_{i,j} = \sum_{t=1}^{k} h_t^{i,j} f_t$ with $\mathrm{lm}(h_t^{i,j} f_t) = \mathrm{lm}(S_{i,j})$ for one value of t and $\mathrm{lm}(h_t^{i,j} f_t) \prec \mathrm{lm}(S_{i,j})$ for all other values of t. Since both s_{ij} and $H^{i,j} = (h_1^{i,j}, \ldots, h_k^{i,j})$ represent $S_{i,j}$ with respect to F, we get that $s_{ij} - H^{i,j}$ is a syzygy of F. Hence $Z = \sum_{1\le i<j\le k} t_{ij}(s_{ij} - H^{i,j})$ is a syzygy of F, so $Q - Z = Q' = (q_1', \ldots, q_k')$ represents f with respect to F. Since the leading terms in $q_1 f_1 + \cdots + q_k f_k$ cancel, we have $\mathrm{lm}(q_1 f_1 + \cdots + q_k f_k) \prec m$. Furthermore, by construction we have $\max\{\mathrm{lm}(q_i f_i)\} \succ \max\{\mathrm{lm}(q_i' f_i)\}$. □

Exercises

1. Prove that $\{x_1^2, x_1x_2 - x_2^2, x_2^3\}$ is a Gröbner basis in Lex for the ideal $\langle x_1^2, x_1x_2 - x_2^2\rangle$.

2. Prove that $\{x_1 + x_2 + x_3, x_2^2 + x_2x_3 + x_3^2, x_3^3 - 1\}$ is a Gröbner basis in Lex for $\langle x_1 + x_2 + x_3, x_1x_2 + x_1x_3 + x_2x_3, x_1x_2x_3 - 1\rangle$.

3. Show that, if $\gcd(\mathrm{lm}(f), \mathrm{lm}(g)) = 1$, then $\mathrm{rem}(S(f,g),(f,g)) = 0$.

3.13 The Buchberger Algorithm

The construction of an algorithm to get a Gröbner basis for an ideal is now easy. The following is an informal description of the so-called Buchberger algorithm.

Bruno Buchberger is an Austrian mathematician, who published his algorithm in the 1960s. It was Buchberger who invented the name Gröbner bases, honouring his thesis adviser, Wolfgang Gröbner.

Start with an ideal $\mathfrak{a} = \langle f_1, \ldots, f_k\rangle$. Calculate all S-polynomials and reduce them with respect to $(f_1, \ldots, f_k)$. If $\mathrm{rem}(S(f_i, f_j), (f_1, \ldots, f_k)) \neq 0$, extend $(f_1, \ldots, f_k)$ with $\mathrm{rem}(S(f_i, f_j), (f_1, \ldots, f_k))$. Calculate all S-polynomials which haven't already been calculated, and reduce them a.s.o. If $\mathrm{rem}(S(f_i, f_j), (f_1, \ldots, f_k)) \neq 0$ we have $\mathrm{lm}(\mathrm{rem}(S(f_i, f_j), (f_1, \ldots, f_k)) \notin \langle \mathrm{lm}(f_1), \ldots, \mathrm{lm}(f_k)\rangle$, so each time we extend the generating set, we also extend the monomial ideal generated by the leading monomials of the generators. Thus termination is guaranteed by Dickson's Lemma. Correctness, i.e., that the final result is a Gröbner basis, follows from Theorem 21.

Note that we have a great deal of freedom when turning this informal description into an actual algorithm. For example, the order in which we calculate the S-polynomials and make the reductions is not essential for the final result. It is, however, very important if we want to make an efficient algorithm. We will return to these questions later.

Further Reading

B. Buchberger, *Ein Algorithmus zum Auffinden der Basiselemente des Restklassenringes nach einem nulldimensionalen Polynomideal*, Univ. of Innsbruck Math. Inst., PhD thesis 1965.

Exercises

1. Determine the reduced Gröbner basis, with respect to Lex and Deglex with $x_1 \succ x_2 \succ x_3$, for the ideal $\langle x_1^2 - x_1x_3^2, x_2^2 - x_2x_3^2\rangle$.

2. Determine the reduced Gröbner basis, with respect to Deglex and Degrevlex with $x_1 \succ x_2 \succ x_3$, for the ideal $\langle x_2^2 + x_1x_3, x_1^2\rangle$.

Chapter 4

Algebraic Sets

We will in this chapter study the connection between solutions of systems of polynomial equations $f_1 = \cdots = f_m = 0, f_i \in k[x_1, \ldots, x_n]$, and ideals in $k[x_1, \ldots, x_n]$. Before making formal definitions and statements, we digress with an informal discussion. Suppose we are given a system of polynomial equations, $f_1 = \cdots = f_m = 0, f_i \in k[x_1, \ldots, x_n]$, where k is a field. Let X be the set of solutions to this system, i.e., $X = \{ (c_1, \ldots, c_n) \mid c_i \in k, f_1(c_1, \ldots, c_n) = \cdots = f_m(c_1, \ldots, c_n) = 0 \}$. Then every linear combination $g_1 f_1 + \cdots + g_m f_m$ of the f_i's is also zero on the points in X. Hence X is the zero set to the whole ideal $\mathfrak{a} = \langle f_1, \ldots, f_m \rangle$. If $h_1, \ldots, h_s$ is another generating set for $\mathfrak{a}$, so $\langle f_1, \ldots, f_m \rangle = \langle h_1, \ldots, h_s \rangle$, then the two systems $f_1 = \cdots = f_m = 0$ and $h_1 = \cdots = h_s = 0$ are equivalent since each h_i is a linear combination of the f_i's, and vice versa. If we enlarge the set X to $Y \supset X$, there are fewer polynomials being zero on the set Y than on X, and if one adds equations to a system, the zero set gets smaller, respectively. If one has two systems of polynomial equations then the set of points solving both systems are exactly those which solve the system consisting of all equations from the two systems. This is the meaning of $\mathfrak{v}(\mathfrak{a} + \mathfrak{b}) = \mathfrak{v}(\mathfrak{a}) \cap \mathfrak{v}(\mathfrak{b})$ below. We encourage the reader to make similar interpretations of the remaining statements in this section. Now we are ready for the formal part. Let X be a subset of k^n, so X consists of vectors $(c_1, \ldots, c_n), c_i \in k$.

4.1 Algebraic Sets and Ideals

An *algebraic set* is a subset of k^n which consists of common zeros of a set of polynomials. We will in this section show how algebraic sets correspond to ideals and how the ideal operations correspond to natural operations on their corresponding algebraic sets.

DEFINITION 1. Let $X \subseteq k^n$. Then

$$\mathfrak{i}(X) = \{\, f \in k[x_1, \ldots, x_n] \mid f(c_1, \ldots, c_n) = 0 \text{ if } (c_1, \ldots, c_n) \in X \,\}$$

is called *the ideal of* X.

The name is justified by the following proposition.

PROPOSITION 2. *For any subset* $X \subseteq k^n$, $\mathfrak{i}(X)$ *is an ideal.*

Proof. Let $f_1, f_2 \in \mathfrak{i}(X)$ and let $g \in k[x_1, \ldots, x_n]$. Then

$$\begin{aligned}
(f_1 + f_2)(c_1, \ldots, c_n) &= f_1(c_1, \ldots, c_n) + f_2(c_1, \ldots, c_n) \\
&= 0 + 0 = 0, \text{ and} \\
(gf_1)(c_1, \ldots, c_n) &= g(c_1, \ldots, c_n) f_1(c_1, \ldots, c_n) \\
&= g(c_1, \ldots, c_n) \cdot 0 = 0
\end{aligned}$$

if $(c_1, \ldots, c_n) \in X$. □

EXAMPLE 3. Let X consist of one point, $X = \{(c_1, \ldots, c_n)\}$. Then we claim that $\mathfrak{i}(X) = \langle x_1 - c_1, \ldots, x_n - c_n \rangle$. Any polynomial $f \in k[x_1, \ldots, x_n]$ is also a polynomial in $x_1 - c_1, \ldots, x_n - c_n$, $f(x_1, \ldots, x_n) = f_1(x_1 - c_1, \ldots, x_n - c_n)$. It follows that $f \in \mathfrak{i}(X)$ if and only if the constant term in f_1 equals zero, i.e., if and only if $f \in \langle x_1 - c_1, \ldots, x_n - c_n \rangle$. □

Let S be a subset of $k[x_1, \ldots, x_n]$.

DEFINITION 4. $\mathfrak{v}(S) = \{\, (c_1, \ldots, c_n) \mid f(c_1, \ldots, c_n) = 0 \text{ if } f \in S \,\}$.

DEFINITION 5 (ALGEBRAIC SET/VARIETY). A subset X of k^n is an *(affine) algebraic set* or a *variety* if $X = \mathfrak{v}(S)$ for some subset $S \subseteq k[x_1, \ldots, x_n]$. When k^n is considered as an algebraic set, it is often denoted $A^n(k)$.

PROPOSITION 6. *(a) If* $X_1 \subseteq X_2 \subseteq k^n$ *then* $\mathfrak{i}(X_2) \subseteq \mathfrak{i}(X_1)$. *If* $S \subseteq T$ *are subsets of* $k[x_1, \ldots, x_n]$ *then* $\mathfrak{v}(T) \subseteq \mathfrak{v}(S)$.

(b) If X *is an algebraic set then there is an ideal* $\mathfrak{a} \subseteq k[x_1, \ldots, x_n]$ *such that* $X = \{\, (c_1, \ldots, c_n) \mid f(c_1, \ldots, c_n) = 0 \text{ if } f \in \mathfrak{a} \,\}$.

(c) For any subset $X \subseteq k^n$, *we have* $X \subseteq \mathfrak{v}(\mathfrak{i}(X))$, *with equality if and only if* X *is an algebraic set.*

(d) $\mathfrak{v}(\langle 0\rangle) = k^n$ *and* $\mathfrak{v}(k[x_1,\ldots,x_n]) = \emptyset$.

(e) If $\mathfrak{a}$ *is an ideal in* $k[x_1,\ldots,x_n]$ *we have* $\mathfrak{v}(\mathfrak{a}) = \mathfrak{v}(\sqrt{\mathfrak{a}})$.

(f) If $\mathfrak{a}$ *and* $\mathfrak{b}$ *are ideals in* $k[x_1,\ldots,x_n]$ *then*

$$\begin{aligned}\mathfrak{v}(\mathfrak{a}\cap\mathfrak{b}) &= \mathfrak{v}(\mathfrak{a})\cup\mathfrak{v}(\mathfrak{b}),\\ \mathfrak{v}(\mathfrak{a}\mathfrak{b}) &= \mathfrak{v}(\mathfrak{a})\cup\mathfrak{v}(\mathfrak{b}),\\ \mathfrak{v}(\mathfrak{a}+\mathfrak{b}) &= \mathfrak{v}(\mathfrak{a})\cap\mathfrak{v}(\mathfrak{b}).\end{aligned}$$

(g) If $\mathfrak{a}$ *is an ideal in* $k[x_1,\ldots,x_n]$ *then* $\mathfrak{a}\subseteq\mathfrak{i}(\mathfrak{v}(\mathfrak{a}))$ *and the inclusion may be strict.*

Proof.

(a) is trivial.

(b) If X is an algebraic set then by definition there is a subset $S \subseteq k[x_1,\ldots,x_n]$ such that $X = \{(c_1,\ldots,c_n) \mid f(c_1,\ldots,c_n) = 0 \text{ if } f\in S\}$. Let $\mathfrak{a} = \langle S\rangle$, the ideal generated by S. Then

$$\mathfrak{a} = \{g_1f_1 + \cdots + g_mf_m \mid g_i \in k[x_1,\ldots,x_n], f_i\in S\}.$$

If $g_1f_1+\cdots+g_mf_m \in \mathfrak{a}$ then $(g_1f_1+\cdots+g_mf_m)(c_1,\ldots,c_n) = g_1(c_1,\ldots,c_n)f_1(c_1,\ldots,c_n)+\cdots+g_m(c_1,\ldots,c_n)f_m(c_1,\ldots,c_n) = 0+\cdots+0 = 0$ if $(c_1,\ldots,c_n)\in X$. Hence $X\subseteq\mathfrak{v}(\mathfrak{a})$ and since $S\subseteq\mathfrak{a}$ gives $\mathfrak{v}(\mathfrak{a})\subseteq\mathfrak{v}(S)$, we have $\mathfrak{v}(\mathfrak{a}) = X$.

(c) The variety $\mathfrak{v}(\mathfrak{i}(X))$ consists of those points $(c_1,\ldots,c_n)\in k^n$ such that $f(c_1,\ldots,c_n) = 0$ for all f such that $f(c_1,\ldots,c_n) = 0$ for all $(c_1,\ldots,c_n)\in X$ so certainly $X\subseteq\mathfrak{v}(\mathfrak{i}(X))$. If $X = \mathfrak{v}(\mathfrak{i}(X))$ then by definition X is an algebraic set since it is of the form $\mathfrak{v}(S)$ for $S = \mathfrak{i}(X)$. Conversely, if X is an algebraic set then by (b) there is an ideal $\mathfrak{b}$ so that $X = \mathfrak{v}(\mathfrak{b})$. Since $\mathfrak{i}(X)$ consists of those polynomials which are zero on all points of X, we have $\mathfrak{b}\subseteq\mathfrak{i}(X)$ and we get $X\subseteq\mathfrak{v}(\mathfrak{i}(X))\subseteq\mathfrak{v}(\mathfrak{b}) = X$, hence $X = \mathfrak{v}(\mathfrak{i}(X))$.

(d) is trivial.

(e) Since $\mathfrak{a}\subseteq\sqrt{\mathfrak{a}}$ we get $\mathfrak{v}(\sqrt{\mathfrak{a}})\subseteq\mathfrak{v}(\mathfrak{a})$. Let $(c_1,\ldots,c_n)\in\mathfrak{v}(\mathfrak{a})$. If $f\in\sqrt{\mathfrak{a}}$ then $f^N\in\mathfrak{a}$ for some N, so $f^N(c_1,\ldots,c_n) = 0$ which gives $f(c_1,\ldots,c_n) = 0$ and $(c_1,\ldots,c_n)\in\mathfrak{v}(\sqrt{\mathfrak{a}})$.

(f) Since $\mathfrak{a}\mathfrak{b}\subseteq\mathfrak{a}\cap\mathfrak{b}\subseteq\mathfrak{a}$ and $\mathfrak{a}\mathfrak{b}\subseteq\mathfrak{a}\cap\mathfrak{b}\subseteq\mathfrak{b}$ we get $\mathfrak{v}(\mathfrak{a})\subseteq\mathfrak{v}(\mathfrak{a}\cap\mathfrak{b})\subseteq\mathfrak{v}(\mathfrak{a}\mathfrak{b})$ and $\mathfrak{v}(\mathfrak{b})\subseteq\mathfrak{v}(\mathfrak{a}\cap\mathfrak{b})\subseteq\mathfrak{v}(\mathfrak{a}\mathfrak{b})$, hence $\mathfrak{v}(\mathfrak{a})\cup\mathfrak{v}(\mathfrak{b})\subseteq\mathfrak{v}(\mathfrak{a}\cap\mathfrak{b})\subseteq\mathfrak{v}(\mathfrak{a}\mathfrak{b})$. Suppose $(c_1,\ldots,c_n)\notin\mathfrak{v}(\mathfrak{a})\cup\mathfrak{v}(\mathfrak{b})$. Then there exist $f_1\in\mathfrak{a}$ and $f_2\in\mathfrak{b}$ such that $f_1(c_1,\ldots,c_n)\neq 0$ and $f_2(c_1,\ldots,c_n)\neq 0$. We have $f_1f_2\in\mathfrak{a}\mathfrak{b}$ but $(f_1f_2)(c_1,\ldots,c_n)\neq 0$, and therefore $(c_1,\ldots,c_n)\neq\mathfrak{v}(\mathfrak{a}\mathfrak{b})$. Thus $\mathfrak{v}(\mathfrak{a}\mathfrak{b})\subseteq\mathfrak{v}(\mathfrak{a})\cup\mathfrak{v}(\mathfrak{b})$ and the first and second equality follows. Since

$\mathfrak{a} \subseteq \mathfrak{a} + \mathfrak{b}$ and $\mathfrak{b} \subseteq \mathfrak{a} + \mathfrak{b}$ we get $\mathfrak{v}(\mathfrak{a} + \mathfrak{b}) \subseteq \mathfrak{v}(\mathfrak{a}) \cap \mathfrak{v}(\mathfrak{b})$. Suppose $(c_1, \ldots, c_n) \in \mathfrak{v}(\mathfrak{a}) \cap \mathfrak{v}(\mathfrak{b})$. Then $f_1(c_1, \ldots, c_n) = 0$ if $f_1 \in \mathfrak{a}$ and $f_2(c_1, \ldots, c_n) = 0$ if $f_2 \in \mathfrak{b}$. If $f \in \mathfrak{a} + \mathfrak{b}$ then $f = f_1 + f_2$ with $f_1 \in \mathfrak{a}$ and $f_2 \in \mathfrak{b}$. Hence $f(c_1, \ldots, c_n) = 0$, so $\mathfrak{v}(\mathfrak{a}) \cap \mathfrak{v}(\mathfrak{b}) \subseteq \mathfrak{v}(\mathfrak{a} + \mathfrak{b})$, and we have proved the third equality.

(g) That $\mathfrak{a} \subseteq \mathfrak{i}(\mathfrak{v}(\mathfrak{a}))$ is immediate from the definitions. Consider the ideal $\mathfrak{a} = \langle x^2 + 1 \rangle$ in $\mathbb{R}[x]$, the polynomial ring in one variable over the real numbers. Then $\mathfrak{v}(\mathfrak{a}) = \emptyset$ since the equation $x^2 + 1 = 0$ has no real solution and we get $\mathfrak{i}(\mathfrak{v}(\mathfrak{a})) = \mathfrak{i}(\emptyset) = \mathbb{R}[x]$ so the inclusion $\mathfrak{a} \subset \mathfrak{i}(\mathfrak{v}(\mathfrak{a}))$ is strict.

□

REMARK. The reason for the strict inclusion in the example in (g) is that there are real polynomials with no real zero. This is no longer true if we consider polynomials with complex coefficients. But there is also another reason for the inclusion to be strict. Consider the ideal $\mathfrak{a} = \langle x^2 \rangle$ in $k[x]$, where k is *any* field. Then $\mathfrak{v}(\mathfrak{a}) = \{0\}$ so $\mathfrak{i}(\mathfrak{v}(\mathfrak{a})) = \langle x \rangle$. The reason for inequality in this case is that the two equations $x^2 = 0$ and $x = 0$ have the same solution. Note that $\sqrt{\langle x^2 \rangle} = \langle x \rangle$.

We now discuss the geometrical meaning of prime ideals in polynomial rings.

DEFINITION 7 (IRREDUCIBLE). An algebraic set $X \subseteq k^n$, k a field, is said to be *irreducible* if X is not the union of two strictly smaller algebraic sets. An algebraic set which is not irreducible is called reducible.

PROPOSITION 8. *Let X be an algebraic set in k^n. Then X is irreducible if and only if $\mathfrak{i}(X)$ is a prime ideal.*

Proof. Suppose that X is reducible, $X = X_1 \cup X_2$ with X_i algebraic sets, $X_i \subseteq X, X_i \neq X, i = 1, 2$. Then $\mathfrak{i}(X) \subseteq \mathfrak{i}(X_1)$, and we claim that the inclusion is strict. We have, according to (c) in the proposition above, that $X_1 = \mathfrak{v}(\mathfrak{i}(X_1))$ and $X = \mathfrak{v}(\mathfrak{i}(X))$, and since $X_1 \neq X$ we must have $\mathfrak{i}(X_1) \neq \mathfrak{i}(X)$. In the same way $\mathfrak{i}(X) \subseteq \mathfrak{i}(X_2), \mathfrak{i}(X) \neq \mathfrak{i}(X_2)$. Take an $f \in \mathfrak{i}(X_1) \backslash \mathfrak{i}(X)$ and a $g \in \mathfrak{i}(X_2) \setminus \mathfrak{i}(X)$. Then $fg \in \mathfrak{i}(X)$ since if $(c_1, \ldots, c_n) \in X_1 \cup X_2$, then either $(c_1, \ldots, c_n) \in X_1$ and then $f(c_1, \ldots, c_n) = 0$, or $(c_1, \ldots, c_n) \in X_2$ and then $g(c_1, \ldots, c_n) = 0$. In any case we have $(fg)(c_1, \ldots, c_n) = f(c_1, \ldots, c_n) g(c_1, \ldots, c_n) = 0$. Thus $\mathfrak{i}(X)$ is not a prime ideal. Now suppose that $\mathfrak{i}(X)$ is not a prime ideal. Choose $f, g \notin \mathfrak{i}(X)$ such that $fg \in \mathfrak{i}(X)$. Let $X_1 = \mathfrak{v}(\mathfrak{i}(X) + \langle f \rangle)$ and $X_2 = \mathfrak{v}(\mathfrak{i}(X) + \langle g \rangle)$. Then it follows that $X_1 \subseteq X$ since $\mathfrak{i}(X) \subseteq \mathfrak{i}(X) + \langle f \rangle$, and $X_1 \neq X$ since $f \notin \mathfrak{i}(X)$, so $f(c_1, \ldots, c_n) \neq 0$ for some $(c_1, \ldots, c_n) \in X$. Similarly $X_2 \subseteq X$ and $X_2 \neq X$. If $(c_1, \ldots, c_n) \in X$

then $(fg)(c_1,\ldots,c_n) = 0$, so either $f(c_1,\ldots,c_n) = 0$ or $g(c_1,\ldots,c_n) = 0$, hence $(c_1,\ldots,c_n) \in X_1 \cup X_2$ and we have the equality $X_1 \cup X_2 = X$. □

We have seen that in the correspondence between algebraic sets and ideals, the inclusions reverse. Smaller algebraic sets corresponds to larger ideals and vice versa. Since ideals satisfy the ACC it will perhaps not come as a surprise that algebraic sets satisfy the *descending chain condition* (DCC), i.e., every strictly descending sequence of algebraic sets is finite.

THEOREM 9. *Let $X_1 \supset X_2 \supset \cdots$ be a strictly descending sequence of algebraic subsets of k^n. Then the sequence is finite. Another way to state this is to say that any nonempty set $\{X_\alpha\}$ of algebraic subsets of k^n has a minimal element, i.e., an element X_{α_0} not containing any other X_α.*

Proof. We get that $\mathfrak{i}(X_1) \subseteq \mathfrak{i}(X_2) \cdots \subseteq$ is an ascending sequence of ideals. If $\mathfrak{i}(X_k) = \mathfrak{i}(X_{k+1})$ we would have $X_k = \mathfrak{v}(\mathfrak{i}(X_k)) = \mathfrak{v}(\mathfrak{i}(X_{k+1})) = X_{k+1}$, so the sequence is strictly ascending and hence finite. The proof of the equivalence between DCC and the minimal condition is completely analogous to the proof of the equivalence between ACC and the maximal condition in Proposition 13 of Section 3.5. □

THEOREM 10. *Every algebraic set in k^n is a finite union of irreducible algebraic sets, $X = X_1 \cup \cdots \cup X_m$, X_i irreducible. If we make this union irredundant, i.e., if no X_i is contained in any $X_j, j \neq i$, then the irreducible components are unique.*

Proof. Let $\{X_\alpha\}$ be the algebraic subsets which are not a finite union of irreducibles. Take a minimal element X_{α_0} in this set. Then X_{α_0} is not irreducible. On the other hand it would contradict the minimality if X_{α_0} were reducible. The only possibility is that $\{X_\alpha\}$ is empty. It remains to show the unicity, so assume that $X = X_1 \cup \cdots \cup X_m = Y_1 \cup \cdots \cup Y_k$ are two irredundant decompositions with X_i and Y_j irreducible. Then $X_1 = X_1 \cap X = X_1 \cap (Y_1 \cup \cdots \cup Y_k) = (X_1 \cap Y_1) \cup \cdots \cup (X_1 \cap Y_k)$. Since X_1 is irreducible we get $X_1 = X_1 \cap Y_j$ for some j, so $X_1 \subseteq Y_j$. In the same way we see that $Y_j \subseteq X_i$ for some i. Since the decomposition is irredundant, we must have $i = 1$ and $X_1 = Y_j$. Continuing like this, we see that $m = k$ and that one decomposition is only a renumbering of the other. □

Exercises

1. Let X be the x_1-axis in $\mathbb{R}^2$, $X = \{\,(t,0) \mid t \in \mathbb{R}\}$. Determine $\mathfrak{i}(X)$ in $\mathbb{R}[x_1, x_2]$.

2. Let X be the parabola $X = \{ (t, t^2) \mid t \in \mathbb{R} \}$. Determine $\mathfrak{i}(X)$ in $\mathbb{R}[x_1, x_2]$.

3. Show that the coordinate axis $\{ (t, 0) \mid t \in k \}$ is an irreducible algebraic set in k^2, and that the union of the coordinate axis $\{ (t, 0) \mid t \in k \} \cup \{ (0, t) \mid t \in k \}$ is a reducible algebraic set in k^2.

4. Show that the parabola $\{ (t, t^2) \mid t \in k \}$ is an irreducible algebraic set in k^2.

5. Show that any finite set in k^n is algebraic. When is it irreducible?

6. Show that if $\{X_\alpha\}$ is a nonempty set of algebraic subsets of k^n, then $\{X_\alpha\}$ has a minimal element, i.e., an element X_{α_0} such that no $X_\alpha, \alpha \neq \alpha_0$ is contained in X_{α_0}.

4.2 Hilbert's Nullstellensatz

We have seen that there is no perfect correspondence between algebraic sets in k^n and ideals in $k[x_1, \ldots, x_n]$. The reason for this is twofold. One reason is that several ideals correspond to the same algebraic set, e.g., $\mathfrak{v}(\mathfrak{a}) = \mathfrak{v}(\sqrt{\mathfrak{a}})$. A second reason is that there are ideals in say $\mathbb{R}[x_1, \ldots, x_n]$ with no real zero. We will in this section show that if we restrict ourselves to radical ideals in $\mathbb{C}[x_1, \ldots, x_n]$ there *is* a perfect correspondence. This proof of the following classical theorem follows an idea by Lars Svensson.

THEOREM 11 (HILBERT'S NULLSTELLENSATZ, WEAK FORM). *Let $\mathfrak{a}$ be an ideal in $\mathbb{C}[x_1, \ldots, x_n]$, $\mathfrak{a} \neq \mathbb{C}[x_1, \ldots, x_n]$. Then $\mathfrak{v}(\mathfrak{a}) \neq \emptyset$.*

Proof. For $n = 1$ the statement is nothing but the Fundamental Theorem of Algebra, Theorem 35 in Section 1.3.3, so we may proceed with induction on n.

Case 1. Assume that $\mathfrak{a} \cap \mathbb{C}[x_i] = \langle 0 \rangle$ for some i. We can suppose that $i = 1$. Let $\mathfrak{a} = \langle f_1, \ldots, f_k \rangle$ and let $\mathfrak{b}$ be the ideal generated by the f_i's in $\mathbb{C}(x_1)[x_2, \ldots, x_n]$, $\mathfrak{b} = \langle f_1, \ldots, f_k \rangle \, \mathbb{C}(x_1)[x_2, \ldots, x_n]$. Then $\mathfrak{b} \neq \langle 1 \rangle$, since if $1 = h_1 f_1 + \cdots + h_k f_k, h_i \in \mathbb{C}(x_1)[x_2, \ldots, x_n]$, there exists $H(x_1) \in \mathbb{C}[x_1]$, such that $H(x_1) h_i \in \mathbb{C}[x_1, x_2, \ldots, x_n]$, so that $H(x_1) = (H(x_1)h_1)f_1 + \cdots + (H(x_1)h_k)f_k \in \mathfrak{a}$, which contradicts the assumption $\mathfrak{a} \cap \mathbb{C}[x_1] = \langle 0 \rangle$. Let $G = \{g_i\}$ be a finite Gröbner basis for $\mathfrak{b}$. There are only finitely many coefficients in $k(x_1)$ involved in all g_i's. That G is a Gröbner basis can be tested by showing that $\mathrm{rem}(S(g_i, g_j), G) = 0$ for all i, j; see Theorem 21 in Section 3.12. This test is a finite procedure, so it involves only finitely many coefficients, including those previously mentioned, $c_i \in \mathbb{C}(x_1)$. Let $c_i = p_i(x_1)/q_i(x_1)$ and let $P(x_1)$ be the product of all p_i's and all q_i's, $P(x_1) \in \mathbb{C}[x_1]$. Since $P(x_1)$ has only finitely many zeros, there is an $m \in \mathbb{Z}$ such that $P(m) \neq 0$. Let $G_m = \{g_i(m, x_2, \ldots, x_n)\} \subseteq \mathbb{C}[x_2, \ldots, x_n]$. Then G_m is a Gröbner basis for $\langle G_m \rangle$ since

$$\operatorname{rem}\Big(S\big(g_i(m, x_2, \ldots, x_n),\ g_j(m, x_2, \ldots, x_n)\big),\ G_m\Big) = 0 \ \ \forall\, i, j.$$

We also get $1 \notin \langle G_m \rangle$, since if $1 \in \langle G_m \rangle$ we would have $g_i(m, x_2, \ldots, x_n) \in \mathbb{C}(x_1)$ for some i, which would give $g_i(x_1, x_2, \ldots, x_n) \in \mathbb{C}[x_1]$, again contradicting the assumption $\mathfrak{a} \cap \mathbb{C}[x_1] = \langle 0 \rangle$. We have reduced the number of variables.

Case 2. Assume that $\mathfrak{a} \cap \mathbb{C}[x_i] \neq \langle 0 \rangle$ for each i. Let $\mathfrak{a} \cap \mathbb{C}[x_1] = \langle f_1(x_1) \rangle$. Since $\deg(f_1) > 0$, there is a $c_1 \in \mathbb{C}$ so that $f_1(x_1) = (x_1 - c_1) f_2(x_1)$. Let $\mathfrak{b} = \mathfrak{a} + \langle x_1 - c_1 \rangle$. Then $\mathfrak{b} \neq \langle 1 \rangle$ since otherwise $1 = f + q(x_1 - c_1)$ for some $f \in \mathfrak{a}$ and some $q \in \mathbb{C}[x_1, \ldots, x_n]$, and then $f_2 = f f_2 + q f_1 \in \mathfrak{a}$, which is a contradiction. Iterating this with $x_2, \ldots, x_n$ we get

$$\mathfrak{a} \subseteq \mathfrak{a} + \langle x_1 - c_1, \ldots, x_n - c_n \rangle \neq \langle 1 \rangle\,.$$

But $\langle x_1 - c_1, \ldots, x_n - c_n \rangle$ is a maximal ideal, which means that we must have $\mathfrak{a} \subseteq \langle x_1 - c_1, \ldots, x_n - c_n \rangle$. Hence

$$(c_1, \ldots, c_n) \in \mathfrak{v}(\langle x_1 - c_1, \ldots, x_n - c_n \rangle) \subseteq \mathfrak{v}(\mathfrak{a}).$$

□

The geometrical meaning of the next corollary is that there is a bijection between points in $\mathbb{C}^n$ and maximal ideals in $\mathbb{C}[x_1, \ldots, x_n]$.

COROLLARY 12. *Every maximal ideal in $\mathbb{C}[x_1, \ldots, x_n]$ is of the form*

$$\langle x_1 - c_1, \ldots, x_n - c_n \rangle$$

for some point $(c_1, \ldots, c_n) \in \mathbb{C}^n$. Conversely, every ideal of the form $\langle x_1 - c_1, \ldots, x_n - c_n \rangle$ is a maximal ideal.

Proof. We have already seen in Exercises 2 and 3 in Section 1.3.2, that for a point $(c_1, \ldots, c_n) \in \mathbb{C}^n$ we have $\mathfrak{i}(\{(c_1, \ldots, c_n)\}) = \langle x_1 - c_1, \ldots, x_n - c_n \rangle$ and that the ideal $\langle x_1 - c_1, \ldots, x_n - c_n \rangle$ is a maximal ideal. Let $\mathfrak{m}$ be a maximal ideal. Then $\mathfrak{v}(\mathfrak{m}) \neq \emptyset$ so there is a point $(c_1, \ldots, c_n) \in \mathfrak{v}(\mathfrak{m})$. But then $\mathfrak{i}(\mathfrak{v}(\mathfrak{m})) \subseteq \mathfrak{i}(\{(c_1, \ldots, c_n)\}) = \langle x_1 - c_1, \ldots, x_n - c_n \rangle$. □

COROLLARY 13 (HILBERT'S NULLSTELLENSATZ, STRONG FORM). *For any ideal* $\mathfrak{a} \subseteq \mathbb{C}[x_1, \ldots, x_n]$ *we have* $\mathfrak{i}(\mathfrak{v}(\mathfrak{a})) = \sqrt{\mathfrak{a}}$.

Proof. We already know from Proposition 6 that $\sqrt{\mathfrak{a}} \subseteq \mathfrak{i}(\mathfrak{v}(\sqrt{\mathfrak{a}})) = \mathfrak{i}(\mathfrak{v}(\mathfrak{a}))$. Suppose $f \in \mathfrak{i}(\mathfrak{v}(\mathfrak{a}))$. Consider the ideal $\mathfrak{a}_1$ generated by $\mathfrak{a}$ and $fy - 1$ in the polynomial ring $\mathbb{C}[x_1, \ldots, x_n, y]$, where y is a new variable. If $(c_1, \ldots, c_n, c) \in \mathfrak{v}(\mathfrak{a}_1) \subseteq \mathbb{C}^{n+1}$ we get that $(c_1, \ldots, c_n) \in \mathfrak{v}(\mathfrak{a})$ since every element in $\mathfrak{a}$ belongs to $\mathfrak{a}_1$. But then $f(c_1, \ldots, c_n) = 0$ so $f(c_1, \ldots, c_n)c - 1 = -1$. Hence $\mathfrak{v}(\mathfrak{a}_1) = \emptyset$ so $\mathfrak{a}_1 = \langle 1 \rangle$. Thus, we can find elements $g_i \in \mathbb{C}[x_1, \ldots, x_n, y]$ so that $1 = g_1 f_1 + \cdots + g_m f_m + g_{m+1}(fy - 1)$, where $f_i \in \mathfrak{a}$. This is an identity in y so it remains true if we set $y = 1/f$, so

$$\begin{aligned} 1 &= g_1(x_1, \ldots, x_n, 1/f) f_1(x_1, \ldots, x_n) + \cdots \\ &\quad + g_m(x_1, \ldots, x_n, 1/f) f_m(x_1, \ldots, x_n) \\ &\quad + g_{m+1}(x_1, \ldots, x_n, 1/f)(f(x_1, \ldots, x_n)/f(x_1, \ldots, x_n) - 1) \\ &= g_1(x_1, \ldots, x_n, 1/f) f_1(x_1, \ldots, x_n) + \cdots \\ &\quad + g_m(x_1, \ldots, x_n, 1/f) f_m(x_1, \ldots, x_n). \end{aligned}$$

Clearing denominators we get that f^N is a linear combination of $f_1, \ldots, f_m$ with coefficients in $\mathbb{C}[x_1, \ldots, x_n]$, so $f^N \in \mathfrak{a}$. We have shown that if $f \in \mathfrak{i}(\mathfrak{v}(\mathfrak{a}))$, then $f^N \in \mathfrak{a}$ for some N, i.e., $\mathfrak{i}(\mathfrak{v}(\mathfrak{a})) \subseteq \sqrt{\mathfrak{a}}$. □

The conclusion of the theorem is the following. If we consider radical ideals in $\mathbb{C}[x_1, \ldots, x_n]$ and algebraic sets in $\mathbb{C}^n$, then the two operations $\mathfrak{v}$ from radical ideals to algebraic sets and $\mathfrak{i}$ from radical ideals to algebraic sets are inverses to each other.

COROLLARY 14. *If* X *is an algebraic set in* $\mathbb{C}^n$ *then* $\mathfrak{v}(\mathfrak{i}(X)) = X$. *If* $\mathfrak{a}$ *is a radical ideal in* $\mathbb{C}[x_1, \ldots x_n]$ *then* $\mathfrak{i}(\mathfrak{v}(\mathfrak{a})) = \mathfrak{a}$.

Proof. If X is an algebraic set then $X = \mathfrak{v}(\mathfrak{a})$ for some ideal $\mathfrak{a}$, according to Proposition 6(b). Hence $X = \mathfrak{v}(\sqrt{\mathfrak{a}})$, according to Proposition 6(e). Thus $\mathfrak{v}(\mathfrak{i}(X)) = \mathfrak{v}(\mathfrak{i}(\sqrt{\mathfrak{a}})) = \mathfrak{v}(\sqrt{\mathfrak{a}}) = X$. If $\mathfrak{a}$ is radical, we get $\mathfrak{i}(\mathfrak{v}(\mathfrak{a})) = \sqrt{\mathfrak{a}} = \mathfrak{a}$. □

4.3 A Dictionary: Algebraic Sets ↔ Radical Ideals

In this section we study the bijection between radical ideals in $\mathbb{C}[x_1, \ldots, x_n]$ and algebraic sets in $\mathbb{C}^n$. We have seen that if $X_1 = \mathfrak{v}(\mathfrak{a}_1)$ and $X_2 = \mathfrak{v}(\mathfrak{a}_2)$ are algebraic sets, then $X_1 \cap X_2$ and $X_1 \cup X_2$ are algebraic sets, since

$\mathfrak{v}(\mathfrak{a}_1) \cap \mathfrak{v}(\mathfrak{a}_2) = \mathfrak{v}(\mathfrak{a}_1 + \mathfrak{a}_2)$ and $\mathfrak{v}(\mathfrak{a}_1) \cup \mathfrak{v}(\mathfrak{a}_2) = \mathfrak{v}(\mathfrak{a}_1 \cap \mathfrak{a}_2)$. We have also seen that a prime ideal is radical, and that the set of radical ideals is closed under intersection, since if $\mathfrak{a}_1 = \sqrt{\mathfrak{a}_1}$ and $\mathfrak{a}_2 = \sqrt{\mathfrak{a}_2}$, then $\mathfrak{a}_1 \cap \mathfrak{a}_2 = \sqrt{\mathfrak{a}_1} \cap \sqrt{\mathfrak{a}_2} = \sqrt{\sqrt{\mathfrak{a}_1} \cap \sqrt{\mathfrak{a}_2}} = \sqrt{\mathfrak{a}_1 \cap \mathfrak{a}_2}$. We now give a "dictionary" for the translation between algebraic sets and radical ideals.

Algebraic sets	*Radical ideals*
Point $(c_1, \dots, c_n) \in \mathbb{C}^n$	Maximal ideal $\langle x_1 - c_1, \dots, x_n - c_n \rangle$
Irreducible set $X = \mathfrak{v}(\mathfrak{p})$	Prime ideal $\mathfrak{p} = \mathfrak{i}(X)$
Union of algebraic sets $X = X_1 \cup X_2$	Intersection of radical ideals $\mathfrak{i}(X) = \mathfrak{i}(X_1) \cap \mathfrak{i}(X_2) = \mathfrak{r}_1 \cap \mathfrak{r}_2 = \mathfrak{r}$
Intersection of algebraic sets $X = X_1 \cap X_2$	Radical of sum of radical ideals $\mathfrak{i}(X) = \sqrt{\mathfrak{i}(X_1) + \mathfrak{i}(X_2)} = \sqrt{\mathfrak{r}_1 + \mathfrak{r}_2} = \mathfrak{r}$

Exercises

1. Show that the set of radical ideals is not closed under addition. Hint: Show that the ideals $\mathfrak{a} = \langle x_1^3 - x_2^2 \rangle$ and $\mathfrak{b} = \langle x_1^5 - x_2^2 \rangle$ are prime ideals in $\mathbb{C}[x_1, x_2]$, but $\mathfrak{a}+\mathfrak{b}$ is not radical.

2. An element in a ring is called *nilpotent* if some power of it equals 0. Show that the set of nilpotent elements in a ring equals the intersection of all prime ideals in the ring.

3. Show that the radical ideals in $\mathbb{C}[x_1, \dots, x_n]$ are exactly those ideals which are intersections of prime ideals.

Chapter 5

Primary Decomposition of Ideals

In the next chapter we will show how to use Gröbner bases to determine the number of solutions of and to solve a system of equations $f_1 = \cdots = f_k = 0, f_i \in \mathbb{C}[x_1, \ldots, x_n]$. If we can write the ideal $\mathfrak{a} = \langle f_1, \ldots, f_k \rangle$ as an intersection of ideals, $\mathfrak{a} = \mathfrak{b} \cap \mathfrak{c}$, then the set of solutions $V(\mathfrak{a})$ will break up as a union of smaller sets, $V(\mathfrak{a}) = V(\mathfrak{b}) \cup V(\mathfrak{c})$; see Proposition 6 in Section 4.1. Two ideals with the same radical determine the same algebraic set since $\mathfrak{v}(\mathfrak{a}) = \mathfrak{v}(\sqrt{\mathfrak{a}})$; see Proposition 6(e) in Section 4.1. When solving equations, however, one sometimes wants to consider the solutions to $x - 1 = 0$ and $(x-1)^2 = 0$ as different, i.e., one wants to take multiplicity into account. We will in this chapter show that any ideal in $\mathbb{C}[x_1, \ldots, x_n]$ is an intersection of so-called primary ideals. A primary ideal has a prime ideal as radical, so its corresponding algebraic set is irreducible. Primary ideals can, however, have multiplicity, so they give a finer description of the solution set. It is equally easy to treat a general Noetherian ring instead of $\mathbb{C}[x_1, \ldots, x_n]$. Thus, in this section, R will denote a Noetherian ring.

DEFINITION 1. An ideal in R is called *irreducible* if it is not an intersection of two strictly larger ideals.

PROPOSITION 2. *Let $\mathfrak{a}$ be an ideal in R. Then $\mathfrak{a}$ is a finite intersection of irreducible ideals.*

Proof. Suppose that there are ideals which are not a finite intersection of irreducible ideals, and call the set of such ideals S. Then S has a maximal element, say $\mathfrak{b}$; see Proposition 13 in Section 3.5. Since $\mathfrak{b}$ is not irreducible, we can write $\mathfrak{b} = \mathfrak{b}_1 \cap \mathfrak{b}_2$ with $\mathfrak{b}_i$ strictly larger than $\mathfrak{b}$. But then $\mathfrak{b}_i$ are finite intersections of irreducible ideals, which gives the contradiction that $\mathfrak{b}$ is a finite intersection of irreducible ideals. □

DEFINITION 3. An ideal $\mathfrak{q}$ is called *primary* if $rs \in \mathfrak{q}, r \notin \mathfrak{q}$ implies that $s^n \in \mathfrak{q}$ for some n.

The following is a more symmetric reformulation of the definition.

LEMMA 4. *An ideal $\mathfrak{q}$ is primary if $ab \in \mathfrak{q}, a, b \notin \mathfrak{q}$ implies that both a and b have some power belonging to $\mathfrak{q}$.*

As for prime ideals and maximal ideals, we can characterize primary ideals in terms of their factor ring.

DEFINITION 5. An element r in a ring is called *nilpotent* if $r^n = 0$ for some n.

PROPOSITION 6. *An ideal $\mathfrak{q}$ in a ring R is primary if and only if every zerodivisor in $R/\mathfrak{q}$ is nilpotent.*

Proof. The condition $rs \in \mathfrak{q}, r \notin \mathfrak{q}$ implies $s^n \in \mathfrak{q}$ becomes $\bar{r}\bar{s} = \bar{0}, \bar{r} \neq \bar{0}$ implies $\bar{s}^n = \bar{0}$ in $R/\mathfrak{q}$. □

LEMMA 7. *If $\mathfrak{q}$ is primary then $\sqrt{\mathfrak{q}}$ is a prime ideal.*

Proof. Let $r = st \in \sqrt{\mathfrak{q}}$. If $s \notin \sqrt{\mathfrak{q}}$ we must show that $t \in \sqrt{\mathfrak{q}}$. If $s \notin \sqrt{\mathfrak{q}}$ then no power of s belongs to $\mathfrak{q}$. Then $t \in \mathfrak{q} \subseteq \sqrt{\mathfrak{q}}$ since $\mathfrak{q}$ is primary. □

If $\mathfrak{q}$ is primary with $\sqrt{\mathfrak{q}} = \mathfrak{p}$ we say that $\mathfrak{q}$ is $\mathfrak{p}$-primary.

LEMMA 8. *An irreducible ideal is primary.*

Proof. Suppose that $\mathfrak{q}$ is irreducible but not primary. Then there are elements $a, b \notin \mathfrak{q}$ such that $ab \in \mathfrak{q}$ but such that no powers of a or b belong to $\mathfrak{q}$. Consider the increasing sequence of ideals $(\mathfrak{q} : \langle b^i \rangle)_{i=1}^{\infty}$. This sequence stabilizes, see Proposition 13 in Section 3.5, so there is an n such that $\mathfrak{q} : \langle b^n \rangle = \mathfrak{q} : \langle b^{n+1} \rangle$. Since $a, b^n \notin \mathfrak{q}$, both $\mathfrak{q} + \langle a \rangle$ and $\mathfrak{q} + \langle b^n \rangle$ are strictly larger than $\mathfrak{q}$. We claim that $(\mathfrak{q} + \langle a \rangle) \cap (\mathfrak{q} + \langle b^n \rangle) = \mathfrak{q}$, which gives us the contradiction, since $\mathfrak{q}$ is irreducible. An element $r \in (\mathfrak{q} + \langle a \rangle) \cap (\mathfrak{q} + \langle b^n \rangle)$ can be written $r = q_1 + r_1 a = q_2 + r_2 b^n$ with $q_i \in \mathfrak{q}$ and $r_i \in R$. Then $rb = q_1 b + r_1 ab = q_2 b + r_2 b^{n+1}$. Since $ab, q_1 b \in \mathfrak{q}$ we get $rb \in \mathfrak{q}$. Since $q_2 b \in \mathfrak{q}$ we get $r_2 b^{n+1} \in \mathfrak{q}$, i.e., $r_2 \in \mathfrak{q} : \langle b^{n+1} \rangle = \mathfrak{q} : \langle b^n \rangle$, so $r_2 b^n \in \mathfrak{q}$. Hence $r = q_2 + r_2 b^n \in \mathfrak{q}$, and we have proved our claim. □

LEMMA 9. *The intersection of two $\mathfrak{p}$-primary ideals is $\mathfrak{p}$-primary.*

Proof. Let $\mathfrak{q}_1$ and $\mathfrak{q}_2$ be $\mathfrak{p}$-primary. Suppose that $ab \in \mathfrak{q}_1 \cap \mathfrak{q}_2$ and that $a \notin \mathfrak{q}_1 \cap \mathfrak{q}_2$. We have to show that some power of b belongs to $\mathfrak{q}_1 \cap \mathfrak{q}_2$. Since $a \notin \mathfrak{q}_1 \cap \mathfrak{q}_2$ we have $a \notin \mathfrak{q}_1$ or $a \notin \mathfrak{q}_2$, say $a \notin \mathfrak{q}_1$. Then $b^m \in \mathfrak{q}_1 \subseteq \sqrt{\mathfrak{q}_1} = \sqrt{\mathfrak{q}_2}$ so $(b^m)^n \in \mathfrak{q}_1 \cap \mathfrak{q}_2$ for some n. □

We are now turning to the question of uniqueness. Let $\mathfrak{a} = \mathfrak{q}_1 \cap \cdots \cap \mathfrak{q}_s$ with $\sqrt{\mathfrak{q}_i} = \mathfrak{p}_i$. We can collect all components in the intersection with radical $\mathfrak{p}_1$. The intersection of those, say $\mathfrak{Q}_1$, is $\mathfrak{p}_1$-primary. We do the same for the other primes. This means that we get $\mathfrak{a} = \mathfrak{Q}_1 \cap \cdots \cap \mathfrak{Q}_t$ with all $\sqrt{\mathfrak{Q}_i}$ different. If we furthermore delete any $\mathfrak{Q}_i$ which contains the intersection of the remaining ones, so the intersection becomes *irredundant*, we say that we have a *primary decomposition* of $\mathfrak{a}$. Primary decompositions are not unique, but there is a great deal of uniqueness.

LEMMA 10. *Suppose $\mathfrak{q}$ is $\mathfrak{p}$-primary and let $\mathfrak{a}$ be an arbitrary ideal.*

1. *If $\mathfrak{a} \not\subseteq \mathfrak{p}$ then $\mathfrak{q} : \mathfrak{a} = \mathfrak{q}$.*
2. *If $\mathfrak{a} \subseteq \mathfrak{p}$ but $\mathfrak{a} \not\subseteq \mathfrak{q}$ then $\mathfrak{q} : \mathfrak{a}$ is $\mathfrak{p}$-primary and contains $\mathfrak{q}$.*
3. *If $\mathfrak{a} \subseteq \mathfrak{q}$ then $\mathfrak{q} : \mathfrak{a} = \langle 1 \rangle$.*

Proof. 1. That $\mathfrak{q} \subseteq \mathfrak{q} : \mathfrak{a}$ is true without restriction on $\mathfrak{a}$. Take $x \in \mathfrak{a} \setminus \mathfrak{p}$. If $c \in \mathfrak{q} : \mathfrak{a}$ then $cx \in \mathfrak{q}$, so $c \in \mathfrak{q}$ since no power of x belongs to $\mathfrak{q}$.

2. First we show that $\mathfrak{q} : \mathfrak{a}$ is primary. Suppose $xy \in \mathfrak{q} : \mathfrak{a}$ and $y \notin \mathfrak{q} : \mathfrak{a}$. Then $ya \notin \mathfrak{q}$ for some $a \in \mathfrak{a}$. Since $xya \in \mathfrak{q}$ we have $x^n \in \mathfrak{q} \subseteq \mathfrak{q} : \mathfrak{a}$ for some n. It remains to show that $\sqrt{\mathfrak{q} : \mathfrak{a}} = \mathfrak{p}$. Since $\mathfrak{q} \subseteq \mathfrak{q} : \mathfrak{a}$ we have $\mathfrak{p} = \sqrt{\mathfrak{q}} \subseteq \sqrt{\mathfrak{q} : \mathfrak{a}}$. Take $a \in \mathfrak{a} \setminus \mathfrak{q}$. If $x \in \mathfrak{q} : \mathfrak{a}$ then $xa \in \mathfrak{q}$ so $x^n \in \mathfrak{q}$ for some n so $x \in \mathfrak{p}$ and hence $\mathfrak{q} : \mathfrak{a} \subseteq \mathfrak{p}$, which gives $\sqrt{\mathfrak{q} : \mathfrak{a}} \subseteq \sqrt{\mathfrak{p}} = \mathfrak{p}$. Hence $\sqrt{\mathfrak{q} : \mathfrak{a}} = \mathfrak{p}$.

3. This is obvious.

□

THEOREM 11. *Let*

$$\mathfrak{a} = \mathfrak{q}_1 \cap \cdots \cap \mathfrak{q}_s = \mathfrak{q}'_1 \cap \cdots \cap \mathfrak{q}'_t$$

be two primary decompositions. Suppose $\sqrt{\mathfrak{q}_i} = \mathfrak{p}_i$ and $\sqrt{\mathfrak{q}'_i} = \mathfrak{p}'_i$. Then $\{\mathfrak{p}_i\} = \{\mathfrak{p}'_i\}$, in particular $s = t$.

Proof. After renumbering and, if necessary, changing the role of $\{\mathfrak{p}_i\}$ and $\{\mathfrak{p}'_i\}$, we can assume that $\mathfrak{p}_1$ is not contained in any of $\mathfrak{p}_i, i > 1$ and not strictly contained in any $\mathfrak{p}'_j$. Suppose $\mathfrak{p}_1 \neq \mathfrak{p}'_j$ for all j. Then $\mathfrak{q}_1$ is not contained in any $\mathfrak{p}'_j$ since that would give $\mathfrak{p}_1 = \sqrt{\mathfrak{q}_1} \subseteq \sqrt{\mathfrak{p}'_j} = \mathfrak{p}'_j$. Hence

$$\mathfrak{a} : \mathfrak{q}_1 = (\mathfrak{q}'_1 \cap \cdots \cap \mathfrak{q}'_t) : \mathfrak{q}_1 = (\mathfrak{q}'_1 : \mathfrak{q}_1) \cap \cdots \cap (\mathfrak{q}'_t : \mathfrak{q}_1) = \mathfrak{q}'_1 \cap \cdots \cap \mathfrak{q}'_t = \mathfrak{a}$$

by Lemma 10. On the other hand

$$\mathfrak{a} : \mathfrak{q}_1 = (\mathfrak{q}_1 \cap \cdots \cap \mathfrak{q}_s) : \mathfrak{q}_1 = (\mathfrak{q}_1 : \mathfrak{q}_1) \cap \cdots \cap (\mathfrak{q}_s : \mathfrak{q}_1) = \langle 1 \rangle \cap \mathfrak{q}_2 \cap \cdots \cap \mathfrak{q}_s \neq \mathfrak{a}$$

which is a contradiction. Hence we can assume that $\mathfrak{p}_1 = \mathfrak{p}'_1$. Now

$$\mathfrak{a} : \mathfrak{q}_1\mathfrak{q}'_1 = (\mathfrak{q}_1 \cap \cdots \cap \mathfrak{q}_s) : \mathfrak{q}_1\mathfrak{q}'_1 = (\mathfrak{q}_1 : \mathfrak{q}_1\mathfrak{q}'_1) \cap \cdots \cap (\mathfrak{q}_s : \mathfrak{q}_1\mathfrak{q}'_1) =$$

$$((\mathfrak{q}_1 : \mathfrak{q}_1) : \mathfrak{q}'_1) \cap \cdots \cap ((\mathfrak{q}_s : \mathfrak{q}_1) : \mathfrak{q}'_1) = (\langle 1 \rangle \cap \mathfrak{q}'_1) \cap (\mathfrak{q}_2 : \mathfrak{q}'_1) \cap \cdots \cap (\mathfrak{q}_s : \mathfrak{q}'_1) =$$

$$\mathfrak{q}_2 \cap \cdots \cap \mathfrak{q}_s.$$

In the same way $\mathfrak{a} : \mathfrak{q}_1\mathfrak{q}'_1 = \mathfrak{q}'_2 \cap \cdots \cap \mathfrak{q}'_t$ and we are ready to use induction on the shortest length of a decomposition. □

DEFINITION 12. The prime ideals that are radicals of some primary decomposition of $\mathfrak{a}$ are said to *belong to* $\mathfrak{a}$ or to be *associated* to $\mathfrak{a}$. The set of associated primes is denoted $\mathrm{Ass}(\mathfrak{a})$.

There might be containment relations amongst the $\mathfrak{p}_i$'s in $\mathrm{Ass}(\mathfrak{a})$. If $\mathfrak{p}_i$ does not contain any $\mathfrak{p}_j, j \neq i$, then $\mathfrak{q}_i$ is called an *isolated* component. Otherwise $\mathfrak{q}_i$ (for geometric reasons) is called an *embedded* component.

EXAMPLE 13. If $\mathfrak{a} = \langle x^2, xy \rangle = \langle x \rangle \cap \langle x^2, y \rangle$ then $\langle x \rangle$ is an isolated component and $\langle x^2, y \rangle$ is an embedded component since $\sqrt{\langle x \rangle} = \langle x \rangle$ is contained in $\langle x, y \rangle = \sqrt{\langle x^2, y \rangle}$. □

THEOREM 14. *The isolated primary components in a primary decomposition are unique.*

Proof. Let $\mathfrak{a} = \mathfrak{q}_1 \cap \cdots \cap \mathfrak{q}_s = \mathfrak{q}'_1 \cap \cdots \cap \mathfrak{q}'_s$ be primary decompositions with $\mathfrak{q}_1$ and $\mathfrak{q}'_1$ isolated components with $\sqrt{\mathfrak{q}_1} = \sqrt{\mathfrak{q}'_1} = \mathfrak{p}_1$. Let $\mathfrak{b} = \mathfrak{q}_2 \cap \cdots \cap \mathfrak{q}_s$ and $\mathfrak{b}' = \mathfrak{q}'_2 \cap \cdots \cap \mathfrak{q}'_s$. We claim that $\mathfrak{b} \not\subseteq \mathfrak{p}_1$, since if $\mathfrak{q}_2 \cap \cdots \cap \mathfrak{q}_s \subseteq \mathfrak{p}_1$ we would have

$$\mathfrak{p}_2 \cdots \mathfrak{p}_s \subseteq \mathfrak{p}_2 \cap \cdots \cap \mathfrak{p}_s = \sqrt{\mathfrak{q}_2} \cap \cdots \cap \sqrt{\mathfrak{q}_s} = \sqrt{\mathfrak{q}_2 \cap \cdots \cap \mathfrak{q}_s} \subseteq \sqrt{\mathfrak{p}_1} = \mathfrak{p}_1$$

which would give $\mathfrak{p}_j \subseteq \mathfrak{p}_1$ for some $j > 1$; see Lemma 64 in Section 1.8. We get

$$\mathfrak{a} : \mathfrak{b} = (\mathfrak{q}_1 \cap \mathfrak{b}) : \mathfrak{b} = (\mathfrak{q}_1 : \mathfrak{b}) \cap (\mathfrak{b} : \mathfrak{b}) = \mathfrak{q}_1 \cap \langle 1 \rangle = \mathfrak{q}_1.$$

On the other hand

$$\mathfrak{a} : \mathfrak{b} = (\mathfrak{q}_1' \cap \mathfrak{b}') : \mathfrak{b} = (\mathfrak{q}_1' : \mathfrak{b}) \cap (\mathfrak{b}' : \mathfrak{b}) = \mathfrak{q}_1' \cap (\mathfrak{b}' : \mathfrak{b}) \subseteq \mathfrak{q}_1',$$

so $\mathfrak{q}_1 \subseteq \mathfrak{q}_1'$. Similarly $\mathfrak{q}_1' \subseteq \mathfrak{q}_1$ so $\mathfrak{q}_1 = \mathfrak{q}_1'$. □

But embedded components however not unique; see Exercise 4 below.

For future reference we characterize the zerodivisors in $k[x_1, \ldots, x_n]/\mathfrak{a}$ by means of a primary decomposition of $\mathfrak{a}$. Let $f \in k[x_1, \ldots, x_n]$ and denote the image of f in $R = k[x_1, \ldots, x_n]/\mathfrak{a}$ by $\bar{f}$. That $\bar{f}$ is a zerodivisor in R means that $\bar{f}\bar{g} = \bar{0}$ for some $\bar{g} \neq \bar{0}$, i.e., that $\langle \bar{0} \rangle : \langle \bar{f} \rangle \neq \langle \bar{0} \rangle$. This is equivalent to $fg \in \mathfrak{a}$ for some $g \notin \mathfrak{a}$, i.e., that $\mathfrak{a} : \langle f \rangle$ is strictly larger than $\mathfrak{a}$.

THEOREM 15. *Let $R = k[x_1, \ldots, x_n]/\mathfrak{a}$ and suppose that $\mathfrak{a} = \mathfrak{q}_1 \cap \cdots \cap \mathfrak{q}_s$ is a primary decomposition with $\sqrt{\mathfrak{q}_i} = \mathfrak{p}_i$. Then the image $\bar{f}$ in R of $f \in k[x_1, \ldots, x_n]$ is a zerodivisor in R if and only if $f \in \cup_{i=1}^{s} \mathfrak{p}_i$.*

Proof. Suppose that $f \notin \cup \mathfrak{p}_i$ and that $fg \in \mathfrak{a}$. Since $f \notin \mathfrak{p}_i$ and $fg \in \mathfrak{a} \subseteq \mathfrak{q}_i$ we get $g \in \mathfrak{q}_i$. This is true for every i so $g \in \cap \mathfrak{q}_i = \mathfrak{a}$. We have shown that if $f \notin \cup \mathfrak{p}_i$ then $\bar{f}$ is a nonzerodivisor in R. Now suppose that $f \in \cup \mathfrak{p}_i, f \notin \mathfrak{a}$, i.e., $\bar{f} \neq \bar{0}$. We shall show that $\bar{f}$ is a zerodivisor in R, i.e., $\bar{f}\bar{g} = \bar{0}$ for some $\bar{g} \neq \bar{0}$. By renumbering we can assume that $f \in \mathfrak{p}_i$ for $i = 1, \ldots, k$ and $f \notin \mathfrak{p}_i$ for $i = k+1, \ldots, s$. Choose $g \in \cap_{i=k+1}^{s} \mathfrak{q}_i \setminus \mathfrak{a}$, which is possible since the decomposition is irredundant. (If $k = s$ we can take $g = 1$.) If $fg \in \mathfrak{a}$ we are ready. For large N we have $f^N \in \mathfrak{q}_i, i = 1, \ldots, k$ since $f \in \mathfrak{p}_i, i = 1, \ldots, k$ and $\sqrt{\mathfrak{q}_i} = \mathfrak{p}_i$. Thus $f^N g \in \cap_{i=1}^{k} \mathfrak{q}_i \cdot \cap_{i=k+1}^{s} \mathfrak{q}_i = \cap_{i=1}^{s} \mathfrak{q}_i$ for all large N. Hence, for some $m \geq 0$ we have $f(f^m g) \in \mathfrak{a}, f^m g \notin \mathfrak{a}$. □

We can get a characterization of ideals primary to a maximal ideal. The following lemma contains the most important ingredient.

LEMMA 16. *Let $\mathfrak{a}$ be an ideal and let $\mathfrak{p}$ be a prime ideal in a Noetherian ring. Then $\sqrt{\mathfrak{a}} = \mathfrak{p}$ if and only if $\mathfrak{p}^n \subseteq \mathfrak{a} \subseteq \mathfrak{p}$ for some n.*

Proof. Suppose $\mathfrak{p}^n \subseteq \mathfrak{a} \subseteq \mathfrak{p}$. Then $\sqrt{\mathfrak{p}^n} \subseteq \sqrt{\mathfrak{a}} \subseteq \sqrt{\mathfrak{p}}$ by Proposition 45.4 in Section 1.3.5. We have $\sqrt{\mathfrak{p}^n} = \sqrt{\mathfrak{p}}$ by Proposition 45.5, and $\sqrt{\mathfrak{p}} = \mathfrak{p}$ by Exercise 5 in Section 1.8. Now suppose $\sqrt{\mathfrak{a}} = \mathfrak{p} = \langle p_1, \ldots, p_r \rangle$. For each i we have $p_i^{m_i} \in \mathfrak{a}$ for some m_i. Let $m = \max\{m_i\}$. We claim that $\mathfrak{p}^n \subseteq \mathfrak{a}$ if $n = mr - r + 1$. The ideal $\mathfrak{p}^n$ is generated by $\{ p_1^{n_1} \cdots p_r^{n_r} \mid n_1 + \cdots + n_r = n \}$. If $n_1 + \cdots + n_r = mr - r + 1$ then $n_i \geq m$ for at least one i, so all generators $p_1^{n_1} \cdots p_r^{n_r}$ for $\mathfrak{p}^n$ belong to $\mathfrak{a}$. Thus $\mathfrak{p}^n \subseteq \mathfrak{a}$. □

PROPOSITION 17. *Let $\mathfrak{a}$ be an ideal and let $\mathfrak{m}$ be a maximal ideal in a Noetherian ring. Then $\mathfrak{a}$ is $\mathfrak{m}$-primary if and only if $\sqrt{\mathfrak{a}} = \mathfrak{m}$. Thus $\mathfrak{a}$ is $\mathfrak{m}$-primary if and only if $\mathfrak{m}^n \subseteq \mathfrak{a} \subseteq \mathfrak{m}$ for some n.*

Proof. If $\mathfrak{a}$ is $\mathfrak{m}$-primary we have $\sqrt{\mathfrak{a}} = \mathfrak{m}$, so suppose for the converse that $\sqrt{\mathfrak{a}} = \mathfrak{m}$. If $\mathfrak{a} = \mathfrak{q}_1 \cap \cdots \cap \mathfrak{q}_k$ is a primary decomposition with $\mathfrak{q}_i$ $\mathfrak{p}_i$-primary, then

$$\sqrt{\mathfrak{a}} = \sqrt{\mathfrak{q}_1 \cap \cdots \cap \mathfrak{q}_k} = \sqrt{\mathfrak{q}_1} \cap \cdots \cap \sqrt{\mathfrak{q}_k} = \mathfrak{p}_1 \cap \cdots \cap \mathfrak{p}_k = \mathfrak{m}.$$

Since $\mathfrak{m}$ is a maximal ideal, we get $k = 1$ and $\mathfrak{p}_1 = \mathfrak{m}$. □

We finish this chapter by showing that for monomial ideals in a polynomial ring $k[x_1, \ldots, x_n]$ we can determine a primary decomposition by the means we have at hand. Let $\mathfrak{a} = \langle m_1, \ldots, m_t \rangle$ be a monomial ideal and suppose $m_1 = x_{i_1}^{n_1} \cdots x_{i_k}^{n_k}$. Then

$$\mathfrak{a} = \left\langle x_{i_1}^{n_1}, m_2, \ldots, m_t \right\rangle \cap \cdots \cap \left\langle x_{i_k}^{n_k}, m_2, \ldots, m_t \right\rangle$$

by Section 2.2. If $m_2 = x_{j_1}^{m_1} \cdots x_{j_l}^{m_l}$, then

$$\left\langle x_{i_s}^{n_s}, m_2, \ldots, m_t \right\rangle = \left\langle x_{i_s}^{n_s}, x_{j_1}^{m_1}, m_3, \ldots, m_t \right\rangle \cap \cdots \cap \left\langle x_{i_s}^{n_s}, x_{j_l}^{m_l}, m_3, \ldots, m_t \right\rangle.$$

Continuing like this, we can write $\mathfrak{a}$ as an intersection of ideals of type $\left\langle x_{i_1}^{n_1}, \ldots, x_{i_k}^{n_k} \right\rangle, n_i > 0$. We leave it as an exercise to show that such an ideal is $\langle x_{i_1}, \ldots, x_{i_k} \rangle$-primary. Collecting all ideals belonging to the same prime ideal and performing their intersection, we obtain an intersection of primary ideals with different primes. Making the intersection irredundant gives us a primary decomposition.

EXAMPLE 18.

$$\mathfrak{a} = \left\langle x^3, x^2y, xy^2z \right\rangle$$

$$= \left\langle x^3, x^2, x \right\rangle \cap \left\langle x^3, x^2, y^2 \right\rangle \cap \left\langle x^3, x^2, z \right\rangle \cap \left\langle x^3, y, x \right\rangle \cap \left\langle x^3, y, y^2 \right\rangle \cap \left\langle x^3, y, z \right\rangle$$

$$= \langle x \rangle \cap \left\langle x^2, y^2 \right\rangle \cap \left\langle x^2, z \right\rangle \cap \langle x, y \rangle \cap \left\langle x^3, y \right\rangle \cap \left\langle x^3, y, z \right\rangle.$$

We have $\left\langle x^3, y \right\rangle \subseteq \left\langle x^3, y, z \right\rangle$ and $\left\langle x^2, y^2 \right\rangle \subseteq \langle x, y \rangle$, so we can delete $\left\langle x^3, y, z \right\rangle$ and $\langle x, y \rangle$. We have $\left\langle x^2, y^2 \right\rangle \cap \left\langle x^3, y \right\rangle = \left\langle x^3, x^2y, x^3y^2, y^2 \right\rangle = \left\langle x^3, x^2y, y^2 \right\rangle$. Thus $\mathfrak{a} = \langle x \rangle \cap \left\langle x^3, x^2y, y^2 \right\rangle \cap \left\langle x^2, z \right\rangle$ is a primary decomposition of $\mathfrak{a}$. The associated prime ideals are $\langle x \rangle$, $\langle x, y \rangle$, and $\langle x, z \rangle$. □

Exercises

1. Show that a monomial ideal in $k[x_1, \ldots, x_n]$ is irreducible if and only if it is of the form $\langle x_{i_1}^{n_1}, \ldots, x_{i_k}^{n_k} \rangle$.

2. Show that an ideal $\mathfrak{q}$ in a Noetherian ring is primary if and only if $\mathfrak{a}\mathfrak{b} \subseteq \mathfrak{q}, \mathfrak{a} \not\subseteq \mathfrak{q}$ implies that $\mathfrak{b}^n \subseteq \mathfrak{q}$ for some n.

3. Show that $\langle x^3, xy, y^4 \rangle$ is primary and that $\langle x^3, xy \rangle$ is not primary in $k[x, y]$.

4. Show that $\langle x^2, xy \rangle = \langle x \rangle \cap \langle x^2, y + \alpha x \rangle$ in $k[x, y]$ are different primary decompositions for every $\alpha \in k$.

5. It can be shown that $\mathfrak{p} = \langle f_1, f_2, f_3 \rangle = \langle y^2 - xz, yz - x^3, z^2 - x^2 y \rangle$ is a prime ideal in $k[x, y, z]$. Show that $\mathfrak{p}^2$ is not a primary ideal and that $\sqrt{\mathfrak{p}^2} = \mathfrak{p}$. Thus powers of prime ideals are not necessarily primary, in particular it is not sufficient that $\sqrt{\mathfrak{a}}$ is prime for $\mathfrak{a}$ to be primary. Hint: Consider $f_2^2 - f_1 f_3$.

6. Show that the ideal $\langle x_1^{n_1}, \ldots, x_k^{n_k} \rangle, n_i > 0$ in $k[x_1, \ldots, x_n]$ is primary to $\langle x_1, \ldots, x_k \rangle$.

7. Make a primary decomposition of $\mathfrak{a} = \langle x^2 y, y^2 z, x z^2 \rangle$.

8. Show that a monomial ideal in $k[x_1, \ldots, x_n]$ is primary if and only if it is of the form $\langle x_{i_1}^{n_1}, \ldots, x_{i_k}^{n_k}, m_1, \ldots, m_t \rangle$, where the m_i's are monomials in $\{x_{i_1}, \ldots, x_{i_k}\}$.

Chapter 6

Solving Systems of Polynomial Equations

We start in the simplest way with a system of polynomial equations in one variable. We allow complex coefficients and look for complex solutions. Thus we are interested in the solutions of

$$f_1(x) = f_2(x) = \cdots = f_k(x) = 0, f_i(x) \in \mathbb{C}[x],$$

i.e., in $\mathfrak{v}(\langle f_1, \ldots, f_k \rangle)$. Now $\langle f_1, \ldots, f_k \rangle = \langle \gcd(f_1, \ldots, f_k) \rangle$, see Theorem 28 in Section 1.3.3, which is a principal ideal. Thus the system $f_1(x) = f_2(x) = \cdots = f_k(x) = 0$ is equivalent to one single equation, $\gcd(f_1, f_2, \ldots, f_k) = 0$. The equation $f(x) = 0$ has n solutions, where $n = \deg(f)$, see Corollary 38 in Section 1.3.3, provided that the solutions are counted with multiplicities. Thus $x^4 - x^2 = x^2(x-1)(x+1) = 0$ has four solutions: $x = 0$ with multiplicity 2 and $x = \pm 1$, each of multiplicity 1. An algebraic way to describe the degree of a polynomial $f(x)$ is $\deg(f) = \dim_{\mathbb{C}} \mathbb{C}[x]/\langle f(x) \rangle$, since

$$\{1 + \langle f(x) \rangle, x + \langle f(x) \rangle, \ldots, x^{n-1} + \langle f(x) \rangle\}$$

is a $\mathbb{C}$-basis for $\mathbb{C}[x]/\langle f(x) \rangle$; see Example 72 in Section 1.9. Thus the number of solutions to $f(x) = 0$ equals $\dim_{\mathbb{C}} \mathbb{C}[x]/\langle f(x) \rangle$.

6.1 Systems with Only One Solution

We now turn to the general (and more interesting) situation of several variables. We have seen that points in $\mathbb{C}^n$ correspond to maximal ideals in $\mathbb{C}[x_1, \ldots, x_n]$, if $P = (c_1, \ldots, c_n)$ then $\mathfrak{i}(\{P\}) = \langle x_1 - c_1, \ldots, x_n - c_n \rangle$;

see Corollary 14 in Section 4.2. Hence if the system $f_1(x_1, \ldots, x_n) = \cdots = f_k(x_1, \ldots, x_n) = 0$ has P as its only solution, i.e., $\mathfrak{v}(\langle f_1, \ldots, f_k \rangle) = \{P\}$, then $\sqrt{\langle f_1, \ldots, f_k \rangle} = \langle x_1 - c_1, \ldots, x_n - c_n \rangle$; see Example 3 in Section 4.1. Let us denote $\langle x_1 - c_1, \ldots, x_n - c_n \rangle$ by $\mathfrak{m}_P$. We have seen that the ideals $\mathfrak{a}$, for which $\sqrt{\mathfrak{a}} = \mathfrak{m}_P$, are exactly those which satisfy $\mathfrak{m}_P^s \subseteq \mathfrak{a} \subseteq \mathfrak{m}_P$ for some s; see Lemma 16 in Chapter 5.

We will next discuss the concept of multiplicity of a solution. We start with some examples. The system $x = y^2 = 0$ corresponds to the intersection of the y-axis $x = 0$ and the x-axis $y = 0$ "counted twice". It is thus natural to say that $(0,0)$ should have multiplicity 2 as a solution to this system. Similarly $(0,0)$ should have multiplicity 4 as a solution to $x^2 = y^2 = 0$, since both the x-axis and the y-axis are double. We note that $\dim_{\mathbb{C}} \mathbb{C}[x,y]/\left\langle x, y^2 \right\rangle = 2$ and $\dim_{\mathbb{C}} \mathbb{C}[x,y]/\left\langle x^2, y^2 \right\rangle = 4$.

DEFINITION 1 (MULTIPLICITY OF A SOLUTION). Let $\mathfrak{a} = \langle f_1, \ldots, f_k \rangle$ be an ideal such that the radical of $\mathfrak{a}$ is $\sqrt{\mathfrak{a}} = \langle x_1 - c_1, \ldots, x_n - c_n \rangle$. Then the multiplicity of the solution $(c_1, \ldots, c_n)$ to the system $f_1 = \cdots = f_k = 0$ is $\dim_{\mathbb{C}} \mathbb{C}[x_1, \ldots, x_n]/\mathfrak{a}$.

We should check that this dimension is finite. We know that $\mathfrak{m}_P^s \subseteq \mathfrak{a}$ for some s, and hence that $\dim_{\mathbb{C}} \mathbb{C}[x_1, \ldots, x_n]/\mathfrak{a} \leq \dim_{\mathbb{C}} \mathbb{C}[x_1, \ldots, x_n]/\mathfrak{m}_P^s$. The latter is easily seen to be finite, since after a coordinate transformation we can assume that $P = (0, 0, \ldots, 0)$, so $\mathfrak{m}_P = \langle x_1, \ldots, x_n \rangle$. The monomials of degree $< s$ constitute a $\mathbb{C}$-basis for $\dim_{\mathbb{C}} \mathbb{C}[x_1, \ldots, x_n]/\langle x_1, \ldots, x_n \rangle^s$, and there are only finitely many such monomials.

Exercises

1. Show the statements above: that $\dim_{\mathbb{C}} \mathbb{C}[x,y]/\left\langle x, y^2 \right\rangle = 2$ and that $\dim_{\mathbb{C}} \mathbb{C}[x,y]/\left\langle x^2, y^2 \right\rangle = 4$.

2. Show that in fact $\dim_{\mathbb{C}} \mathbb{C}[x_1, \ldots, x_n]/\mathfrak{m}_P^s = \binom{n+s-1}{n}$.

6.2 Systems with Finitely Many Solutions

We will now study systems of equations with finitely many solutions. Since points are considered to be zero dimensional, such systems are often called zero-dimensional systems.

6.2.1 Decomposition of the Ideal

Now suppose that $f_1 = \cdots = f_k = 0$ has finitely many solutions $P_1, \ldots, P_s$. Let $\mathfrak{a} = \langle f_1, \ldots, f_k \rangle$. Then $\mathfrak{v}(\mathfrak{a}) = \{P_1, \ldots, P_s\}$ so $\sqrt{\mathfrak{a}} = \mathfrak{m}_{P_1} \cap \cdots \cap \mathfrak{m}_{P_s}$;

see Proposition 6 in Section 4.1. We have proved that any ideal $\mathfrak{a}$ has a primary decomposition, $\mathfrak{a} = \mathfrak{q}_1 \cap \cdots \cap \mathfrak{q}_t$ with $\sqrt{\mathfrak{q}_i} = \mathfrak{p}_i$, a prime ideal. Then

$$\sqrt{\mathfrak{a}} = \sqrt{\mathfrak{q}_1 \cap \cdots \cap \mathfrak{q}_t} = \sqrt{\mathfrak{q}_1} \cap \cdots \cap \sqrt{\mathfrak{q}_t} = \mathfrak{p}_1 \cap \cdots \cap \mathfrak{p}_t.$$

Thus $\mathfrak{p}_1 \cap \cdots \cap \mathfrak{p}_t = \mathfrak{m}_{P_1} \cap \cdots \cap \mathfrak{m}_{P_s}$, which gives

$$\mathfrak{v}(\mathfrak{p}_1 \cap \cdots \cap \mathfrak{p}_t) = \mathfrak{v}(\mathfrak{p}_1) \cup \cdots \cup \mathfrak{v}(\mathfrak{p}_t) = \{P_1, \ldots, P_s\} =$$

$$\mathfrak{v}(\mathfrak{m}_{P_1}) \cup \cdots \cup \mathfrak{v}(\mathfrak{m}_{P_s}) = \mathfrak{v}(\mathfrak{m}_{P_1} \cap \cdots \cap \mathfrak{m}_{P_s}).$$

We have that $\mathfrak{v}(\mathfrak{p}_i)$ is an irreducible set since $\mathfrak{p}_i$ is a prime ideal. The only irreducible subsets of $\{P_1, \ldots, P_s\}$ are the one-point subsets $\{P_j\}$, so we get that $\mathfrak{v}(\mathfrak{p}_i) = \{P_j\}$ for some j, i.e.

$$\mathfrak{p}_i = \mathfrak{i}(\mathfrak{v}(\mathfrak{p}_i)) = \mathfrak{i}(\{P_j\}) = \mathfrak{m}_{P_j}.$$

We have proved the first part of the following theorem.

THEOREM 2. *Suppose that $f_1 = \cdots = f_k = 0$ has a finite set of solutions $\{P_1, \ldots, P_s\}$. Then $\langle f_1, \ldots, f_k \rangle = \mathfrak{q}_1 \cap \cdots \cap \mathfrak{q}_s$, with $\sqrt{\mathfrak{q}_i} = \mathfrak{m}_{P_i}$ for $i = 1, \ldots, s$. The ideals $\mathfrak{q}_i$ are unique.*

Proof. All primary components are isolated, so they are unique; see Chapter 5, Proposition 14. □

6.2.2 Decomposition of the Ring

Next we will see that writing the ideal $\mathfrak{a}$ as an intersection of simpler ideals can be used to decompose the ring $\mathbb{C}[x_1, \ldots, x_n]/\mathfrak{a}$ as a product of simpler rings; we claim that

$$\mathbb{C}[x_1, \ldots, x_n]/\mathfrak{a} \simeq \mathbb{C}[x_1, \ldots, x_n]/\mathfrak{q}_1 \times \cdots \times \mathbb{C}[x_1, \ldots, x_n]/\mathfrak{q}_s$$

if $\mathfrak{a} = \mathfrak{q}_1 \cap \cdots \cap \mathfrak{q}_s$.

LEMMA 3. *Let $\mathfrak{q}_i, i = 1, \ldots, s$ be ideals in $\mathbb{C}[x_1, \ldots, x_n]$ with $\sqrt{\mathfrak{q}_i} = \mathfrak{m}_i$, where $\mathfrak{m}_i$ are different maximal ideals for $i = 1, \ldots, s$. Then*

$$\mathfrak{q}_1 \cap \cdots \cap \mathfrak{q}_{s-1} + \mathfrak{q}_s = \langle 1 \rangle .$$

Proof. We first note that $\mathfrak{m}_i + \mathfrak{m}_s = \langle 1 \rangle$ if $i \neq s$, since $\mathfrak{m}_i + \mathfrak{m}_s$ strictly contains $\mathfrak{m}_i$, and the only ideal which strictly contains a maximal ideal is $\langle 1 \rangle$. This gives that $\mathfrak{q}_i + \mathfrak{q}_s = \langle 1 \rangle$ since

$$\langle 1 \rangle = \sqrt{\langle 1 \rangle} = \sqrt{\mathfrak{m}_i + \mathfrak{m}_s} = \sqrt{\sqrt{\mathfrak{q}_i} + \sqrt{\mathfrak{q}_s}} = \sqrt{\mathfrak{q}_i + \mathfrak{q}_s}$$

(here we have used Proposition 45 in Section 1.3.5) and the only ideal $\mathfrak{a}$ with radical $\langle 1\rangle$ is $\mathfrak{a} = \langle 1\rangle$, since a power of 1 belongs to $\mathfrak{a}$. Now

$$\langle 1\rangle = \langle 1\rangle^{s-1} = (\mathfrak{q}_1 + \mathfrak{q}_s)\cdots(\mathfrak{q}_{s-1} + \mathfrak{q}_s) \subseteq \mathfrak{q}_1\cdots\mathfrak{q}_{s-1} + \mathfrak{q}_s \subseteq \cap_{i=1}^{s-1}\mathfrak{q}_i + \mathfrak{q}_s.$$

□

We are now ready to prove our claim.

THEOREM 4. *If* $\mathfrak{a} = \cap_{i=1}^{s}\mathfrak{q}_i \subseteq \mathbb{C}[x_1,\ldots,x_n]$, *with* $\sqrt{\mathfrak{q}_i}$ *different maximal ideals for* $i = 1,\ldots,s$, *then*

$$\mathbb{C}[x_1,\ldots,x_n]/\mathfrak{a} \simeq \mathbb{C}[x_1,\ldots,x_n]/\mathfrak{q}_1 \times\cdots\times \mathbb{C}[x_1,\ldots,x_n]/\mathfrak{q}_s.$$

Proof. We define a map

$$\phi : \mathbb{C}[x_1,\ldots,x_n] \to \mathbb{C}[x_1,\ldots,x_n]/\mathfrak{q}_1 \times\cdots\times \mathbb{C}[x_1,\ldots,x_n]/\mathfrak{q}_s$$

by $\phi(f) = (f+\mathfrak{q}_1,\ldots,f+\mathfrak{q}_s)$. It is easy to see that ϕ is a homomorphism. The kernel equals

$$\{\, f \mid f+\mathfrak{q}_i = \mathfrak{q}_i, i = 1,\ldots,s \,\} = \{\, f \mid f \in \mathfrak{q}_i, i = 1,\ldots,s \,\} = \cap_{i=1}^{s}\mathfrak{q}_i = \mathfrak{a},$$

hence ϕ gives an isomorphism $\psi : \mathbb{C}[x_1,\ldots,x_n]/\mathfrak{a} \to \mathrm{im}(\phi)$, defined by $\psi(f+\mathfrak{a}) = (f+\mathfrak{q}_1,\ldots,f+\mathfrak{q}_s))$, according to Theorem 61 in Section 1.7. It remains to show that ϕ is surjective. We show this by induction on s. If $s = 2$ there are $q_i \in \mathfrak{q}_i, i = 1,2$, such that $q_1 + q_2 = 1$ since $\mathfrak{q}_1 + \mathfrak{q}_2 = \langle 1\rangle$. Hence $\phi(q_1) = (0+\mathfrak{q}_1, 1+\mathfrak{q}_2)$ and $\phi(q_2) = (1+\mathfrak{q}_1, 0+\mathfrak{q}_2)$ so $\phi(f_2q_1+f_1q_2) = (f_1+\mathfrak{q}_1, f_2+\mathfrak{q}_2)$ and ϕ is surjective. By induction we have

$$\mathbb{C}[x_1,\ldots,x_n]/\cap_{i=1}^{s-1}\mathfrak{q}_i \simeq \mathbb{C}[x_1,\ldots,x_n]/\mathfrak{q}_1 \times\cdots\times \mathbb{C}[x_1,\ldots,x_n]/\mathfrak{q}_{s-1},$$

so for each $(s-1)$-tuple $(f_1,\ldots,f_{s-1})$ there is an f such that $f+\mathfrak{q}_i = f_i+\mathfrak{q}_i, i = 1,\ldots,s-1$. By Lemma 3, $\mathfrak{q}_1\cap\cdots\cap\mathfrak{q}_{s-1}+\mathfrak{q}_s = \langle 1\rangle$, so there exist $a \in \mathfrak{q}_1\cap\cdots\cap\mathfrak{q}_{s-1}$ and $b \in \mathfrak{q}_s$ such that $a+b = 1$. Let $(f_1,\ldots,f_s)$ be any s-tuple of polynomials. By induction there is an f such that $f+\mathfrak{q}_i = f_i+\mathfrak{q}_i$ for $i = 1,\ldots,s-1$. Since $a \in \mathfrak{q}_i$ for $i < s$, we have

$$f_sa + fb + \mathfrak{q}_i = fb + \mathfrak{q}_i = f(1-a) + \mathfrak{q}_i = f + \mathfrak{q}_i = f_i + \mathfrak{q}_i$$

if $i < s$. Since $b \in \mathfrak{q}_s$ we have

$$f_sa + fb + \mathfrak{q}_s = f_sa + \mathfrak{q}_s = f_s(1-b) + \mathfrak{q}_s = f_s + \mathfrak{q}_s.$$

Hence $\phi(f_sa + fb) = (f_1+\mathfrak{q}_1,\ldots,f_s+\mathfrak{q}_s)$ so ϕ is surjective. □

COROLLARY 5. *If $f_1 = \cdots = f_k = 0$ has a finite number of solutions then the number of solutions, counted with multiplicity, equals*

$$\dim_{\mathbb{C}} \mathbb{C}[x_1,\ldots,x_n]/\langle f_1,\ldots,f_k\rangle .$$

Proof. If $\langle f_1,\ldots,f_k\rangle = \mathfrak{q}_1 \cap \cdots \cap \mathfrak{q}_s$ then $\dim_{\mathbb{C}} \mathbb{C}[x_1,\ldots,x_n]/\langle f_1,\ldots,f_k\rangle = \dim_{\mathbb{C}}(\mathbb{C}[x_1,\ldots,x_n]/\mathfrak{q}_1 \times \cdots \times \mathbb{C}[x_1,\ldots,x_n]/\mathfrak{q}_s) = \dim_{\mathbb{C}} \mathbb{C}[x_1,\ldots,x_n]/\mathfrak{q}_1 + \cdots + \dim_{\mathbb{C}} \mathbb{C}[x_1,\ldots,x_n]/\mathfrak{q}_s$; see Theorem 77 in Section 1.9. □

We have seen that if $f_1 = f_2 = \cdots = f_k = 0$ has finitely many solutions then $\dim_{\mathbb{C}} \mathbb{C}[x_1,\ldots,x_n]/\mathfrak{a}$, where $\mathfrak{a} = \langle f_1, f_2,\ldots,f_k\rangle$, equals the number of solutions. In particular this dimension is finite. We will now show that the converse to this statement is true, i.e., if $\dim_{\mathbb{C}} \mathbb{C}[x_1,\ldots,x_n]/\mathfrak{a} < \infty$, where $\mathfrak{a} = \langle f_1, f_2,\ldots,f_k\rangle$, then the number of solutions to $f_1 = f_2 = \cdots = f_k = 0$ is finite. We know that $\dim_{\mathbb{C}} \mathbb{C}[x_1,\ldots,x_n]/\mathfrak{a} = \dim_{\mathbb{C}} \mathbb{C}[x_1,\ldots,x_n]/l(\mathfrak{a})$, where $l(\mathfrak{a})$ is the ideal of leading monomials of $\mathfrak{a}$ (in some ordering); see Proposition 15 in Section 3.6. We first characterize those monomial ideals $l(\mathfrak{a})$ for which

$$\dim_{\mathbb{C}} \mathbb{C}[x_1,\ldots,x_n]/l(\mathfrak{a}) < \infty.$$

PROPOSITION 6. *For a monomial ideal $\mathfrak{b}$ we have*

$$\dim_{\mathbb{C}} \mathbb{C}[x_1,\ldots,x_n]/\mathfrak{b} < \infty$$

if and only if for each i there is an m_i such that $x_i^{m_i} \in \mathfrak{b}$.

Proof. Suppose that for each i there is an m_i such that $x_i^{m_i} \in \mathfrak{b}$, and let $N = \max\{m_i\}$. Then each monomial of degree greater than $n(N-1)$ belongs to $\mathfrak{b}$, because a monomial of degree greater than $n(N-1)$ must contain at least a power N of some variable, so there are only finitely many monomials outside $\mathfrak{b}$. On the other hand, if $\dim_{\mathbb{C}} \mathbb{C}[x_1,\ldots,x_n]/\mathfrak{b} < \infty$, then for each i there is an m_i such that $x_i^{m_i} \in \mathfrak{b}$, because otherwise all infinitely many powers of x_i would be outside $\mathfrak{b}$. □

Given any ordering, we have

$$\dim_{\mathbb{C}} \mathbb{C}[x_1,\ldots,x_n]/\mathfrak{a} = \dim_{\mathbb{C}} \mathbb{C}[x_1,\ldots,x_n]/l(\mathfrak{a}),$$

where the initial ideal is calculated in the given ordering. Thus, although Gröbner bases, and thus initial ideals, depend heavily on the ordering, $\dim_{\mathbb{C}} \mathbb{C}[x_1,\ldots,x_n]/l(\mathfrak{a})$ does not. Thus the property that each variable has a power in $l(\mathfrak{a})$ does not depend on the ordering of the monomials we have used (though the actual power might depend on the ordering). We now investigate the consequences for the Gröbner basis if we choose

Lex with $x_1 \succ x_2 \succ \cdots \succ x_n$. Suppose $\dim_{\mathbb{C}} \mathbb{C}[x_1, \ldots, x_n]/\mathfrak{a} < \infty$ and that $\mathfrak{b} = l(\mathfrak{a})$ in Lex. Then there is an $m_n \in \mathbb{N}$ such that $x_n^{m_n} \in \mathfrak{b}$, hence there is an f in the Gröbner basis with $\mathrm{lm}(f) = x_n^{m_n}$. But the only monomials that are smaller than $x_n^{m_n}$ in this ordering are $x_n^i, i < m_n$. Hence $f \in \mathbb{C}[x_n], \deg(f) = m_n$, so there are only at most m_n different possible values of x_n in a solution to the system. By choosing Lex with say $x_n \succ x_1 \succ x_2 \succ \cdots \succ x_{n-1}$ we see in the same way that there are only finitely many possibilities for x_{n-1} in a solution to the system a.s.o.. We have proven the following theorem.

THEOREM 7. *Let $\mathfrak{a} = \langle f_1, \ldots, f_k \rangle$. Then there are finitely many solutions to $f_1 = \cdots = f_k = 0$ if and only if* $\dim_{\mathbb{C}} \mathbb{C}[x_1, \ldots, x_n]/\mathfrak{a} < \infty$. *The number of solutions, counted with multiplicity, is* $\dim_{\mathbb{C}} \mathbb{C}[x_1, \ldots, x_n]/\mathfrak{a}$.

6.3 Solving Zero-dimensional Systems

We now turn to the question of how to actually get the solutions to a system of polynomial equations. We can reduce the problem to solving polynomial equations in one variable. In one variable we have no help from Gröbner bases. One method comes from the reasoning above. For each variable x_i we can get an equation, $g_i(x_i) = 0$, that the x_i-coordinate c_i must satisfy for any solution $(c_1, c_2, \ldots, c_n)$ to the system. If $\deg(g_i) = m_i$ we have $m_1 m_2 \cdots m_n$ candidate solutions and we can plug them into the system to see if they are genuine solutions. This is theoretically possible, but practically impossible in many cases as it could mean calculating n different Gröbner bases in Lex. Theoretical results and overwhelming practical evidence show how it is a very time-consuming task to calculate Gröbner bases in Lex. It is almost always many times slower than a corresponding calculation in Degrevlex. We will now describe a method where one calculation in Lex suffices, but we will later indicate that there are even faster methods. Hence consider the system $f_1 = f_2 = \cdots = f_k = 0$, where $f_i \in \mathbb{C}[x_1, \ldots, x_n]$, and suppose there are finitely many solutions (this will be checked automatically during the calculations). Choose Lex with $x_1 \succ \cdots \succ x_n$. As above, the Gröbner basis contains an element $g_n(x_n)$ in one variable. If we consider the reduced Gröbner basis there is only one such polynomial, otherwise the polynomials of higher degree could be reduced by those of lower degree. Denote the solutions to $g_n(x_n) = 0$ by $c_1, \ldots, c_{m_n}$. There is also an element g_{n-1} with $\mathrm{lm}(g_{n-1}) = x_{n-1}^{m_{n-1}}$ in the Gröbner basis. The only monomials smaller than $x_{n-1}^{m_{n-1}}$ that can occur in $l(\mathfrak{a})$ are $x_{n-1}^j x_n^k$ with $j < m_{n-1}, k < m_n$. Hence there can be only finitely many elements $G_1(x_{n-1}, x_n), \ldots, G_l(x_{n-1}, x_n)$ in the Gröbner basis corresponding to such monomials. Plug in $x_n = c_1$ in each G_i. Then we get a finite set of polynomials $G_i(x_{n-1}, c_1)$ in one variable. The system $G_1(x_{n-1}, c_1) = \cdots = G_l(x_{n-1}, c_1)$ is equivalent to one equation $\gcd(G_1(x_{n-1}, c_1), \ldots, G_l(x_{n-1}, c_1)) = 0$. This equation has only finitely many solutions $c_{1,1}, \ldots, c_{1,m_{n-1}}$. Repeat this process for $c_2, \ldots, c_{m_n}$. There is also an element g_{n-2} with $\mathrm{lm}(g_{n-2}) = x_{n-2}^{m_{n-2}}$ in the Gröbner basis. The only monomials smaller than $x_{n-2}^{m_{n-2}}$ that can occur in $l(\mathfrak{a})$ are $x_{n-2}^j x_{n-1}^k x_n^l$ with $j < m_{n-2}, k < m_{n-1}, l < m_n$. Hence there can be only finitely many elements $H_1(x_{n-2}, x_{n-1}, x_n), \ldots, H_m(x_{n-2}, x_{n-1}, x_n)$ in the Gröbner basis corresponding to such monomials. Plug in $x_n = c_1, x_{n-1} = c_{1,1}$ in each H_i. Then we get a finite set of polynomials $H_i(x_{n-2}, c_{1,1}, c_1)$ in one variable. The system

$$H_1(x_{n-2}, c_{1,1}, c_1) = \cdots = H_m(x_{n-2}, c_{1,1}, c_1)$$

is equivalent to one equation

$$\gcd(H_1(x_{n-2}, c_{1,1}, c_1), \ldots, H_m(x_{n-2}, c_{1,1}, c_1)) = 0.$$

This equation has only finitely many solutions $c_{2,1}, \ldots, c_{2,m_2}$. Repeat this process for each pair $(c_{j,i}, c_j)$. It is clear that induction gives us all solutions to the system.

EXAMPLE 8. Consider the system

$$\begin{aligned} x_1^2 + x_2^2 + x_3^2 - 3 &= 0 \\ x_1 x_2 x_3 - 1 &= 0 \\ x_1 x_3 + x_2 - 2 &= 0. \end{aligned}$$

The ideal

$$\langle x_1^2 + x_2^2 + x_3^2 - 3, x_1 x_2 x_3 - 1, x_1 x_3 + x_2 - 2 \rangle$$

has in Lex with $x_1 \succ x_2 \succ x_3$ the reduced Gröbner basis

$$\{x_1 + x_2 x_3 - 1/2 x_3^5 + 2x_3^3 - 7/2 x_3, x_2^2 - 2x_2 + 1,$$

$$x_2 x_3^2 - x_2 + 1/2 x_3^4 - 2x_3^2 + 3/2, x_3^6 - 3x_3^4 + 3x_3^2 - 1\}.$$

From the last equation we get $x_3 = \pm 1$. If we substitute $x_3 = 1$ in the first three equations, we get $x_1 + x_2 - 2 = x_2^2 - 2x_2 + 1 = 0$, so $x_2 = x_1 = 1$. If we substitute $x_3 = -1$, we get $x_1 - x_2 + 2 = x_2^2 - 2x_2 + 1 = 0 = 0$, so $x_2 = 1, x_1 = -1$. Hence the system has the two solutions $(1, 1, 1)$ and $(-1, 1, -1)$. (One can show that they both have multiplicity 4.) □

Exercises

1. Solve the system of equations

$$\begin{aligned} x^3 + x^2y - xy - y^2 &= 0 \\ xy - x - y + 1 &= 0 \\ x^2 - y^2 &= 0 \end{aligned}$$

by calculating the Gröbner basis of

$$\langle x^3 + x^2y - xy - y^2, xy - x - y + 1, x^2 - y^2 \rangle$$

in Lex.

† 2. Solve the system

$$\begin{aligned} xy - z &= 0 \\ xz - y &= 0 \\ yz - x &= 0 \end{aligned}$$

3. Solve the system

$$\begin{aligned} x^3 + xy + y^2 - z^2 &= 0 \\ x^2 - xz + y + 2z - 2 &= 0 \\ x^4 + xyz + yz + z^2 - 2 &= 0 \\ x^2 - y - 2z + 1 &= 0. \end{aligned}$$

4. The ideal

$$\mathfrak{a} = \langle a + b + c + d + 1, ab + bc + cd + d + a, abc + bcd + cd + ad + ab,$$
$$abcd + bcd + cda + dab + abc, abcd - 1 \rangle$$

has ideal of leading monomials $l(\mathfrak{a}) = \langle a, b^2, bc^2, bcd, bd^2, c^3, cd^2, d^3 \rangle$ in Degrevlex. Determine the number of solutions to

$$\begin{aligned} a + b + c + d + 1 &= 0 \\ ab + bc + cd + d + a &= 0 \\ abc + bcd + cd + ad + ab &= 0 \\ abcd + bcd + cda + dab + abc &= 0 \\ abcd - 1 &= 0 \end{aligned}$$

5. Determine the number of solutions to

$$\begin{aligned} x_1x_2 - 1 &= 0 \\ x_2x_3 - x_1 &= 0 \\ x_3^2 - x_2 - 1 &= 0 \end{aligned}$$

6. A point $(a_1, \ldots, a_n)$ is called a *critical point* for $f(x_1, \ldots, x_n)$ if

$$\frac{\partial f}{\partial x_i}(a_1, \ldots, a_n) = 0$$

for all i. Determine the critical points for $f(x, y, z) = x^3 + 2xy^2 + y^3 + yz + z^2$.

6.4 Systems of Higher Dimension

For zero-dimensional systems it is clear what one should mean by solving the system, at least if all solutions have multiplicity 1. A solution is then simply a complete list of points satisfying the equations. For systems with infinitely many solutions the situation is more difficult. It is not even clear what one should mean by "solving" such a system. There is no natural easiest way to present the solutions. Despite that, one can sometimes use Gröbner bases to get an easier presentation of the solutions than given by the system itself. As for zero-dimensional systems, the Lex ordering is best suited to present the solutions.

EXAMPLE 9. Consider the system

$$\begin{aligned} x^2 - y^2 + x + y - z &= 0 \\ x^2 + 2y^2 - 2x + y - z &= 0 \\ x^3 - x^2z - xy^2 - 2y^2z + x^2 + xy + xz - yz + z^2 &= 0. \end{aligned}$$

A Gröbner basis for

$$\langle x^2 - y^2 + x + y - z, x^2 + 2y^2 - 2x + y - z,$$
$$x^3 - x^2z - xy^2 - 2y^2z + x^2 + xy + xz - yz + z^2\rangle$$

is $\{x - y^2, y^4 + y - z\}$. Thus the solutions are given by

$$\begin{aligned} z &= t^4 + t \\ y &= t \\ x &= t^2. \end{aligned}$$

□

Exercise

1. The method of Lagrange multipliers tells that if a maximum (or minimum) of $f(x_1, \ldots, x_n)$ subject to the condition $g(x_1, \ldots, x_n) = 0$ exists, it is attained at a point where $\frac{\partial}{\partial x_i}(f + \lambda g) = 0$ for some λ.

   (a) Determine $\max(xy^2z^3)$ when $x + y + z - 1$ and $x, y, z > 0$.

   (b) Determine the point on the surface $xy + xz + yz = 1$ that is closest to the origin.

Chapter 7

Applications of Gröbner Bases

Solving systems of polynomial equations, which we treated in the previous chapter, is one of the most important applications of Gröbner bases. In this chapter we will consider some other applications within mathematics. We stress that it is constructiveness we are interested in; we want to show that certain operations may actually be performed constructively (by hand or with a computer).

We recall some important facts that we will use. Let $\mathfrak{a}$ be an ideal in $k[x_1, \ldots, x_n]$ and let G be a Gröbner basis for $\mathfrak{a}$. Any element f has a unique normal form with respect to G; $\mathrm{NF}(f)$ is a linear combination, with coefficients in k, of monomials outside the ideal of leading monomials $l(\mathfrak{a})$. We have for all f that $f - \mathrm{NF}(f) \in \mathfrak{a}$. In many problems we need a specific ordering of the monomials. Usually we divide the variables into two groups, say $\{x_1, \ldots, x_m\}$ and $\{y_1, \ldots, y_n\}$ in $k[x_1, \ldots, x_m, y_1, \ldots, y_n]$, and we want all monomials (except the monomial 1) in the first group to be larger than any monomial in the last group, or similarly, that all x_i are larger than any monomial in the y_j's. This is achieved by the Lex ordering with $x_1 \succ \cdots \succ x_m \succ y_1 \cdots \succ y_n$ but there are several other orderings satisfying our need; see Section 3.1.

7.1 Membership Problems

7.1.1 Ideal Membership

We have already touched on the problem of how to judge whether an element belongs to a given ideal. We recall the method. Let $\mathfrak{a} = \langle f_1, \ldots, f_r \rangle$ be an ideal in $k[x_1, \ldots, x_n]$ and let $g \in k[x_1, \ldots, x_n]$. Determine a Gröbner basis G of $\mathfrak{a}$. We recall that each element in $k[x_1, \ldots, x_n]$ has a unique normal form with respect to G, and that $\mathrm{NF}(f) = 0$ if and only if $f \in \mathfrak{a}$.

EXAMPLE 1. Suppose we want to show that

$$y^3 + z^3 \in \langle x^2, xy + y^2, xz + z^2 \rangle = \mathfrak{a}.$$

A Gröbner basis for $\mathfrak{a}$ in Degrevlex is $\{x^2, xy + y^2, xz + z^2, y^3, z^3\}$. Since y^3 and z^3 belong to the Gröbner basis we have $\mathrm{NF}(y^3 + z^3) = 0$. Thus $y^3 + z^3$ belongs to $\mathfrak{a}$. We note that, if we remember the calculation of the Gröbner basis, we can actually get an expression

$$y^3 + z^3 = f_1 \cdot x^2 + f_2 \cdot (xy + y^2) + f_3 \cdot (xz + z^2).$$

The S-polynomial

$$S(x^2, xy + y^2) = y \cdot x^2 - x \cdot (xy + y^2) = -xy^2 = -y \cdot (xy + y^2) + 1 \cdot y^3,$$

so

$$y^3 = y \cdot x^2 + (-x + y)(xy + y^2).$$

Similarly

$$z^3 = z \cdot x^2 + (-x + z)(xz + z^2)$$

so

$$y^3 + z^3 = (y + z)x^2 + (y - x)(xy + y^2) + (z - x)(xz + z^2).$$

□

Exercise

1. Determine the smallest power of z which belongs to the ideal $\mathfrak{a} = \langle x^2 + xz, xy + y^2, yz + z^2 \rangle$. Express this smallest power of z as a linear combination of the generators of $\mathfrak{a}$.

7.1.2 Radical Membership

If we assume that we are in a polynomial ring over $\mathbb{C}$, we can answer the following question: Does $g \in \sqrt{\langle f_1, \ldots, f_r \rangle}$? Note that, according to Corollary 13 in Section 4.2, we can also formulate the question like this: Is every solution to $f_1 = \cdots = f_r = 0$ also a solution to $g = 0$? By using a trick we will reduce this problem to an ideal membership problem in a ring with an extra variable.

PROPOSITION 2. *Let $\langle f_1, \ldots, f_r \rangle$ be an ideal in $\mathbb{C}[x_1, \ldots, x_n]$. Then an element $g \in \mathbb{C}[x_1, \ldots, x_n]$ belongs to $\sqrt{\langle f_1, \ldots, f_r \rangle}$ if and only if the ideal $\mathfrak{b} = \langle f_1, \ldots, f_r, 1 - yg \rangle$ in $\mathbb{C}[x_1, \ldots, x_n, y]$ equals $\mathbb{C}[x_1, \ldots, x_n, y]$, i.e., if and only if $1 \in \mathfrak{b}$. This is true if and only if the reduced Gröbner basis of $\mathfrak{b}$ is $\{1\}$.*

Proof. This is exactly what we proved in Hilbert's Nullstellensatz; see Corollary 13 in Section 4.2.

□

EXAMPLE 3. Let us show, using this method, that $y \in \sqrt{\langle x^2, xy + y^2 \rangle}$. Consider the ideal $\mathfrak{b} = \left\langle x^2, xy + y^2, 1 - yz \right\rangle$ in $\mathbb{C}[x, y, z]$. A Gröbner basis for $\mathfrak{b}$ is $\{1\}$. □

Exercise

1. Determine $\sqrt{\mathfrak{a}}$ if $\mathfrak{a} = \left\langle x^2 + xz, xy + y^2, yz + z^2 \right\rangle$ in $\mathbb{C}[x, y, z]$.

7.1.3 Subalgebra Membership

Let $f_1, \ldots, f_r \in k[x_1, \ldots, x_n]$. The subalgebra $k[f_1, \ldots, f_r]$ of $k[x_1, \ldots, x_n]$ generated by $f_1, \ldots, f_r$ is the set of polynomials $f(f_1, \ldots, f_r)$, where f is a polynomial in r variables. Hence $k[f_1, \ldots, f_r]$ consists of all polynomials which are the result of a substitution $y_1 = f_1, \ldots, y_r = f_r$ in some polynomial $f(y_1, \ldots, y_r) \in k[y_1, \ldots, y_r]$. Another way to phrase this is to say that $k[f_1, \ldots, f_r]$ is the image of the homomorphism $\phi : k[y_1, \ldots, y_r] \longrightarrow k[x_1, \ldots, x_n]$ defined by $\phi(y_i) = f_i(x_1, \ldots, x_n)$. Thus

$$k[x^2] = \left\{ a_0 + a_2 x^2 + a_4 x^4 + \cdots + a_{2k} x^{2k} \;\middle|\; a_i \in k \right\},$$

and

$$k[x^2, y^3] = \{\sum a_{ij} x^{2i} y^{3j}\}[a_{ij} \in k].$$

It is clear that $k[f_1, \ldots, f_r]$ is a ring.

Let $f_1, \ldots, f_r, g \in k[x_1, \ldots, x_n]$. The subalgebra membership problem is to decide whether or not $g \in k[f_1, \ldots, f_r]$. We will solve this and furthermore, if $g \in k[f_1, \ldots, f_r]$, determine a polynomial $h(y_1, \ldots, y_r)$ such that $g = h(f_1, \ldots, f_r)$.

PROPOSITION 4. *Let $f_1, \ldots, f_r, g \in k[x_1, \ldots, x_n]$ and consider the ideal $\mathfrak{b} = \langle y_1 - f_1, \ldots, y_r - f_r \rangle$ in $k[x_1, \ldots, x_n, y_1, \ldots, y_r]$. We use an ordering of the monomials in $k[x_1, \ldots, x_n, y_1, \ldots, y_r]$ such that each x_i is larger than every monomial in $\{y_1, \ldots, y_r\}$. Then $g \in k[f_1, \ldots, f_r]$ if and only if the normal form of g with respect to a Gröbner basis of $\mathfrak{b}$ belongs to $k[y_1, \ldots, y_r]$. If so, and the normal form of g is $h(y_1, \ldots, y_r)$, then $g = h(f_1, \ldots, f_r)$.*

Proof. Consider the homomorphism

$$\phi : k[x_1, \ldots, x_n, y_1, \ldots, y_r] \longrightarrow k[x_1, \ldots, x_n]$$

defined by

$$\phi(f(x_1, \ldots, x_n, y_1, \ldots, y_r)) = f(x_1, \ldots, x_n, f_1, \ldots, f_r).$$

It is clear that $\ker(\phi) = \mathfrak{b}$. Suppose $g \in k[f_1, \ldots, f_r]$, say $g = h(f_1, \ldots, f_r)$. Then $g - h(y_1, \ldots, y_r) \in \mathfrak{b}$, which gives that $\mathrm{NF}(g) = \mathrm{NF}(h(y_1, \ldots, y_r))$. But every polynomial in $k[y_1, \ldots, y_r]$ will, in the ordering we have chosen, reduce to polynomials in $k[y_1, \ldots, y_r]$, so $\mathrm{NF}(g) \in k[y_1, \ldots, y_r]$. Since $g - h(y_1, \ldots, y_r) \in \mathfrak{b}$ we get $g = \phi(g) = \phi(h(y_1, \ldots, y_r)) = h(f_1, \ldots, f_r)$. Conversely, if $\mathrm{NF}(g) = h(y_1, \ldots, y_r)$ we have $g - h(y_1, \ldots, y_r) \in \mathfrak{b}$, so $\phi(g - \mathrm{NF}(g)) = g - h(f_1, \ldots, f_r) = 0$. □

EXAMPLE 5. We will show that $x^8 - x^2 \in k[x^4 + x^2, x^3]$. Consider the ideal $\mathfrak{b} = \langle y_1 - x^4 - x^2, y_2 - x^3 \rangle$ in $k[x, y_1, y_2]$. The Gröbner basis for $\mathfrak{b}$ in the Lex ordering with $x \succ y_1 \succ y_2$ is

$$G = \{x^2 + xy_2 - y_1, xy_1 + xy_2^2 - y_1y_2 - y_2,$$

$$xy_2^3 - xy_2 + y_1^2 - y_1y_2^2 - 2y_2^2, y_1^3 - 3y_1y_2^2 - y_2^4 - y_2^2\}.$$

The normal form of $x^8 - x^2$ with respect to G is $y_1^2 - 2y_2^2 - y_1$. Thus

$$x^8 - x^2 = (x^4 + x^2)^2 - 2(x^3)^2 - (x^4 + x^2) \in k[x^4 + x^2, x^3].$$

□

A polynomial $f(x_1, x_2, \ldots, x_n)$ is called *symmetric* if any permutation of the variables leaves f fixed. Examples of symmetric polynomials are the following, called the elementary symmetricfunctions:

$$\begin{aligned}
\sigma_1 &= x_1 + x_2 + \cdots + x_n \\
\sigma_2 &= \sum_{1 \le i < j \le n} x_i x_j \\
\sigma_3 &= \sum_{1 \le i < j < k \le n} x_i x_j x_k \\
&\vdots \\
\sigma_n &= x_1 x_2 \cdots x_n
\end{aligned}$$

It is well known and not so hard to prove that every symmetric polynomial is a polynomial in the elementary symmetric functions.

EXAMPLE 6. The polynomial $g = x^2y + x^2z + y^2x + y^2z + z^2x + z^2y$ is symmetric. We will write g as a polynomial in the elementary symmetric functions $\sigma_1 = x + y + z, \sigma_2 = xy + xz + yz, \sigma_3 = xyz$. Consider the ideal

$$\mathfrak{b} = \langle y_1 - (x + y + z), y_2 - (xy + xz + yz), y_3 - xyz \rangle$$

in $k[x, y, z, y_1, y_2, y_3]$ equipped with the lexicographical ordering induced by $x \succ y \succ z \succ y_1 \succ y_2 \succ y_3$. A Gröbner basis for $\mathfrak{b}$ is

$$\{x + y + z - y_1, y^2 + yz - yy_1 + z^2 - zy_1 + y_2, z^3 - z^2y_1 + zy_2 - y_3\},$$

so the normal form of g is $y_1y_2 - 3y_3$. Hence $g = \sigma_1\sigma_2 - 3\sigma_3$. □

Exercises

1. Write $x^3y + x^3z + y^3x + y^3z + z^3x + z^3y$ as a polynomial in the elementary symmetric functions.
2. Show that $x^{15} \in \mathbb{Z}_2[x^4, x^6 + x^7]$.
3. Show that $k[x^3, x^4, x^5]$ consists of all polynomials with coefficient 0 both for x and x^2.

7.2 Calculation in Factor Rings of Polynomial Rings

Let $R = k[x_1, \ldots, x_n]/\mathfrak{a}$. We will consider the problem of how to actually perform ring operations in R. The first thing we want is a good representation of the elements in R. This is achieved by taking the normal

Table 7.1: Multiplication table using normal forms.

$\cdot$	1	x	y	z	xz	y^2	z^2	z^3
1	1	x	y	z	xz	y^2	z^2	z^3
x	x	$-xz$	$-y^2$	xz	$-z^3$	$-z^3$	z^3	0
y	y	$-y^2$	y^2	$-z^2$	$-z^3$	z^3	$-z^3$	0
z	z	xz	$-z^2$	z^2	z^3	z^3	z^3	0
xz	xz	$-z^3$	$-z^3$	z^3	0	0	0	0
y^2	y^2	$-z^3$	z^3	z^3	0	0	0	0
z^2	z^2	z^3	$-z^3$	z^3	0	0	0	0
z^3	z^3	0	0	0	0	0	0	0

form with respect to a Gröbner basis, i.e., we represent the image $\bar{f}$ of f in R by $\mathrm{NF}(f)$. Thus $\bar{f} + \bar{g} = \overline{f+g}$ is represented by $\mathrm{NF}(f+g)$ which actually equals $\mathrm{NF}(f) + \mathrm{NF}(g)$, since both $\mathrm{NF}(f)$ and $\mathrm{NF}(g)$, and hence $\mathrm{NF}(f) + \mathrm{NF}(g)$, are linear combinations of monomials outside $l(\mathfrak{a})$. The element $\bar{f}\bar{g} = \overline{fg}$ is represented by $\mathrm{NF}(fg)$ which equals $\mathrm{NF}(\mathrm{NF}(f)\,\mathrm{NF}(g))$ since

$$\mathrm{NF}(fg) - \mathrm{NF}(\mathrm{NF}(f)\,\mathrm{NF}(g)) =$$

$$\mathrm{NF}(fg) - fg + fg - f\,\mathrm{NF}(g) + f\,\mathrm{NF}(g) - \mathrm{NF}(f)\,\mathrm{NF}(g) +$$

$$\mathrm{NF}(f)\,\mathrm{NF}(g) - \mathrm{NF}(\mathrm{NF}(f)\,\mathrm{NF}(g)) =$$

$$(\mathrm{NF}(fg) - fg) + f(g - \mathrm{NF}(g)) + \mathrm{NF}(g)(f - \mathrm{NF}(f)) +$$

$$\mathrm{NF}(f)\,\mathrm{NF}(g) - \mathrm{NF}(\mathrm{NF}(f)\,\mathrm{NF}(g)) \in \mathfrak{a},$$

because $h - \mathrm{NF}(h) \in \mathfrak{a}$ for all h, which we use on $h = fg, h = g, h = f$, and $h = \mathrm{NF}(f)\,\mathrm{NF}(g)$.

EXAMPLE 7. Consider the ideal $\mathfrak{a} = \langle x^2 + xz, xy + y^2, yz + z^2 \rangle \subseteq k[x, y, z]$. A Gröbner basis for $\mathfrak{a}$ in Degrevlex with $x \succ y \succ z$ is

$$\left\{x^2 + xz, xy + y^2, yz + z^2, xz^2 - z^3, y^3 - z^3, z^4\right\}.$$

The elements in $k[x, y, z]/\mathfrak{a}$ can thus be represented by linear combinations of $\{1, x, y, z, xz, y^2, z^2, z^3\}$ (the monomials outside $l(\mathfrak{a})$), and $k[x, y, z]/\mathfrak{a}$ is thus an eight-dimensional vector space over k. We give the multiplication table for $k[x, y, z]/\mathfrak{a}$; see Table 7.1. □

Exercise

1. Construct a multiplication table for

$$k[x,y,z]/\langle x^2 - z^2, y^2 - z^2, xy + xz + yz\rangle .$$

7.3 Elimination

Let $\mathfrak{b}$ be an ideal in $k[x_1,\ldots,x_m,y_1,\ldots,y_n]$. The problem of elimination is to determine the ideal $\mathfrak{b}\cap k[y_1,\ldots,y_n]$. We will now see how this can be solved by means of Gröbner basis techniques.

PROPOSITION 8. *Let $\mathfrak{b}$ be an ideal in $k[x_1,\ldots,x_m,y_1,\ldots,y_n]$ equipped with an ordering such that every x_i is larger than any monomial in the y_j's. If G is a Gröbner basis for $\mathfrak{b}$, then $G\cap k[y_1,\ldots,y_n]$ is a Gröbner basis for $\mathfrak{b}\cap k[y_1,\ldots,y_n]$.*

Proof. It is clear that $G\cap k[y_1\ldots,y_n]\subseteq \mathfrak{b}\cap k[y_1,\ldots,y_n]$. Let $f\in\mathfrak{b}\cap k[y_1,\ldots,y_n]$. Then $\mathrm{NF}(f,G)=0$. But the reductions giving $\mathrm{NF}(f,G)=0$ will only involve elements from $G\cap k[y_1,\ldots,y_n]$, hence

$$\mathrm{NF}(f,G\cap k[y_1,\ldots,y_n])=0$$

so $G\cap k[y_1,\ldots,y_n]$ is a Gröbner basis for $\mathfrak{b}\cap k[y_1,\ldots,y_n]$. □

EXAMPLE 9. A subalgebra $k[f_1,\ldots,f_r]$ of $k[x_1,\ldots,x_n]$ is the image of the homomorphism $\phi: k[y_1,\ldots,y_r]\longrightarrow k[x_1,\ldots,x_n]$ where $\phi(y_i)=f_i$. Hence $k[f_1,\ldots,f_r]\simeq k[y_1,\ldots,y_r]/\ker(\phi)$ and $\ker(\phi)$ is the intersection of the ideal $\langle y_1-f_1,\ldots,y_r-f_r\rangle$ in $k[x_1,\ldots,x_n,y_1,\ldots,y_r]$ with $k[y_1,\ldots,y_r]$, i.e., $\ker(\phi)=\langle y_1-f_1,\ldots,y_r-f_r\rangle\cap k[y_1,\ldots,y_r]$. Hence we can determine the presentation of $k[f_1,\ldots,f_r]$ as a factor ring of $k[y_1,\ldots,y_r]$. Consider for example $k[x^3,x^5,x^7]$. We determine

$$\mathfrak{b}=\langle y_1-x^3,y_2-x^5,y_3-x^7\rangle\cap k[y_1,y_2,y_3]$$

by calculating a Gröbner basis for $\mathfrak{b}$ in Lex with $x\succ y_1\succ y_2\succ y_3$. The result is

$$\{x^3-y_1, x^2y_1-y_2, xy_3-y_1y_2, xy_1^2-y_3, y_3^2-y_1^3y_2, y_2y_3-y_1^4,$$
$$y_1y_3-y_2^2, y_2^3-y_1^5\}.$$

Hence

$$k[x^3,x^5,x^7]\simeq k[y_1,y_2,y_3]/\langle y_3^2-y_1^3y_2, y_2y_3-y_1^4, y_1y_3-y_2^2, y_2^3-y_1^5\rangle .$$

Note that we do not get a minimal generating set for the ideal since

$$y_2^3 - y_1^5 = -y_2(y_1y_3 - y_2^2) + y_1(y_2y_3 - y_1^4),$$

so $y_2^3 - y_1^5$ is not needed as a generator. □

Exercise

1. Write $k[x^3, x^7, x^8]$ as a factor ring $k[y_1, y_2, y_3]/\mathfrak{a}$. Determine a minimal set of generators for $\mathfrak{a}$.

7.4 Ideal Operations

Given two ideals $\mathfrak{a} = \langle f_1, \ldots, f_r\rangle$ and $\mathfrak{b} = \langle g_1, \ldots, g_s\rangle$ we want to get a generating set for $\mathfrak{a} + \mathfrak{b}$, $\mathfrak{a} \cdot \mathfrak{b}$, $\mathfrak{a} \cap \mathfrak{b}$, and $\mathfrak{a} : \mathfrak{b}$. For the sum and the product there are no problems: $\mathfrak{a} + \mathfrak{b} = \langle f_1, \ldots, f_r, g_1, \ldots, g_s\rangle$ and $\mathfrak{a} \cdot \mathfrak{b} = \langle f_1g_1, \ldots, f_rg_s\rangle$. The intersection and quotient of ideals are, however, not equally easy.

7.4.1 Intersection of Ideals

We will reduce the problem of calculating the intersection of ideals to an elimination problem by adding a new variable. Hence we can use Proposition 8 to solve the problem of getting a generating set for the intersection of two ideals.

PROPOSITION 10. *Let $\mathfrak{a} = \langle f_1, \ldots, f_r\rangle$ and $\mathfrak{b} = \langle g_1, \ldots, g_s\rangle$ be ideals in $k[x_1, \ldots, x_n]$. Consider the ideal*

$$\mathfrak{c} = \langle yf_1, \ldots, yf_r, (1-y)g_1, \ldots, (1-y)g_s\rangle$$

in $k[x_1, \ldots, x_n, y]$. Then $\mathfrak{a} \cap \mathfrak{b} = \mathfrak{c} \cap k[x_1, \ldots, x_n]$.

Proof. Suppose $f \in \mathfrak{a} \cap \mathfrak{b}$. Then $yf \in y\mathfrak{a}$ and $(1-y)f \in (1-y)\mathfrak{b}$ so

$$f = yf + (1-y)f \in y\mathfrak{a} + (1-y)\mathfrak{b} = \mathfrak{c}.$$

Suppose $f \in \mathfrak{c} \cap k[x_1, \ldots, x_n]$. Then

$$f =$$

$$h_1(x_1, \ldots, x_n, y)yf_1(x_1, \ldots, x_n) + \cdots + h_r(x_1, \cdots, x_n, y)yf_r(x_1, \ldots, x_n) +$$
$$h_{r+1}(x_1, \ldots, x_n, y)(1-y)g_1(x_1, \ldots, x_n) + \cdots +$$

$$h_{r+s}(x_1,\ldots,x_n,y)(1-y)g_s(x_1,\ldots,x_n).$$

Since $f \in k[x_1,\ldots,x_n]$ we get

$$f = h_{r+1}(x_1,\ldots,x_n,0)g_1(x_1,\ldots,x_n) + \cdots +$$

$$h_{r+s}(x_1,\ldots,x_n,0)g_s(x_1,\ldots,x_n) \in \mathfrak{b}$$

and

$$f = h_1(x_1,\ldots,x_n,1)f_1(x_1,\ldots,x_n) + \cdots +$$

$$h_r(x_1,\ldots,x_n,1)f_r(x_1,\ldots,x_n) \in \mathfrak{a}.$$

□

EXAMPLE 11. Let $\mathfrak{a} = \langle x^2+y^2, xy\rangle$ and $\mathfrak{b} = \langle x^2-y^2\rangle$. Let

$$\mathfrak{c} = \langle x^2t+y^2t, xyt, tx^2-ty^2-x^2+y^2\rangle.$$

A Gröbner basis for $\mathfrak{c}$ in Lex with $t \succ x \succ y$ is

$$\{2x^2t-x^2-y^2, xyt, 2y^2t+x^2+y^2, x^3-xy^2, x^2y-y^3\}.$$

Hence $\mathfrak{a} \cap \mathfrak{b} = \langle x^3-xy^2, x^2y-y^3\rangle$. □

Exercise

1. Determine $\langle x^3+x, xy+y^2\rangle \cap \langle y^3+y, xy+x^2\rangle$ in $k[x,y]$.

7.4.2 Ideal Quotient

Let $\mathfrak{a}$ be an ideal and $\mathfrak{b} = \langle g_1,\ldots,g_s\rangle$; see Proposition 45.3 in Section 1.3.5. Then $\mathfrak{a} : \mathfrak{b} = \cap_{i=1}^{s} \mathfrak{a} : \langle g_i\rangle$. Hence it is sufficient to consider the case when $\mathfrak{b}$ is principal, since we know how to perform intersections.

PROPOSITION 12. *If* $\mathfrak{a} \cap \langle g\rangle = \langle h_1,\ldots,h_t\rangle$ *then* $\mathfrak{a} : \langle g\rangle = \langle h_1/g,\ldots,h_t/g\rangle$.

Proof. Let $f \in \mathfrak{a} : \langle g\rangle$. Then $fg \in \mathfrak{a} \cap \langle g\rangle$. Let $h' \in \mathfrak{a} \cap \langle g\rangle$. Then $h' = gh$ for some h and $h \in \mathfrak{a} : \langle g\rangle$. Thus $h' \in \mathfrak{a}\cap\langle g\rangle$ if and only if $h'/g \in \mathfrak{a} : \langle g\rangle$. □

EXAMPLE 13. Let $\mathfrak{a} = \langle x^2+y^2, xy\rangle$ and $\mathfrak{b} = \langle x^2-y^2\rangle$. Since $\mathfrak{a} \cap \mathfrak{b} = \langle x^3-xy^2, x^2y-y^3\rangle$ we get $\mathfrak{a} : \mathfrak{b} = \langle x, y\rangle$. □

Exercise

1. Determine $\langle x^3+x, xy+y^2\rangle : \langle y^3+y, xy+x^2\rangle$ in $k[x,y]$.

7.5 Supplementary Exercises

1. Show that $x^5y - 2x^4y^2 \in \langle xy^3 - 2x^3y + 3x^2y^2, x^3y^2 + 5x^2y^3 \rangle$.

2. Show that $x^{11} + x^{15} \in \mathbb{Z}_2[x^4, x^6 + x^7, x^{10}]$.

3. Show that

$$\sqrt{\langle x^2 - x + xz - z, xy - y + y^2, y + yz + z^2 + 2z + 1 \rangle} =$$

$$\langle x - 1, y, z + 1 \rangle .$$

4. Let $\mathfrak{b} = \langle xy - y, xz - x, yz - z \rangle$. Determine $\mathfrak{b} \cap k[x, y]$ and $\mathfrak{b} \cap k[x]$.

5. Let $R = k[x_1, \ldots, x_n]/\langle g_1, \ldots, g_s \rangle = k[x_1, \ldots, x_n]/\mathfrak{a}$. Define a homomorphism $\phi : k[y_1, \ldots, y_r] \longrightarrow R$ by $\phi(y_i) = f_i + \mathfrak{a}$. Show that $\ker(\phi) = (\langle y_1 - f_1, \ldots, y_r - f_r \rangle + \mathfrak{a}) \cap k[y_1, \ldots, y_r]$.

6. Show that $R = \mathbb{Q}[x]/\langle x^3 + x + 1 \rangle$ is a field and a vector space of dimension 3 over $\mathbb{Q}$. Show that any element $\alpha \in R$ is a root of a polynomial equation of degree at most 3 with coefficients in $\mathbb{Q}$. (Hint: Since $\dim_{\mathbb{Q}} R = 3$, we get that $1, \alpha, \alpha^2, \alpha^3$ are linearly dependent.) Determine an equation for (the image of) $x + 1$.

7. Write $k[x^4 + x^2, x^3]$ as $k[y_1, y_2]/\mathfrak{a}$.

8. Write $k[x^4, x^3y, xy^3, y^4]$ as a factor ring of $k[y_1, y_2, y_3, y_4]$.

9. Show that $\langle y^2 - xz, yz - x^3, z^2 - x^2y \rangle$ is a prime ideal in $k[x, y, z]$. Hint: Write $k[t^3, t^4, t^5]$ as a factor ring of $k[x, y, z]$.

Chapter 8

Homogeneous Algebras

We will in this chapter study the class of ideals in a polynomial ring which are generated by homogeneous elements (and their factor rings). We will show that this class is closed under the usual operations on ideals and that the primary components of such an ideal can be chosen to be generated by homogeneous elements. In the second part of the chapter we will study the connection between an arbitrary ideal and its "homogenization", which is an ideal generated by homogeneous elements in a ring with one more variable. Finally we will study Gröbner bases of ideal, generated by homogeneous elements.

8.1 Homogeneous Ideals and Algebras

Let $A = k[x_1, \ldots, x_n]$. An element $f \in A$ is called *homogeneous* if f is a linear combination (with coefficients in k) of monomials of the same degree. Thus $x^3 + 2xy^2$ is homogeneous but $x^3 + 2xy^2 - xy$ is not. The homogeneous elements of degree i in A constitute together with $\{0\}$ a vector space A_i over k with the set of monomials of degree i as a basis. Furthermore $A_iA_j \subseteq A_{i+j}$, i.e., a product of two homogeneous elements of degree i and j, respectively, is homogeneous of degree $i + j$. Every element $f \in A \setminus \{0\}$ can be uniquely written as $f = f_0 + f_1 + \cdots + f_k$, where $f_i \in A_i$. The f_i's are called the *homogeneous components* of f. Another way to express this is to say that A is the direct sum of the A_i's, $A = \oplus_{i \geq 0} A_i$; see Lemma 79 in Section 1.9. An ideal $\mathfrak{a} \subseteq A$ is called homogeneous if $\mathfrak{a}$ has a generating set consisting of homogeneous elements.

EXAMPLE 1. If $\mathfrak{a}$ is generated by monomials, then $\mathfrak{a}$ is homogeneous. □

LEMMA 2. *An ideal $\mathfrak{a}$ is homogeneous if and only if the following condition holds. If $F \in \mathfrak{a}$ then every homogeneous component of F belongs to $\mathfrak{a}$.*

Proof. Suppose that $\mathfrak{a} = \langle f_1, \dots, f_k \rangle$ with f_i homogeneous for all i and let $F \in \mathfrak{a}$. Then $F = g_1 f_1 + \cdots + g_k f_k$. Let F_j be the homogeneous component of degree j of F. Then $F_j = g_1^{(j-\deg f_1)} f_1 + \cdots + g_k^{(j-\deg f_k)} f_k$ where $g_i^{(j-\deg f_i)}$ is the homogeneous component of degree $(j - \deg f_i)$ of g_i. Hence $F_j \in \mathfrak{a}$. For the converse let $\mathfrak{a}$ be generated by $h_1, \dots, h_k$. Each homogeneous component $h_i^{(j)}$ of each h_i belongs to $\mathfrak{a}$. Then $\mathfrak{a}$ is generated by the $h_i^{(j)}$'s. □

REMARK. Another way to express the statement of the lemma is the following. Let $\mathfrak{a}$ be a homogeneous ideal and let $\mathfrak{a}_i$ be the k-vector space of homogeneous polynomials of degree i in $\mathfrak{a}$. Then $\mathfrak{a} = \oplus_{i \geq 0} \mathfrak{a}_i$.

DEFINITION 3 (HOMOGENEOUS ALGEBRA). A ring $R = k[x_1, \dots, x_n]/\mathfrak{a}$ is called a *homogeneous algebra* if $\mathfrak{a}$ is homogeneous.

We will now see that there exists in a homogeneous algebra a direct sum decomposition in homogeneous components, much like in the polynomial ring.

LEMMA 4. *If $R = k[x_1, \dots, x_n]/\mathfrak{a}$ is a homogeneous algebra then any element in R can be written $\bar{f} = \bar{f}_0 + \cdots + \bar{f}_k$, where f_i is homogeneous of degree i or zero, and where $\bar{f}_i$ is unique. Equivalently, $R = \oplus R_i$ where R_i is the image of A_i in the natural map $A \longrightarrow A/\mathfrak{a} = R$. On the other hand, if $R = k[x_1, \dots, x_n]/\mathfrak{a}$ and $R = \oplus R_i$ where R_i is the image of A_i, then $\mathfrak{a}$ is homogeneous.*

Proof. Since any $f \in A$ can be written $f = f_0 + \cdots + f_k$ with $f_i \in A_i$, then any element $\bar{f} \in R$ can be written $\bar{f} = \bar{f}_0 + \cdots + \bar{f}_k$, $f_i \in A_i$. Suppose $\bar{f} = \bar{f}_0 + \cdots + \bar{f}_k = \bar{g}_0 + \cdots + \bar{g}_k = \bar{g}$, so $\overline{f_0 - g_0} + \cdots + \overline{f_k - g_k} = \bar{0}$. This means that $f - g = (f_0 - g_0) + \cdots + (f_k - g_k) \in \mathfrak{a}$. But $f_i - g_i$ is the homogeneous component of degree i of $f - g$, so $f_i - g_i \in \mathfrak{a}$ since $\mathfrak{a}$ is homogeneous. Hence $\overline{f_i - g_i} = \bar{0}$, so $\bar{f}_i = \bar{g}_i$. Now suppose that $R = \oplus R_i$ where $R_i = \operatorname{im} A_i$. This means that $\overline{f_0 + \cdots + f_k} = \bar{f}_0 + \cdots + \bar{f}_k = \bar{0}$ if and only if $\bar{f}_i = \bar{0}$ for every i, i.e., $f_0 + \cdots + f_k \in \mathfrak{a}$ if and only if $f_i \in \mathfrak{a}$ for all i, i.e., if and only if $\mathfrak{a}$ is homogeneous. □

COROLLARY 5. *Let $\mathfrak{a} \subseteq \mathfrak{b}$ be homogeneous ideals in $k[x_1, \dots, x_n]$ and let $R = k[x_1, \dots, x_n]/\mathfrak{a} = \oplus_{i \geq 0} R_i$, $S = k[x_1, \dots, x_n]/\mathfrak{b} = \oplus_{i \geq 0} S_i$. Then $S_i \simeq \phi(R_i)$, where $\phi : R \to S$ is the homomorphism defined by $\phi(f + \mathfrak{a}) = f + \mathfrak{b}$.*

Proof. That $\phi(R_i) = S_i$ is clear. If $f_1 + \mathfrak{a} = f_2 + \mathfrak{a}$ then $f_1 - f_2 \in \mathfrak{a} \subseteq \mathfrak{b}$, so $f_1 + \mathfrak{b} = f_2 + \mathfrak{b}$. Hence ϕ is well defined, and it is easy to check that ϕ is a homomorphism. □

The set of homogeneous ideals behave well under ideal operations.

LEMMA 6. *If $\mathfrak{a}$ and $\mathfrak{b}$ are homogeneous ideals in $k[x_1, \ldots, x_n]$ then $\mathfrak{a} + \mathfrak{b}$, $\mathfrak{ab}$, $\mathfrak{a} \cap \mathfrak{b}$, $\mathfrak{a} : \mathfrak{b}$, and $\sqrt{\mathfrak{a}}$ are homogeneous.*

Proof. If $\mathfrak{a}$ and $\mathfrak{b}$ are generated by homogeneous elements then $\mathfrak{a} + \mathfrak{b}$ and $\mathfrak{ab}$ are generated by homogeneous elements. Suppose that any element in $\mathfrak{a}$ has all its homogeneous components in $\mathfrak{a}$ and that the same is true for $\mathfrak{b}$. Then any element in $\mathfrak{a} \cap \mathfrak{b}$ has all its homogeneous components in $\mathfrak{a} \cap \mathfrak{b}$, so $\mathfrak{a} \cap \mathfrak{b}$ is homogeneous by Lemma 2. If $\mathfrak{b} = \langle f_1, \ldots, f_k \rangle$ with f_i homogeneous then $\mathfrak{a} : \mathfrak{b} = (\mathfrak{a} : \langle f_1 \rangle) \cap \cdots \cap (\mathfrak{a} : \langle f_k \rangle)$, hence it is sufficient to assume that $\mathfrak{b}$ is principal. If $\mathfrak{a} \cap \langle f \rangle = \langle g_1, \ldots, g_m \rangle$ with g_i homogeneous then $\mathfrak{a} : \langle f \rangle = \langle g_1/f, \ldots, g_m/f \rangle$, cf. Proposition 12 in Section 7.4.2, and g_i/f is homogeneous. Finally, let $F = F_{i_1} + \cdots + F_{i_m} \in \sqrt{\mathfrak{a}}, F_{i_j} \in A_{i_j}, i_1 < \cdots < i_m$. Then $F^n \in \mathfrak{a}$ for some n. The homogeneous component of lowest degree in F^n is $F_{i_1}^n$. Since $\mathfrak{a}$ is homogeneous we have $F_{i_1}^n \in \mathfrak{a}$, so $F_{i_1} \in \sqrt{\mathfrak{a}}$. Hence $F - F_{i_1} = F_{i_2} + \cdots + F_{i_m} \in \sqrt{\mathfrak{a}}$ and we are ready to use induction on m. □

The following criteria are useful when checking whether a homogeneous ideal is prime or primary.

LEMMA 7. *Suppose that $\mathfrak{a}$ is homogeneous.*

1. *If $ab \in \mathfrak{a}, a \notin \mathfrak{a}$ implies that $b \in \mathfrak{a}$ for* **homogeneous** *elements a, b, then $\mathfrak{a}$ is a prime ideal.*
2. *If $ab \in \mathfrak{a}, a \notin \mathfrak{a}$ implies that $b^n \in \mathfrak{a}$ for some n for* **homogeneous** *elements a, b, then $\mathfrak{a}$ is a primary ideal.*

Proof. 1. Suppose that $fg \in \mathfrak{a}$ where $f = f_0 + \cdots + f_k$ and $g = g_0 + \cdots + g_m$ with $f_i, g_i \in A_i$. Suppose that $f \notin \mathfrak{a}$. Then $f_i \notin \mathfrak{a}$ for some i, let i_0 be the smallest such i. The homogeneous component of degree i_0 of fg is $f_0 g_{i_0} + \cdots + f_{i_0} g_0$. Since $fg \in \mathfrak{a}$ and $\mathfrak{a}$ is homogeneous we get that this component belongs to $\mathfrak{a}$. Since all terms except possibly $f_{i_0} g_0$ belong to $\mathfrak{a}$, we get that $f_{i_0} g_0$, and thus g_0, belongs to $\mathfrak{a}$. Then $f(g - g_0) \in \mathfrak{a}$ and we can repeat the same argument to get $g_1 \in \mathfrak{a}$. We are ready to use induction on m.

2. Suppose $fg \in \mathfrak{a}$ where $f = f_{i_1} + \cdots + f_{i_k}$ and $g = g_{j_1} + \cdots + g_{j_l}$ with f_{i_j}, g_{j_i} homogeneous components. Suppose $f \notin \mathfrak{a}$. If $f_{i_1} \in \mathfrak{a}$ then $(f - f_{i_1})g \in \mathfrak{a}, f - f_{i_1} \notin \mathfrak{a}$, so we can just as well assume that $f_{i_1} \notin \mathfrak{a}$. If we can show that for each i there is an m_j such that $g_{j_i}^{m_i} \in \mathfrak{a}$ if $m = \max\{m_i\}$. The homogeneous component of lowest degree in fg is $f_{i_1}g_{j_1}$. Since $fg \in \mathfrak{a}$ and $\mathfrak{a}$ is homogeneous we have $f_{i_1}g_{j_1} \in \mathfrak{a}$. Since $f_{i_1} \notin \mathfrak{a}$ we have $g_{j_1}^{m_1} \in \mathfrak{a}$ for some m_1. Now suppose we have proved $g_{j_1}^{m_1}, \ldots, g_{j_s}^{m_s} \in \mathfrak{a}$. Let $M = \max\{m_{j_1}, \ldots, m_{j_s}\}$. Consider

$$F = f(g_{j_{s+1}} + \cdots + g_{j_l})^n = f(g - (g_{j_1} + \cdots + g_{j_s}))^n =$$

$$f\sum_{i=0}^{n} \binom{n}{i} (-1)^{n-i} g^i (g_{j_1} + \cdots + g_{j_s})^{n-i} =$$

$$(-1)^n f(g_{j_1} + \cdots + g_{j_s})^n + f\sum_{i=1}^{n} \binom{n}{i} (-1)^{n-i} g^i (g_{j_1} + \cdots + g_{j_s})^{n-i}.$$

If we choose $n = s(M-1)+1$, the first term belongs to $\langle g_{j_1}^M, \ldots, g_{j_s}^M \rangle \subseteq \mathfrak{a}$. The remaining terms belong to $\mathfrak{a}$ since $fg \in \mathfrak{a}$, thus $F \in \mathfrak{a}$. The homogeneous component of lowest degree in F is $f_{i_1}g_{j_{s+1}}$ and this component belongs to $\mathfrak{a}$ since $\mathfrak{a}$ is homogeneous. Thus $g_{j_{s+1}}^{m_{j_{s+1}}} \in \mathfrak{a}$ for some $m_{j_{s+1}}$ and we are through by induction.

□

REMARK. There is only one homogeneous maximal ideal in $k[x_1, \ldots, x_n]$, namely $\mathfrak{M} = \langle x_1, \ldots, x_n \rangle$ consisting of all homogeneous elements of positive degree. All homogeneous ideals, except $\langle 1 \rangle$, lie in $\mathfrak{M}$.

NOTATION 8. For any ideal $\mathfrak{a}$ in $k[x_1, \ldots, x_n]$ we denote by $\mathfrak{a}^*$ the largest homogeneous ideal in $\mathfrak{a}$, or similarly, the ideal generated by all homogeneous elements in $\mathfrak{a}$.

We will now investigate to what extent properties of an ideal $\mathfrak{a}$ are preserved in $\mathfrak{a}^*$.

LEMMA 9. *1. If $\mathfrak{a} \subseteq \mathfrak{b}$ then $\mathfrak{a}^* \subseteq \mathfrak{b}^*$.*

2. $\sqrt{\mathfrak{a}^*} = \sqrt{\mathfrak{a}}^*$.

3. If $\mathfrak{p}$ is prime then $\mathfrak{p}^*$ is prime.

4. If $\mathfrak{q}$ is primary then $\mathfrak{q}^*$ is primary.

Proof. 1. This is obvious.

2. Since $\mathfrak{a}^* \subseteq \mathfrak{a}$ we have $\sqrt{\mathfrak{a}^*} \subseteq \sqrt{\mathfrak{a}}$ but $\sqrt{\mathfrak{a}^*}$ is homogeneous, see Lemma 6, and $\sqrt{\mathfrak{a}}^*$ is the largest homogeneous ideal in $\sqrt{\mathfrak{a}}$, so $\sqrt{\mathfrak{a}^*} \subseteq \sqrt{\mathfrak{a}}^*$. Let $F \in \sqrt{\mathfrak{a}}^*$. Since $\sqrt{\mathfrak{a}}^*$ is homogeneous, all homogeneous components of F belong to $\sqrt{\mathfrak{a}}^*$. Hence all homogeneous components F_{i_j} of F have a power $F_{i_j}^{m_{i_j}}$ in $\mathfrak{a}$. If F has k homogeneous components, $F = F_{i_1} + \cdots + F_{i_k}$, and $m = \max\{m_{i_j}\}$, then $F^{k(m-1)+1} \in \langle F_{i_1}^m, \dots, F_{i_k}^m \rangle$; this is a homogeneous ideal and is thus contained in $\mathfrak{a}^*$, so $F \in \sqrt{\mathfrak{a}^*}$.
3. By Lemma 7.1 it is sufficient to check that if f, g are homogeneous with $fg \in \mathfrak{p}^*, f \notin \mathfrak{p}^*$ then $g \in \mathfrak{p}^*$. We have $fg \in \mathfrak{p}$ and $f \notin \mathfrak{p}$ since $f \in \mathfrak{p}$ would imply $f \in \mathfrak{p}^*$ since f is homogeneous. Thus $g \in \mathfrak{p}$, so $g \in \mathfrak{p}^*$ since g is homogeneous.
4. By Lemma 7.2 it is enough to check that if f, g are homogeneous, $fg \in \mathfrak{q}^*, f \notin \mathfrak{q}^*$ then $g^n \in \mathfrak{q}^*$ for some n. We have $fg \in \mathfrak{q}$ and $f \notin \mathfrak{q}$ as above. Thus $g^n \in \mathfrak{q}$ for some n. Since g^n is homogeneous, we have $g^n \in \mathfrak{q}^*$.

□

THEOREM 10. *If $\mathfrak{a}$ is a homogeneous ideal then $\mathfrak{a}$ has a primary decomposition $\mathfrak{a} = \mathfrak{q}_1 \cap \cdots \cap \mathfrak{q}_s$ with $\mathfrak{q}_i$ homogeneous for all i.*

Proof. Let $\mathfrak{q}_1 \cap \cdots \cap \mathfrak{q}_s$ be any primary decomposition of $\mathfrak{a}$. Since $\mathfrak{q}_i^* \subseteq \mathfrak{q}_i$, we have $\mathfrak{q}_1^* \cap \cdots \cap \mathfrak{q}_s^* \subseteq \mathfrak{q}_1 \cap \cdots \cap \mathfrak{q}_s$. On the other hand, since $\mathfrak{a} \subseteq \mathfrak{q}_i$ for all i we get $\mathfrak{a} \subseteq \mathfrak{q}_i^*$ for all i since $\mathfrak{a}$ is homogeneous, hence $\mathfrak{a} \subseteq \mathfrak{q}_1^* \cap \cdots \cap \mathfrak{q}_s^*$. □

COROLLARY 11. *If $\mathfrak{a}$ is homogeneous then the isolated components and the associated primes of $\mathfrak{a}$ are homogeneous. The embedded components can be chosen to be homogeneous.*

Proof. The isolated components are unique; see Proposition 14 in Chapter 5. Since there is a primary decomposition with homogeneous primary ideals $\mathfrak{q}_i$, we get that $\mathfrak{p}_i = \sqrt{\mathfrak{q}_i}$ are homogeneous; see Lemma 6. Also the associated primes are unique. □

The conclusion of the following theorem is the following important fact. Let $A = k[x_1, \dots, x_n]/\mathfrak{a}$ be a homogeneous algebra. Then either there is a homogeneous nonzero element $s \in A$ such that s annihilates *every* homogeneous element of positive degree in A, or there is a homogeneous nonzerodivisor in A (this nonzerodivisor can be chosen to be of degree 1 if k is infinite).

THEOREM 12. *Let $\mathfrak{a} \neq \langle 1 \rangle$ be a homogeneous ideal in $k[x_1, \dots, x_n]$. If the homogeneous maximal ideal $\mathfrak{M}$ is associated to $\mathfrak{a}$, there is a homogeneous element $f \in k[x_1, \dots, x_n] \setminus \mathfrak{a}$ such that $f\mathfrak{M} \subseteq \mathfrak{a}$. If $\mathfrak{M}$ is not associated to $\mathfrak{a}$,* *there is a homogeneous element $f \in \mathfrak{M}$ such that $\mathfrak{a} : \langle f \rangle = \mathfrak{a}$. Equivalently, if $\mathfrak{M}$ belongs to $\mathfrak{a}$, there is a homogeneous $f \notin \mathfrak{a}$ such that $\bar{f} \cdot \overline{\mathfrak{M}} = \bar{0}$ in $R = k[x_1, \ldots, x_n]/\mathfrak{a}$; and if $\mathfrak{M}$ does not belong to $\mathfrak{a}$ there is a homogeneous $f \in \mathfrak{M}$ such that $\bar{f}$ is a nonzerodivisor in R. Furthermore, if the field k is infinite and $\mathfrak{M}$ does not belong to $\mathfrak{a}$, there is a homogeneous element f of degree 1 such that $\mathfrak{a} : \langle f \rangle = \mathfrak{a}$ (equivalently, there exists a homogeneous element f of degree 1 such that $\bar{f}$ is a nonzerodivisor in R).*

Proof. Let $\mathfrak{a} = \mathfrak{q}_1 \cap \cdots \cap \mathfrak{q}_s$ be a primary decomposition with homogeneous primary ideals and assume $\sqrt{\mathfrak{q}_1} = \mathfrak{M}$. Take an element $g \in \mathfrak{q}_2 \cap \cdots \cap \mathfrak{q}_s, g \notin \mathfrak{q}_1$. This is possible since $\mathfrak{q}_2 \cap \cdots \cap \mathfrak{q}_s \not\subseteq \mathfrak{q}_1$. We can assume that g is homogeneous since both $\mathfrak{q}_2 \cap \cdots \cap \mathfrak{q}_s$ and $\mathfrak{q}_1$ are homogeneous. Then $\mathfrak{a} : \langle g \rangle = \mathfrak{q}_1 \cap \cdots \cap \mathfrak{q}_s : \langle g \rangle = (\mathfrak{q}_1 : \langle g \rangle) \cap \cdots \cap (\mathfrak{q}_s : \langle g \rangle)$, and since $g \in \mathfrak{q}_i$ if $i > 1$ we have $\mathfrak{q}_i : \langle g \rangle = \langle 1 \rangle$ if $i > 1$, so $\mathfrak{a} : \langle g \rangle = \mathfrak{q}_1 : \langle g \rangle$ which is $\mathfrak{M}$-primary. We consider the set of ideals $\mathfrak{a} : \langle g \rangle$ with g homogeneous such that $\mathfrak{a} : \langle g \rangle$ is $\mathfrak{M}$-primary. We have just proved that this set is nonempty, so this set has a maximal element, say $\mathfrak{a} : \langle f \rangle$; see Proposition 13 in Section 3.5. We claim that $\mathfrak{a} : \langle f \rangle$ is a prime ideal. To prove that $\mathfrak{a} : \langle f \rangle$ is a prime ideal, it suffices to prove that if x and y are homogeneous elements such that $xy \in \mathfrak{a} : \langle f \rangle, x \notin \mathfrak{a} : \langle f \rangle$, then $y \in \mathfrak{a} : \langle f \rangle$; see Lemma 7.1. Suppose $xy \in \mathfrak{a} : \langle f \rangle, x \notin \mathfrak{a} : \langle f \rangle$ and x, y homogeneous. Then $y \in \mathfrak{a} : \langle fx \rangle$ and $\mathfrak{a} : \langle fx \rangle \neq \langle 1 \rangle$ since $fx \notin \mathfrak{a}$. Since $\mathfrak{a} : \langle fx \rangle$ is homogeneous (Lemma 6), we have $\mathfrak{a} : \langle fx \rangle \subseteq \mathfrak{M}$. Since

$$\mathfrak{M} \subseteq \sqrt{\mathfrak{a} : \langle x \rangle} \subseteq \sqrt{\mathfrak{a} : \langle fx \rangle} \subseteq \sqrt{\mathfrak{M}} = \mathfrak{M},$$

so $\sqrt{\mathfrak{a} : \langle fx \rangle} = \mathfrak{M}$, so $\mathfrak{a} : \langle fx \rangle$ is $\mathfrak{M}$-primary by Proposition 17 in Chapter 5. But $\mathfrak{a} : \langle fx \rangle = \mathfrak{a} : \langle f \rangle$ since $\mathfrak{a} : \langle f \rangle \subseteq \mathfrak{a} : \langle fx \rangle$, fx is homogeneous, $\mathfrak{a} : \langle fx \rangle$ is $\mathfrak{M}$-primary, and $\mathfrak{a} : \langle f \rangle$ is maximal. Thus $y \in \mathfrak{a} : \langle f \rangle$ and $\mathfrak{a} : \langle f \rangle$ is a prime ideal. But then $\mathfrak{a} : \langle f \rangle = \sqrt{\mathfrak{a} : \langle f \rangle} = \mathfrak{M}$, which gives $f\mathfrak{M} \subseteq \mathfrak{a}$. For the factor ring R this means $\bar{f} \cdot \overline{\mathfrak{M}} = \bar{0}$. Now suppose that $\mathfrak{M}$ does not belong to $\mathfrak{a}$. By Theorem 15 in Chapter 5 it suffices to show that there is a homogeneous element in $\mathfrak{M}$ which does not belong to $\mathfrak{p}_1 \cup \cdots \cup \mathfrak{p}_s$. Suppose the contrary, that $\mathfrak{M} \subseteq \mathfrak{p}_1 \cup \cdots \cup \mathfrak{p}_s$. By deleting some $\mathfrak{p}_i$ if necessary, we can assume that $\mathfrak{M} \not\subseteq \cup_{i \neq j} \mathfrak{p}_i$ for every j. Take for each j a homogeneous element $x_j \in \mathfrak{M}, x_j \notin \cup_{i \neq j} \mathfrak{p}_i$. Let $\deg x_1 = d_1$ and $\deg x_2 \cdots x_s = d_2$. Consider the element $x_1^{d_2} + (x_2 \cdots x_s)^{d_1}$. This is a homogeneous element (of degree $d_1 d_2$). Since $x_1 \notin \cup_{i \neq 1} \mathfrak{p}_i$ and all $\mathfrak{p}_i$'s are prime ideals, we have $x_1^{d_2} \notin \cup_{i \neq 1} \mathfrak{p}_i$. Since $x_1 \in \cup_{i=1}^{s} \mathfrak{p}_i$ we must have $x_1 \in \mathfrak{p}_1$, and thus $x_1^{d_1} \in \mathfrak{p}_1$. As above $x_2 \in \mathfrak{p}_2, \ldots, x_s \in \mathfrak{p}_s$. Thus $x_2 \cdots x_s \in \mathfrak{p}_i$ if $i > 1$, so $(x_2 \cdots x_s)^{d_1} \in \mathfrak{p}_i$ if $i > 1$. As above $x_2, \ldots, x_s \notin \mathfrak{p}_1$, so $(x_2 \cdots x_s)^{d_1} \notin \mathfrak{p}_1$ and $(x_2 \cdots x_s)^{d_1} \in \mathfrak{p}_i$ if $i > 1$. We claim that the homogeneous element $y = x_1^{d_2} + (x_2 \cdots x_s)^{d_1}$

does not belong to any $\mathfrak{p}_i$, since if $y \in \mathfrak{p}_1$ then $y - x_1^{d_2} = (x_2 \cdots x_s)^{d_1} \in \mathfrak{p}_1$, a contradiction, and if $y \in \mathfrak{p}_i$ for some $i > 1$ then $y - (x_2 \cdots x_s)^{d_1} = x_1^{d_2} \in \mathfrak{p}_i$, a contradiction. Finally suppose that k is an infinite field. For each i we have that the elements of $\mathfrak{p}_i$ of degree 1 constitute a k-subspace V_i of the space V of all elements of degree 1, and that $V_i \neq V$. If all elements of degree 1 belong to some $\mathfrak{p}_i$, we get $V = \cup_{i=1}^{s} V_i$. But a vector space over an infinite field cannot be a union of a finite number of proper subspaces. □

Exercises

1. Show that a vector space over an infinite field cannot be a union of a finite number of proper subspaces. Show that a vector space over a finite field *can* be a union of a finite number of proper subspaces.

2. Let $R = \mathbb{Z}_2[x, y]/\langle x^2y + xy^2 \rangle$. Show that there is no nonzerodivisor in R which is homogeneous of degree 1. Determine a homogeneous nonzerodivisor of degree 2.

8.2 Homogenizing and Dehomogenizing

For each polynomial (or ideal) in $k[x_1, \ldots, x_n]$ there is a natural way that to construct a homogeneous polynomial (or homogeneous ideal) in a polynomial ring $k[x_1, \ldots, x_n, y]$ with one more variable.

8.2.1 Homogenizing Polynomials

Let $f = f_0 + \cdots + f_k \in k[x_1, \ldots, x_n] = A$, f_i homogeneous of degree i, $f_k \neq 0$.

DEFINITION 13 (HOMOGENIZATION OF A POLYNOMIAL). The *homogenization* of f with respect to y is

$$h(f) = f_0 y^k + f_1 y^{k-1} + \cdots + f_{k-1} y + f_k \in k[x_1, \ldots, x_n, y].$$

Thus for example $h(1 + x_1x_2 + x_3^3) = y^3 + x_1x_2y + x_3^3$. It is practical to have a formula for homogenizing.

LEMMA 14. *We have* $h(f(x_1, \ldots, x_n)) = y^{\deg f} f(x_1/y, \ldots, x_n/y)$.

Proof. If $f = f_0 + \cdots + f_k, f_k \neq 0$, we have

$$h(f) = f_0 y^k + \cdots + f_k = y^k (f_0 + f_1/y + \cdots + f_k/y^k)$$

and it is easily checked that $f_i/y^i = f_i(x_1/y. \ldots, x_n/y)$. □

Homogenization works well with multiplication.

LEMMA 15. *We have $h(fg) = h(f)h(g)$.*

Proof. We have $\deg(fg) = \deg f + \deg g$, so

$$h(fg) = y^{\deg f+\deg g} fg(x_1/y, \ldots, x_n/y) =$$

$$y^{\deg f+\deg g} f(x_1/y, \ldots, x_n/y)g(x_1/y, \ldots, x_n/y) =$$

$$y^{\deg f} f(x_1/y, \ldots, x_n/y)y^{\deg g} g(x_1/y, \ldots, x_n/y) = h(f)h(g).$$

□

Homogenization works less well with sums, unless $\deg f = \deg g = \deg(f+g)$.

EXAMPLE 16. If $f = x_1, g = x_1^2$ then $h(f) = f, h(g) = g$, and $h(f+g) = x_1 y + x_1^2 \neq h(f) + h(g)$. If $f = 1 + x_1x_2, g = x_1 - x_1x_2$, we have $h(f) = y^2 + x_1x_2, h(g) = x_1y - x_1x_2, h(f+g) = y + x_1 \neq h(f) + h(g)$. □

LEMMA 17. *We have*

$$h(f+g) = y^{\deg(f+g)-\deg f} h(f) + y^{\deg(f+g)-\deg g} h(g).$$

In particular, if $\deg f = \deg g = \deg(f+g)$, we have $h(f+g) = h(f)+h(g)$.

Proof. We have

$$h(f+g) = y^{\deg(f+g)}(f(x_1/y, \ldots, x_n/y) + g(x_1/y, \ldots, x_n/y)) =$$

$$y^{\deg(f+g)-\deg f} y^{\deg f} f(x_1/y, \ldots, x_n/y)+$$

$$y^{\deg(f+g)-\deg g} y^{\deg g} g(x_1/y, \ldots, x_n/y) =$$

$$y^{\deg(f+g)-\deg f} h(f) + y^{\deg(f+g)-\deg g} h(g).$$

□

Exercise

1. Let $f(x) \in k[x]$. Show that f is irreducible if and only if $h(f)$ is irreducible.

8.2.2 Homogenizing Ideals

DEFINITION 18 (HOMOGENIZATION OF AN IDEAL). Let $\mathfrak{a}$ be an ideal in $A = k[x_1, \ldots, x_n]$. Then $h(\mathfrak{a})$ is the homogeneous ideal in $A[y]$ which is generated by all $h(f), f \in \mathfrak{a}$.

For principal ideals $\mathfrak{a} = \langle f \rangle$ it is easy to determine $h(\mathfrak{a})$.

PROPOSITION 19. *If* $\mathfrak{a} = \langle f \rangle$ *then* $h(\mathfrak{a}) = \langle h(f) \rangle$.

Proof. It is clear that $\langle h(f) \rangle \subseteq h(\langle f \rangle)$. If $g \in \langle f \rangle$ then $g = fg_1$ for some g_1 and $h(g) = h(fg_1) = h(f)h(g_1) \in \langle h(f) \rangle$ by Lemma 15, hence $h(\langle f \rangle) \subseteq \langle h(f) \rangle$. □

EXAMPLE 20. If $\mathfrak{a} = \langle f_1, \ldots, f_k \rangle$ then $\langle h(f_1), \ldots, h(f_k) \rangle \subseteq h(\mathfrak{a})$ but there might be strict inequality. If $\mathfrak{a} = \langle x_1x_2 - 1, x_1^2 - x_2 \rangle$ then $x_2^2 - x_1y \in h(\mathfrak{a})$ but $x_2^2 - x_1y \notin \langle x_1x_2 - y^2, x_1^2 - yx_2 \rangle$. □

Exercises

1. Show that if $\mathfrak{a} = \langle x_1x_2 - 1, x_1^2 - x_2 \rangle$, then $x_2^2 - x_1y \in h(\mathfrak{a})$ but $x_2^2 - x_1y \notin \langle x_1x_2 - y^2, x_1^2 - yx_2 \rangle$.

8.2.3 Dehomogenizing Polynomials

There is an opposite operation to homogenization: for each homogeneous polynomial we can define a polynomial in a polynomial ring with one less variable. In fact there are just as many ways to dehomogenize a homogeneous polynomial as there are variables.

DEFINITION 21 (DEHOMOGENIZATION OF A POLYNOMIAL). If the polynomial $f(x_1, \ldots, x_n, y) \in k[x_1, \ldots, x_n, y]$ is homogeneous, then the polynomial $a(f) = f(x_1, \ldots, x_n, 1)$ is the *dehomogenization* of f with respect to y.

If f and g are homogeneous of the same degree, $f + g$ is homogeneous (or 0) and $a(f + g) = a(f) + a(g)$, and for all homogeneous f, g we have that fg is homogeneous and $a(fg) = a(f)a(g)$.

8.2.4 Dehomogenizing Ideals

We can define the operation of dehomogenization not only on homogeneous polynomials, but also on homogeneous ideals.

DEFINITION 22 (DEHOMOGENIZATION OF AN IDEAL). If the ideal $\mathfrak{a}$ in the ring $k[x_1,\ldots,x_n,y]$ is homogeneous, then $a(\mathfrak{a})$ is the ideal in $k[x_1,\ldots,x_n]$, generated by the dehomogenizations of all homogeneous elements in $\mathfrak{a}$.

The map

$$d : k[x_1,\ldots,x_n,y] \longrightarrow k[x_1,\ldots,x_n]$$

defined by $d(f(x_1,\ldots,x_n,y)) = f(x_1,\ldots,x_n,1)$ is a homomorphism, so if $\mathfrak{a}$ is an ideal in $k[x_1,\ldots,x_n,y]$ then $d(\mathfrak{a}) = \{\, d(f) \mid f \in \mathfrak{a}\,\}$ is an ideal in $k[x_1,\ldots,x_n]$. We claim that if $\mathfrak{a}$ is homogeneous then $d(\mathfrak{a})$ is generated by the images of homogeneous elements in $\mathfrak{a}$. Let $f = f_0 + \cdots + f_k \in \mathfrak{a}$, where f_i is homogeneous of degree i. Then $d(f) = d(f_0) + d(f_1) + \cdots + d(f_k)$ and $f_i \in \mathfrak{a}$ since $\mathfrak{a}$ is homogeneous, so $d(f_i) \in a(\mathfrak{a})$. Thus it is clear that $d(\mathfrak{a}) = a(\mathfrak{a})$.

8.2.5 Homogenization versus Dehomogenization

The operations of homogenization and dehomogenization are almost inverse to each other. More exactly, if $f \in k[x_1,\ldots,x_n] = A$ then $a(h(f)) = f$, but if g is a homogeneous polynomial in $A[y]$ then $h(a(f)) = y^{-m}f$ where y^m is the largest power of y which divides f. In particular, if y does not divide f then $h(a(f)) = f$. For an ideal $\mathfrak{a}$ in A we get $a(h(\mathfrak{a})) = \mathfrak{a}$.

LEMMA 23. *Let $\mathfrak{a}$ be an ideal in A and let f be a homogeneous element in $A[y]$. Then $f \in h(\mathfrak{a})$ if and only if $f = y^m h(g)$ for some $g \in \mathfrak{a}$ and some m.*

Proof. If $f = y^m h(g)$ then obviously $f \in h(\mathfrak{a})$. If $f \in h(\mathfrak{a})$ then $f = \sum g_i h(f_i)$ for some homogeneous $g_i \in A[y]$ and some $f_i \in \mathfrak{a}$. We can suppose that $\deg(g_i h(f_i)) = \deg(f)$ for all i, because all other terms must cancel. Let y^{m_i} be the largest power of y that divides g_i and let $g_i = y^{m_i} g_i'$. Then

$$\sum g_i h(f_i) = \sum y^{m_i} g_i' h(f_i) = \sum y^{m_i} h(a(g_i'))h(f_i) = \sum y^{m_i} h(a(g_i') f_i).$$

If $\min\{m_i\} = m$ this sum equals $y^m \sum y^{m_i - m} h(a(g_i')f_i) = y^m \sum h(a(g_i')f_i)$ and $a(g_i')f_i \in \mathfrak{a}$. □

LEMMA 24. *If $\mathfrak{a}$ is an ideal in $k[x_1,\ldots,x_n]$ then $h(\mathfrak{a}) : \langle y^m \rangle = h(\mathfrak{a})$ for each m.*

Proof. That $h(\mathfrak{a}) \subseteq h(\mathfrak{a}) : \langle y^m\rangle$ is clear. If $y^m f \in h(\mathfrak{a})$ for some homogeneous $f \in k[x_1,\ldots,x_n]$ then $y^m f = y^n h(g)$ for some $g \in \mathfrak{a}$ and $n \geq m$ since y does not divide $h(g)$. Thus $y^{m-n} f = h(g)$ and $y^{m-n} f \in h(\mathfrak{a})$, so $f = y^{n-m}(y^{m-n} f) \in h(\mathfrak{a})$. □

For ideals we have:

LEMMA 25. *If $\mathfrak{a}$ is a homogeneous ideal in $k[x_1, \ldots, x_n, y]$ then $\mathfrak{a} \subseteq h(a(\mathfrak{a}))$ and $y^m h(a(\mathfrak{a})) \subseteq \mathfrak{a}$ for some positive integer m.*

Proof. Suppose f is a homogeneous element in $\mathfrak{a}$. Then $a(f) \in a(\mathfrak{a})$, hence $y^{-k} f = h(a(f)) \in h(a(\mathfrak{a}))$ for some k, so $f = y^k h(a(f)) \in h(a(\mathfrak{a}))$ for some k. Thus $\mathfrak{a} \subseteq h(a(\mathfrak{a}))$. Now suppose $\mathfrak{a} = \langle f_1, \ldots, f_r \rangle$. We have $h(a(f_i)) = y^{-m_i} f_i$. Let $m = \max\{m_i\}$. Then $y^m h(a(\mathfrak{a})) \subseteq \mathfrak{a}$. □

LEMMA 26. *Let $\mathfrak{a}$ and $\mathfrak{b}$ be two homogeneous ideals in $k[x_1, \ldots, x_n, y]$. Then $a(\mathfrak{a}) = a(\mathfrak{b})$ if and only if $\mathfrak{a} : \langle y^m \rangle = \mathfrak{b} : \langle y^m \rangle$ for some integer m.*

Proof. We have for each $m \geq 1$ that $a(\mathfrak{a} : \langle y^m \rangle) = a(\mathfrak{a}) : a(\langle y^m \rangle) = a(\mathfrak{a}) : \langle 1 \rangle = a(\mathfrak{a})$. (Here we have used that $a(\mathfrak{a} : \mathfrak{b}) = a(\mathfrak{a}) : a(\mathfrak{b})$; see the exercises below.) If $\mathfrak{a} : \langle y^m \rangle = \mathfrak{b} : \langle y^m \rangle$ we get

$$a(\mathfrak{a}) = a(\mathfrak{a}) : \langle 1 \rangle = a(\mathfrak{a}) : a(\langle y^m \rangle) = a(\mathfrak{a} : \langle y^m \rangle)$$

$$= a(\mathfrak{b} : \langle y^m \rangle) = a(\mathfrak{b}) : a(\langle y^m \rangle) = a(\mathfrak{b}) : \langle 1 \rangle = a(\mathfrak{b}).$$

Now assume that $a(\mathfrak{a}) = a(\mathfrak{b})$. Since $\mathfrak{a} : \langle y^i \rangle \subseteq \mathfrak{a} : \langle y^{i+1} \rangle$ we have by Proposition 13 in Section 3.5 that $\mathfrak{a} : \langle y^m \rangle = \mathfrak{a} : \langle y^{m+1} \rangle = \cdots$ for some m, and we can assume also that $\mathfrak{b} : \langle y^m \rangle = \mathfrak{b} : \langle y^{m+1} \rangle = \cdots$ by choosing m large enough. Since

$$a(\mathfrak{a} : \langle y^m \rangle) = a(\mathfrak{a}) = a(\mathfrak{b}) = a(\mathfrak{b} : \langle y^m \rangle)$$

we have $h(a(\mathfrak{a} : \langle y^m \rangle)) = h(a(\mathfrak{b} : \langle y^m \rangle))$. By Lemma 25 we have that $y^m (h(a(\mathfrak{a} : \langle y^m \rangle))) \subseteq \mathfrak{a} : \langle y^m \rangle$ for m large enough. Thus it follows that $h(a(\mathfrak{a} : \langle y^m \rangle)) \subseteq \mathfrak{a} : \langle y^m \rangle$. Since for any ideal $\mathfrak{b}$ it is true that $\mathfrak{b} \subseteq h(a(\mathfrak{b}))$, we have $\mathfrak{a} : \langle y^m \rangle \subseteq h(a(\mathfrak{a} : \langle y^m \rangle))$, so we get $h(a(\mathfrak{a} : \langle y^m \rangle)) = \mathfrak{a} : \langle y^m \rangle$ and therefore

$$\mathfrak{a} : \langle y^m \rangle = h(a(\mathfrak{a} : \langle y^m \rangle)) = a(\mathfrak{a} : \langle y^m \rangle) = a(\mathfrak{a}) =$$

$$a(\mathfrak{b}) = a(\mathfrak{b} : \langle y^m \rangle) = h(a(\mathfrak{b} : \langle y^m \rangle)) = \mathfrak{b} : \langle y^m \rangle .$$

□

PROPOSITION 27. *1. If $\mathfrak{p}$ is a prime ideal in A then $h(\mathfrak{p})$ is a prime ideal in $A[y]$.*

2. *If $\mathfrak{q}$ is $\mathfrak{p}$-primary in A then $h(\mathfrak{q})$ is $h(\mathfrak{p})$-primary in $A[y]$.*
3. *If $\mathfrak{a} = \mathfrak{q}_1 \cap \cdots \cap \mathfrak{q}_s$ is a primary decomposition then $h(\mathfrak{a}) = h(\mathfrak{q}_1) \cap \cdots \cap h(\mathfrak{q}_s)$ is a primary decomposition.*

Proof. 1. We use Lemma 7.1. Suppose $fg \in h(\mathfrak{p}), f \notin h(\mathfrak{p})$ with f, g homogeneous elements in $A[y]$. Then $a(f)a(g) = a(fg) \in \mathfrak{p}$ since $\mathfrak{p} = a(h(\mathfrak{p}))$. We have $a(f) \notin \mathfrak{p}$, since $a(f) \in \mathfrak{p}$ would give $h(a(f)) \in h(\mathfrak{p})$ and f is a multiple of $h(a(f))$ and $f \notin h(\mathfrak{p})$. Since $a(f) \notin \mathfrak{p}$ we have $a(g) \in \mathfrak{p}$. Then $g = y^m h(a(g))$ for some m, so $g \in h(\mathfrak{p})$ since $h(a(g)) \in h(\mathfrak{p})$.

2. We use Lemma 7.2. Suppose $fg \in h(\mathfrak{q}), f \notin h(\mathfrak{q})$ where f, g are homogeneous in $A[y]$. Then $a(f)a(g) = a(fg) \in \mathfrak{q}$ and $a(f) \notin \mathfrak{q}$ as above, so $(a(g))^k \in \mathfrak{q}$ for some k. Since $(a(g))^k = a(g^k)$ we get $g^k = y^m h(a(g^k)) \in h(\mathfrak{q})$.

3. We have $h(\mathfrak{a}) = h(\mathfrak{q}_1 \cap \cdots \cap \mathfrak{q}_s)$. It is not hard to see that $h(\mathfrak{q}_1 \cap \cdots \cap \mathfrak{q}_s) = h(\mathfrak{q}_1) \cap \cdots \cap h(\mathfrak{q}_s)$ (we leave that as an exercise; see below), so $h(\mathfrak{a})$ is expressed as an intersection of primary ideals. Now $h(\mathfrak{a}) \neq h(\mathfrak{b})$ if $\mathfrak{a} \neq \mathfrak{b}$ since if $h(\mathfrak{a}) = h(\mathfrak{b})$ we get $\mathfrak{a} = a(h(\mathfrak{a})) = a(h(\mathfrak{b})) = \mathfrak{b}$. Thus $h(\sqrt{\mathfrak{q}_i}) \neq h(\sqrt{\mathfrak{q}_j})$ if $i \neq j$. It only remains to show that $\cap_{i \neq j} h(\mathfrak{q}_i) \neq \cap_{1=1}^{s} h(\mathfrak{q}_i)$ for every j. Since $\cap_{i \neq j} \mathfrak{q}_i \neq \mathfrak{a}$ we get $\cap_{i \neq j} h(\mathfrak{q}_i) = h(\cap_{i \neq j} \mathfrak{q}_i) \neq h(\mathfrak{a}) = \cap_{i=1}^{s} h(\mathfrak{q}_i)$.

□

PROPOSITION 28. *1. If $\mathfrak{p}$ is a homogeneous prime ideal in A which does not contain y, then $a(\mathfrak{p})$ is a prime ideal in A.*

2. If $\mathfrak{q}$ is homogeneous $\mathfrak{p}$-primary in $A[y]$ and $\mathfrak{q}$ does not contain any power of y, then $a(\mathfrak{q})$ is $a(\mathfrak{p})$-primary in A.

3. If $\mathfrak{a} = \mathfrak{q}_1 \cap \cdots \cap \mathfrak{q}_s$ is a homogeneous primary decomposition and no power of y belongs to $\mathfrak{a}$, then $a(\mathfrak{a}) = \cap_j a(\mathfrak{q}_j)$ is a primary decomposition, where the intersection is taken over those primary components which do not contain any power of y.

Proof. 1. Suppose $fg \in a(\mathfrak{p}), f \notin a(\mathfrak{p})$. Then $y^m h(fg) = y^m h(f)h(g) \in \mathfrak{p}$ for some positive integer m. We have $h(f) \notin \mathfrak{p}$ since $f = a(h(f)) \notin a(\mathfrak{p})$, and we have assumed that $y \notin \mathfrak{p}$. Hence $h(g) \in \mathfrak{p}$, so $g = a(h(g)) \in a(\mathfrak{p})$.

2. Suppose $fg \in a(\mathfrak{q}), f \notin a(\mathfrak{q})$. As above, we get $y^m h(f)h(g) \in \mathfrak{q}$ and $h(f) \notin \mathfrak{q}$. We have assumed that $y^m \notin \mathfrak{q}$, so $h(g) \in \sqrt{\mathfrak{q}}$. Thus $g = a(h(g)) \in a(\sqrt{\mathfrak{q}}) = \sqrt{a(\mathfrak{q})}$. (The last equality is left as an exercise; see below.)

3. We have $a(\mathfrak{q}_1 \cap \cdots \cap \mathfrak{q}_s) = a(\mathfrak{q}_1) \cap \cdots \cap a(\mathfrak{q}_s)$ (exercise; see below). If $y^m \in \mathfrak{q}_i$ then $1 = a(y^m) \in a(\mathfrak{q}_i)$, so we can delete $a(\mathfrak{q}_i)$. We can thus assume that no $\mathfrak{q}_i$ contains a power of y, hence $a(\mathfrak{q}_i)$ is primary as we have just shown. It remains to show that the decomposition is irredundant, i.e., $\cap_{i \neq t} a(\mathfrak{q}_i) \neq \cap a(\mathfrak{q}_i)$ for each t, $1 \leq t \leq s$. Suppose

that $\cap_{i\neq t}a(\mathfrak{q}_i) = \cap a(\mathfrak{q}_i)$, so that $a(\cap_{i\neq t}\mathfrak{q}_i) = a(\cap\mathfrak{q}_i)$. Since $\mathfrak{q}_1 \cap \cdots \cap \mathfrak{q}_s$ is a primary decomposition, we have $\cap_{i\neq t}\mathfrak{q}_i \not\subseteq \mathfrak{q}_t$, so we get $\cap_{i\neq t} \cap \mathfrak{q}_i : \langle y^m \rangle \not\subseteq \mathfrak{q}_t$. Since no power of y does belong to $\mathfrak{q}_t$, we have $\cap\mathfrak{q}_i : \langle y^j \rangle \subseteq \cap_{i\neq t}\mathfrak{q}_i : \langle y^j \rangle = \cap_{i\neq t}\mathfrak{q}_i$ for all j. Thus $\cap_{i\neq t}\mathfrak{q}_i : \langle y^j \rangle \neq \cap\mathfrak{q}_i : \langle y^j \rangle$ for all j, and the conclusion follows by Lemma 26.

□

Exercises

1. Let g be a homogeneous polynomial in $k[x_1, \ldots, x_n, y]$. Prove that $h(a(f)) = y^{-m}f$ where y^m is the largest power of y which divides f.

2. Show that $a(h(\mathfrak{a})) = \mathfrak{a}$ if $\mathfrak{a}$ is an ideal in A.

3. Let $\mathfrak{a}, \mathfrak{b}$ be ideals in $k[x_1, \ldots, x_n]$. Show that

   (a) $h(\mathfrak{a} + \mathfrak{b}) = h(\mathfrak{a}) + h(\mathfrak{b})$.

   (b) $h(\mathfrak{a}\mathfrak{b}) = h(\mathfrak{a})h(\mathfrak{b})$.

   (c) $h(\mathfrak{a} \cap \mathfrak{b}) = h(\mathfrak{a}) \cap h(\mathfrak{b})$.

   (d) $h(\mathfrak{a} : \mathfrak{b}) = h(\mathfrak{a}) : h(\mathfrak{b})$.

   (e) $h(\sqrt{\mathfrak{a}}) = \sqrt{h(\mathfrak{a})}$.

4. Let $\mathfrak{a}, \mathfrak{b}$ be ideals in $k[x_1, \ldots, x_n, y]$. Show that

   (a) $a(\mathfrak{a} + \mathfrak{b}) = a(\mathfrak{a}) + a(\mathfrak{b})$.

   (b) $a(\mathfrak{a}\mathfrak{b}) = a(\mathfrak{a})a(\mathfrak{b})$.

   (c) $a(\mathfrak{a} \cap \mathfrak{b}) = a(\mathfrak{a}) \cap a(\mathfrak{b})$.

   (d) $a(\mathfrak{a} : \mathfrak{b}) = a(\mathfrak{a}) : a(\mathfrak{b})$.

   (e) $a(\sqrt{\mathfrak{a}}) = \sqrt{a(\mathfrak{a})}$.

8.3 Gröbner Bases for Homogeneous Ideals

The set of homogeneous ideals constitute an important class of ideals. We will now investigate what could be said about their Gröbner bases. We will also show that we can get a Gröbner basis for an arbitrary ideal from the Gröbner basis of its homogenization.

LEMMA 29. *Let $\mathfrak{a}$ be a homogeneous ideal in $k[x_1, \ldots, x_n]$. Then the reduced Gröbner basis of $\mathfrak{a}$ in any ordering consists of homogeneous elements.*

Proof. Let $\mathfrak{a} = \langle f_1, \ldots, f_k \rangle$, f_i homogeneous. The S-polynomial of two homogeneous polynomials is homogeneous, and if we reduce a homogeneous polynomial with respect to a set of homogeneous polynomials, the result is a homogeneous polynomial. Thus, using the Buchberger algorithm, we stay all the time inside the set of homogeneous polynomials. □

We recall the concept of grading an ordering from Lemma 2 in Section 3.1. If $\prec$ is an arbitrary monomial ordering, the grading $\prec_g$ of $\prec$ is the monomial ordering defined by $m_1 \prec_g m_2$ if $\deg m_1 < \deg m_2$ or if $\deg m_1 = \deg m_2$ and $m_1 \prec m_2$. For example, Deglex is the grading of Lex.

PROPOSITION 30. *Let $\mathfrak{a}$ be a homogeneous ideal. Then the reduced Gröbner basis in an ordering $\prec$ and in its grading $\prec_g$ coincide.*

Proof. We know by Lemma 29 that the two reduced Gröbner bases consist of homogeneous elements. We will show, by induction on i, that the elements of degree i coincide. Let k be the least degree of a homogeneous element in $\mathfrak{a}$. Then the elements of degree k in $l_{\prec}(\mathfrak{a})$ and $l_{\prec_g}(\mathfrak{a})$ coincide, since $\prec$ and $\prec_g$ coincide in a fixed degree. Take one element f_i for each generator m_i of $l(\mathfrak{a})_k$. Then, when reducing the f_i's, we can only use elements of degree k, so the result is the same in the two orderings. Now suppose that the two reduced Gröbner bases coincide in degree $< i$. Take one homogeneous element $g_i \in \mathfrak{a}$ for each generator of $l(\mathfrak{a})_i$ which does not belong to the ideal generated by all elements of degree $< i$ in $\mathfrak{a}$. Since we can only use elements of degree at most i when reducing the g_i's, and the two orderings coincide in degree i, and their Gröbner bases coincide in degree $< i$, the result of a reduction becomes the same in the two orderings. □

DEFINITION 31 (GOOD EXTENSION OF AN ORDERING). Let $\prec$ be an ordering of the monomials in $A = k[x_1, \ldots, x_n]$, and $\prec_1$ an ordering of the monomials in $k[x_1, \ldots, x_n, y]$. We call $\prec_1$ a *good extension* of $\prec$ if $\mathrm{lm}_{\prec}(f) = \mathrm{lm}_{\prec_1}(h(f))$ for any $f \in A$.

If $\prec$ is any ordering on A and $\prec_1$ is the product ordering of $\prec$ and the ordering on the powers of y, then $\prec_1$ is a good extension of $\prec$. An example is Lex with $x_1 \succ \cdots \succ x_n \succ y$ a good extension of Lex with $x_1 \succ \cdots \succ x_n$. Another example is $\prec_1$=Degrevlex with $x_1 \succ \cdots \succ x_n \succ y$ from $\prec$=Degrevlex with $x_1 \succ \cdots \succ x_n$.

The following proposition is a rather immediate consequence of the definitions.

PROPOSITION 32. *Let $\mathfrak{a}$ be an ideal in $k[x_1, \ldots, x_n]$ and let $h(\mathfrak{a})$ be its homogenization in $k[x_1, \ldots, x_n, y]$. Suppose that $\prec_1$ is a good extension to $k[x_1, \ldots, x_n, y]$ of an ordering $\prec$ on $k[x_1, \ldots, x_n]$. If $G = \{g_1, \ldots, g_s\}$ is the reduced Gröbner basis for $h(\mathfrak{a})$ in $\prec_1$, then $a(G) = \{a(g_1), \ldots, a(g_s)\}$ is the reduced Gröbner basis for $\mathfrak{a}$ in $\prec$.*

More interesting is the following proposition.

PROPOSITION 33. *Let $\mathfrak{a}$ be an ideal in $k[x_1, \ldots, x_n]$ and let $h(\mathfrak{a})$ be its homogenization in $k[x_1, \ldots, x_n, y]$. Suppose that $\prec_1$ is a good extension to $k[x_1, \ldots, x_n, y]$ of an ordering $\prec$ on $k[x_1, \ldots, x_n]$. If $G = \{g_1, \ldots, g_s\}$ is the reduced Gröbner basis for $\mathfrak{a}$ in $\prec$, then $h(G) = \{h(g_1), \ldots, h(g_s)\}$ is the reduced Gröbner basis for $h(\mathfrak{a})$ in $\prec_1$.*

Proof. Since $h(G) \subseteq h(\mathfrak{a})$ it suffices to show that

$$l(h(\mathfrak{a})) = l(\langle \text{lm}(h(g_1)), \ldots, \text{lm}(h(g_s)) \rangle).$$

Since $h(\mathfrak{a})$ is a homogeneous ideal it suffices to show that if $f \in h(\mathfrak{a})$ is homogeneous, then $\text{lm}(f) \in \langle \text{lm}(h(g_1)), \ldots, \text{lm}(h(g_s)) \rangle$. If f is homogeneous then, by Lemma 23, $f = y^m h(g)$ for some $g \in \mathfrak{a}$, and we get

$$\text{lm}(f) = y^m \,\text{lm}(h(g)) = y^m \,\text{lm}(g) \in \langle \text{lm}(g_1), \ldots, \text{lm}(g_s) \rangle =$$

$$l(\langle \text{lm}(h(g_1)), \ldots, \text{lm}(h(g_s)) \rangle).$$

We now show that $h(G)$ is reduced. It is clear that $\text{lm}(h(g_i)) = \text{lm}(g_i)$ are monic and constitute a minimal set of generators for $l(h(\mathfrak{a})) = l(\mathfrak{a})$. Let $y^k m$ be a monomial occurring in $h(g_i)$, where m is a monomial in $k[x_1, \ldots, x_n]$. Then m occurs in g_i and no $\text{lm}(g_j)$ divides m since G is reduced. Hence no $\text{lm}(h(g_i)) = \text{lm}(g_i)$ divides $y^k m$. □

Experience indicates that it is often much easier to calculate Gröbner bases of homogeneous ideals. Thus it would be desirable to use homogenization to calculate the Gröbner basis for a nonhomogeneous ideal $\mathfrak{a}$. It is trivial to homogenize the generators of $\mathfrak{a}$. We have seen that, in general, if $\mathfrak{a} = \langle f_1, \ldots, f_k \rangle$ then $h(\mathfrak{a}) \neq \langle h(f_1), \ldots, h(f_k) \rangle$. The following proposition shows that it suffices to homogenize the generators.

PROPOSITION 34. *Let $\mathfrak{a} = \langle f_1, \ldots, f_k \rangle$ be an ideal in $A = k[x_1, \ldots, x_n]$ and let $\prec$ be any ordering on A. Let $\prec_1$ be a good extension of $\prec$ to $A[y]$. If $G = \{g_1, \ldots, g_s\}$ is a Gröbner basis of $\langle h(f_1), \ldots, h(f_k) \rangle$ in $\prec_1$, then $\{a(g_1), \ldots, a(g_s)\}$ is a Gröbner basis of $\mathfrak{a}$ (in general not reduced even if G is reduced).*

Proof. We have $h(\mathfrak{a}) : \langle y^m \rangle = h(\mathfrak{a})$ according to Lemma 24. Since

$$a(\langle h(f_1), \ldots, h(f_k) \rangle) = \langle a(h(f_1)), \ldots, a(h(f_k)) \rangle = \langle f_1, \ldots, f_k \rangle = \mathfrak{a},$$

Lemma 26 gives that $\langle h(f_1), \ldots, h(f_k) \rangle : \langle y^m \rangle = h(\mathfrak{a})$ for some m. Thus for any element $g \in h(\mathfrak{a})$, we have $y^m g \in \langle h(f_1), \ldots, h(f_k) \rangle$ for some m. Let $g' \in \mathfrak{a}$. Then $y^m h(g') \in \langle h(f_1), \ldots, h(f_k) \rangle$ for some m, so there is a j whereby $\mathrm{lm}(g_j)$ divides $\mathrm{lm}(y^m h(g')) = y^m \mathrm{lm}(h(g') = y^m \mathrm{lm}(g')$. Thus $a(\mathrm{lm}(g_j)) = \mathrm{lm}(a(g_j))$ divides $\mathrm{lm}(g')$, so $l(\mathfrak{a}) \subseteq \langle \mathrm{lm}(a(g_1)), \ldots, \mathrm{lm}(a(g_s)) \rangle$. In fact we have equality, since the other inclusion is trivial. We have found elements $a(g_j) \in \mathfrak{a}$, such that their leading monomials generate $l(\mathfrak{a})$. Thus $\{a(g_j)\}$ is a Gröbner basis by definition. □

REMARK. Even if the obtained Gröbner basis $\{a(g_j)\}$ is not reduced, there is no difficulty in obtaining a reduced Gröbner basis. Just pick the elements in the Gröbner basis whose initial forms constitute a minimal generating set for $l\,\langle a(g_j) \rangle$.

EXAMPLE 35. Let

$$\mathfrak{a} = \langle x_1 + x_2 + x_3 + x_4 + x_5, x_1x_2 + x_2x_3 + x_3x_4 + x_4x_5 + x_5x_1,$$
$$x_1x_2x_3 + x_2x_3x_4 + x_3x_4x_5 + x_4x_5x_1 + x_5x_1x_2,$$
$$x_1x_2x_3x_4 + x_2x_3x_4x_5 + x_3x_4x_5x_1 + x_4x_5x_1x_2 + x_5x_1x_2x_3, x_1x_2x_3x_4x_5 - 1\rangle$$

and let

$$\mathfrak{b} = \langle x_1 + x_2 + x_3 + x_4 + x_5, x_1x_2 + x_2x_3 + x_3x_4 + x_4x_5 + x_5x_1,$$
$$x_1x_2x_3 + x_2x_3x_4 + x_3x_4x_5 + x_4x_5x_1 + x_5x_1x_2,$$
$$x_1x_2x_3x_4 + x_2x_3x_4x_5 + x_3x_4x_5x_1 + x_4x_5x_1x_2 + x_5x_1x_2x_3, x_1x_2x_3x_4x_5 - y^5\rangle.$$

The initial ideal of $\mathfrak{b}$ in Degrevlex with $x_1 \prec x_2 \prec x_3 \prec x_4 \prec x_5 \prec y$ is

$$\{x_1, x_2^2, x_2x_3^2, x_2x_3x_4^2, x_2x_3x_4x_5^2, x_2x_3x_5^5, x_2x_4^4x_5, x_2x_4^3x_5^2, x_2x_4^3y^5, x_2x_4^2x_5^3y^5,$$
$$x_2x_4x_5^6, x_2x_5^5y^5, x_3^4x_5^2, x_3^3x_4, x_3^3x_5^3, x_3^3y^5, x_3^2x_4^3, x_3^2x_4^2x_5, x_3^2x_4^2y^5, x_3^2x_4x_5^5, x_3^2x_5^6,$$
$$x_3^2x_5^5y^5, x_3x_4^4x_5, x_3x_4^3x_5^2, x_3x_4^3y^5, x_3x_4^2x_5^4, x_3x_4^2x_5^3y^5, x_3x_4x_5^6, x_3x_5^7y^5, x_4^5x_5^2,$$
$$x_4^4x_5^3, x_4^4y^5, x_4^3x_5^4, x_4^2x_5^6, x_4x_5^7y^5, x_5^8y^5\},$$

so the reduced Gröbner basis has 38 elements. If we dehomogenize, we get

$$l(\mathfrak{a}) = \{x_1, x_2^2, x_2x_3^2, x_2x_3x_4^2, x_2x_3x_4x_5^2, x_2x_3x_5^5, x_2x_4^3, x_2x_4^2x_5^3, x_2x_5^5, x_3^3,$$
$$x_3^2x_4^2, x_3^2x_4x_5^5, x_3^2x_5^6, x_3x_4^2x_5^3, x_3x_4x_5^6, x_3x_5^7, x_4^5x_5^2, x_4^4, x_4^3x_5^4, x_4^2x_5^6, x_4x_5^7, x_5^8\},$$

so the reduced Gröbner basis has 24 elements. □

Exercises

1(a) Show that if $\succ$ is any degree ordering on $A = k[x_1, \ldots, x_n]$, then the product ordering $\succ_1$ of $\succ$ with the ordering of the monomials in y, i.e., $x_1^{e_1} \cdots x_n^{e_n} y^e \succ_1 x_1^{f_1} \cdots x_n^{f_n} y^f$ if $x_1^{e_1} \cdots x_n^{e_n} \succ x_1^{f_1} \cdots x_n^{f_n}$ or if $x_1^{e_1} \cdots x_n^{e_n} = x_1^{f_1} \cdots x_n^{f_n}$ and $e > f$, is a good extension of $\succ$.

(b) Show that Degrevlex on $A[y]$ with $x_1 \succ \cdots \succ x_n \succ y$ is a good extension of Degrevlex on A with $x_1 \succ \cdots \succ x_n$.

2. Prove Proposition 32.

3. Determine the reduced Gröbner basis in Degrevlex with $x_1 \succ x_2 \succ x_3$ of $\langle x_1x_2 - 1, x_2x_3 - x_1, x_3^2 - x_2 - 1\rangle$ by determining the reduced Gröbner basis in Degrevlex with $x_1 \succ x_2 \succ x_3 \succ y$ of

$$\langle x_1x_2 - y^2, x_2x_3 - x_1y, x_3^2 - x_2y - y^2\rangle .$$

4. Show that if $\succ_1$ is a good extension of $\succ$ then the grading $\succ_{1,g}$ of $\succ_1$ is a good extension of the grading $\succ_g$ of $\succ$. In particular, if $\succ$ is a degree ordering and $\succ_1$ is a good extension of $\succ$, then $\succ_{1,g}$ is also a good extension of $\succ$. Conclude that Deglex with $x_1 \succ \cdots \succ x_n \succ y$ is a good extension of Deglex with $x_1 \succ \cdots \succ x_n$.

5. Determine the reduced Gröbner basis in Deglex with $x \succ y \succ z$ of $\langle xy - z, xz - y, yz - x\rangle$ by determining the reduced Gröbner basis in Deglex with $x \succ y \succ z \succ t$ of $\langle xy - zt, xz - yt, yx - xt\rangle$.

6. Let $\mathfrak{a} = \langle f_1, \ldots, f_k\rangle$ be an ideal in $k[x_1, \ldots, x_n]$. Show that if $G = \{g_1, \ldots, g_s\}$ is a Gröbner basis of $\langle h(f_1), \ldots, h(f_k)\rangle$ in Degrevlex with $x_1 \succ \cdots \succ x_n \succ y$, then $\{h(a(g_1)), \ldots, h(a(g_s))\}$ is a Gröbner basis of $h(\mathfrak{a})$ (not necessarily reduced even if G is reduced).

Chapter 9

Projective Varieties

We have seen that extending the field from $\mathbb{R}$ to $\mathbb{C}$ greatly simplified the correspondence between geometric objects (affine varieties) and algebraic objects (ideals in or factor rings of polynomial rings). We are now going to introduce a new kind of geometric objects, projective varieties, which will make many geometric statements more clear or "symmetric". (We give some examples to show what we mean below.) This will be achieved by adding to the affine space k^n points "at infinity". We start by describing the projective line $P^1(\mathbb{R})$, which is obtained by adding one point ∞ to $\mathbb{R}$. We give the elements in $P^1(\mathbb{R})$, which we call points, coordinates in the following way. Define an equivalence relation on the set $\mathbb{R}^2 \setminus \{(0,0)\}$ by $(x,y) \sim (x_1,y_1)$ if $(x,y) = (tx_1, ty_1)$ for some $t \neq 0$. Then $P^1(\mathbb{R})$ is the set of equivalence classes. We will write points in $P^1(\mathbb{R})$ as $(x : y)$, where (x,y) is a representative for the class $(x : y)$. Thus $(1 : 2) = (3 : 6)$. If one representative (x,y) for a class has $y \neq 0$, then every representative for this class has second coordinate $\neq 0$, and there is a unique representative for the class with second coordinate 1. We identify $r \in \mathbb{R}$ with the class $(r : 1)$. The only remaining class is $(1 : 0)$, which we call the point at infinity. We are now ready for the definition of $P^n(k)$.

DEFINITION 1 (THE PROJECTIVE SPACE). Let k be a field. Then $P^n(k)$ consists of the equivalence classes of elements in $k^{n+1} \setminus \{(0,0,\ldots,0)\}$, where the equivalence is defined by

$$(x_1,\ldots,x_{n+1}) \sim (y_1,\ldots,y_{n+1}) \text{ if } (x_1,\ldots,x_{n+1}) = (ty_1,\ldots,ty_{n+1})$$

for some $t \neq 0$.

We write the equivalence classes $(x_1 : \cdots : x_{n+1})$ where $(x_1,\ldots,x_{n+1})$ is a representative for the class. We identify $(x_1,\ldots,x_n) \in k^n$ with $(x_1 :$

$\cdots : x_n : 1) \in P^n(k)$, and the remaining points in $P^n(k)$, i.e., those with last coordinate 0, are called the points at infinity.

As an example $P^2(\mathbb{R})$ is $\mathbb{R}^2$ extended with one point $(a : b : 0)$ at infinity for each line through the origin with direction $(a, b) \neq (0, 0)$. Note that, since $(a : b : 0) = (-a : -b : 0)$, each line contributes only one point at infinity, travelling along a line one approaches the same point at infinity in both directions.

EXAMPLE 2. Let $f(x, y) \in \mathbb{R}[x, y]$ and consider the affine curve $\mathfrak{v}(f) = \{(a, b) \in \mathbb{R}^2; f(a, b) = 0\}$. Let $h(f) = z^{\deg f} f(x/z, y/z) \in \mathbb{R}[x, y, z]$ be the homogenization of f. Since $h(f)$ is homogeneous, we have that

$$h(f)(ta, tb, tc) = t^{\deg f} h(f)(a, b, c),$$

so if $h(f)$ is zero on a representative (a, b, c) for a point $(a : b : c)$ in $P^2(\mathbb{R})$, then $h(f)$ is zero on every representative for $(a : b : c)$. Thus the statement that $h(f)$ is zero on a point in $P^2(\mathbb{R})$ is well defined. We call the set of points in $P^2(\mathbb{R})$ for which $h(f)$ is zero the projectivization of the curve $\mathfrak{v}(f)$. □

EXAMPLE 3. A line l in $\mathbb{R}^2$ has an equation $Ax + By + C = 0$ with $(A, B) \neq (0, 0)$. The projectivization of l is the set of points in $P^2(\mathbb{R})$ satisfying the equation $Ax + By + Cz = 0$. It is natural to make the following more symmetric definition. A projective line in $P^2(\mathbb{R})$ is the set of points in $P^2(\mathbb{R})$ satisfying an equation $Ax + By + Cz = 0$ with $(A, B, C) \neq (0, 0, 0)$. Besides the projectivizations of affine lines there is only one more, namely $z = 0$. This line consists of all points at infinity, and is thus called "the line at infinity". If $Ax + By + C = 0$ and $A_1x + B_1y + C_1 = 0$ are two nonparallel affine lines intersecting in the point (a, b), then the projective lines $Ax + By + Cz = 0$ and $A_1x + B_1y + C_1z = 0$ intersect in the projective point $(a : b : 1)$. If $Ax + By + C = 0$ and $Ax + By + C_1 = 0$ are two parallel lines, and thus do not intersect, then the projective lines $Ax + By + Cz = 0$ and $Ax + By + C_1z = 0$ intersect in the point $(B : -A : 0)$ at infinity. The intersection between $Ax + By + Cz = 0$ and $z = 0$ is also $(B : -A : 0)$ (if $(A, B) \neq (0, 0)$). Thus *any* pair of projective lines intersect in one projective point. □

EXAMPLE 4. The affine hyperbola $xy = 1$ in $\mathbb{R}^2$ corresponds to the projective curve $xy = z^2$. We have added the two points $(0 : 1 : 0)$ and $(1 : 0 : 0)$ corresponding to the two directions of the asymptotes to the affine points. The affine parabola $y = x^2$ corresponds to the projective curve $yz = x^2$, which has one point $(0 : 1 : 0)$ at infinity. The affine circle $x^2 + y^2 = 1$ corresponds to the projective curve $x^2 + y^2 = z^2$

which has no point at infinity. Note that if we introduce new variables by $X = x, Y = z - y, Z = z + y$, the equation becomes $X^2 = YZ$. If we now forget that we have used one variable for homogenization and treat all variables equally, we see that there is no difference between the three projective curves. The affine curves just describe different aspects of the same projective curve. □

The theory of projective sets runs in parallel to the theory of affine algebraic sets. Let $P = (c_1 : \cdots : c_{n+1}) \in P^n(k)$. If f is a homogeneous polynomial such that $f(c_1, \ldots, c_{n+1}) = 0$, then $f(tc_1, \ldots, tc_{n+1}) = 0$ for all $t \in k$. We say that f vanishes on P or that P is a zero of f.

DEFINITION 5 (IDEAL OF A SET). If $X \subseteq P^n(k)$ then $\mathfrak{I}(X)$ denotes the homogeneous ideal, generated by all homogeneous polynomials, which vanishes for every $c \in X$.

DEFINITION 6 (PROJECTIVE SET). If $\mathfrak{a}$ is a homogeneous ideal of the polynomial ring $k[x_1, \ldots, x_{n+1}]$, then the set of all projective points that are zeros of all polynomials in $\mathfrak{a}$ is denoted $\mathfrak{V}(\mathfrak{a})$ and is called the *projective set* or *variety* defined by $\mathfrak{a}$.

EXAMPLE 7. If $\mathfrak{a} = \langle x_1, \ldots, x_{n+1} \rangle$ then $\mathfrak{V}(\mathfrak{a}) = \emptyset$. Therefore $\langle x_1, \ldots, x_{n+1} \rangle$ is often called *the irrelevant maximal ideal.* More generally, if $\mathfrak{a}$ is a homogeneous ideal with $\sqrt{\mathfrak{a}} = \langle x_1, \ldots, x_{n+1} \rangle$ then $\mathfrak{V}(\mathfrak{a}) = \emptyset$, since for each i we have $x_i^{n_i} \in \mathfrak{a}$ for some positive n_i, hence $(0, 0, \ldots, 0)$ is the only common zero, and $(0, 0, \ldots, 0)$ does not give a point in $P^n(k)$. □

PROPOSITION 8. *(a) If $X_1 \subseteq X_2 \subseteq P^n(k)$ then $\mathfrak{I}(X_2) \subseteq \mathfrak{I}(X_1)$. If $\mathfrak{a} \subseteq \mathfrak{b}$ are homogeneous ideals in $k[x_1, \ldots, x_{n+1}]$ then $\mathfrak{V}(\mathfrak{b}) \subseteq \mathfrak{V}(\mathfrak{a})$.*

(b) For any subset $X \subseteq P^n(k)$ we have $X \subseteq \mathfrak{V}(\mathfrak{I}(X))$, with equality if and only if X is a projective set.

(c) If $\mathfrak{a}$ is a homogeneous ideal in $k[x_1, \ldots, x_{n+1}]$ we have $\mathfrak{V}(\mathfrak{a}) = \mathfrak{V}(\sqrt{\mathfrak{a}})$.

(d) If $\mathfrak{a}$ and $\mathfrak{b}$ are homogeneous ideals in $k[x_1, \ldots, x_{n+1}]$ then

$$\begin{aligned} \mathfrak{V}(\mathfrak{a} \cap \mathfrak{b}) &= \mathfrak{V}(\mathfrak{a}) \cup \mathfrak{V}(\mathfrak{b}), \\ \mathfrak{V}(\mathfrak{a}\mathfrak{b}) &= \mathfrak{V}(\mathfrak{a}) \cup \mathfrak{V}(\mathfrak{b}), \\ \mathfrak{V}(\mathfrak{a} + \mathfrak{b}) &= \mathfrak{V}(\mathfrak{a}) \cap \mathfrak{V}(\mathfrak{b}). \end{aligned}$$

(e) If $\mathfrak{a}$ is a homogeneous ideal in $k[x_1, \ldots, x_{n+1}]$ then $\mathfrak{a} \subseteq \mathfrak{I}(\mathfrak{V}(\mathfrak{a}))$ and the inclusion may be strict.

Proof. Parallel to the proof of Proposition 6 in Section 4.1. □

DEFINITION 9 (IRREDUCIBLE). A projective set $X \subseteq P^n(k)$, k a field, is said to be *irreducible* if X is not the union of two strictly smaller projective sets. A projective set which is not irreducible is called reducible.

PROPOSITION 10. *Let X be a projective set in $P^n(k)$. Then X is irreducible if and only if $\mathfrak{I}(X)$ is a prime ideal.*

Proof. Parallel to the proof of Proposition 8 in Section 4.1. □

THEOREM 11. *Let $X_1 \supset X_2 \supset \cdots$ be a strictly descending sequence of projective subsets of $P^n(k)$. Then the sequence is finite. Another way to state this is to say that any nonempty set $\{X_\alpha\}$ of projective subsets of $P^n(k)$ has a minimal element, i.e., an element X_{α_0} not containing some other X_α.*

Proof. Parallel to the proof of Theorem 9 in Section 4.1. □

THEOREM 12. *Every projective set in $P^n(k)$ is a finite union of irreducible projective sets, $X = X_1 \cup \cdots \cup X_m$, X_i irreducible. If we make this union irredundant, i.e., if no X_i is contained in some $X_j, j \neq i$, then the irreducible components are unique.*

Proof. Parallel to the proof of Theorem 10 in Section 4.1. □

THEOREM 13 (HILBERT'S PROJECTIVE NULLSTELLENSATZ, WEAK FORM). *Let $\mathfrak{a}$ be a homogeneous ideal in $\mathbb{C}[x_1, \ldots, x_{n+1}]$, with $\sqrt{\mathfrak{a}}$ strictly contained in $\langle x_1, \ldots, x_{n+1}\rangle$. Then $\mathfrak{V}(\mathfrak{a}) \neq \emptyset$.*

Proof. Consider the *affine* algebraic set $\mathfrak{v}(\mathfrak{a})$ in $A^{n+1}(\mathbb{C})$. As an affine algebraic set, $\mathfrak{v}(\mathfrak{a})$ is called the *affine cone* of the projective set $\mathfrak{V}(\mathfrak{a})$. The affine Nullstellensatz, Theorem 11 in Section 4.2, shows that $\mathfrak{v}(\mathfrak{a})$ is nonempty. Since $\sqrt{\mathfrak{a}} \neq \langle x_1, \ldots, x_{n+1}\rangle$, $\mathfrak{v}(\mathfrak{a})$ does not consist of only $(0, 0, \ldots, 0)$, so $\mathfrak{V}(\mathfrak{a})$ is nonempty. □

COROLLARY 14 (STRONG PROJECTIVE NULLSTELLENSATZ). *For any homogeneous ideal $\mathfrak{a} \subseteq \mathbb{C}[x_1, \ldots, x_{n+1}]$ with $\sqrt{\mathfrak{a}}$ strictly contained in the ideal $\langle x_1, \ldots, x_{n+1}\rangle$ we have $\mathfrak{I}(\mathfrak{V}(\mathfrak{a})) = \sqrt{\mathfrak{a}}$.*

Proof. As in the proof of the affine case, Corollary 13 in Section 4.2, we have $\sqrt{\mathfrak{a}} \subseteq \mathfrak{I}(\mathfrak{V}(\sqrt{\mathfrak{a}})) = \mathfrak{I}(\mathfrak{V}(\mathfrak{a}))$. Suppose that $f = f_0 + \cdots + f_s \in \mathfrak{I}(\mathfrak{V}(\mathfrak{a}))$, f_i homogeneous of degree i. Then $f(x_1, \ldots, x_{n+1}) = 0$ for every $(x_1 : \cdots : x_{n+1}) \in \mathfrak{I}(\mathfrak{V}(\mathfrak{a}))$, so $f(tx_1, \ldots, tx_{n+1}) = f_0 + tf_1(x_1, \ldots, x_{n+1}) + \cdots + t^s f_s(x_1, \ldots, x_{n+1}) = 0$ for every $t \in \mathbb{C}$. Since $\mathbb{C}$ is infinite, this polynomial in t must equal the zero polynomial, hence $f_i(x_1, \ldots, x_{n+1}) = 0$ for every i. Thus $\mathfrak{I}(\mathfrak{V}(\mathfrak{a})) = \mathfrak{i}(\mathfrak{v}(\mathfrak{a}))$ and the claim follows from the affine case. □

COROLLARY 15. *If X is a projective set in $P^n(\mathbb{C})$ then $\mathfrak{V}(\mathfrak{I}(X)) = X$. If $\mathfrak{a} = \sqrt{\mathfrak{a}}$ and is strictly contained in $\langle x_1, \ldots x_{n+1}\rangle$ then $\mathfrak{I}(\mathfrak{V}(\mathfrak{a})) = \mathfrak{a}$.*

Proof. Parallel to the proof of Corollary 14 in Section 4.2. □

Exercises

1. Prove Proposition 8. Use the result that the set of homogeneous ideals are closed under the usual ideal theoretic operations.

2. Prove Proposition 10 using Lemma 7 in Section 8.1.

9.1 Projective Closure of an Algebraic Set

Let V be an algebraic set in $A^n(\mathbb{C})$. Then V can be considered as a subset of $P^n(\mathbb{C})$ via the map $\phi : A^n(\mathbb{C}) \to P^n(\mathbb{C})$ defined by $\phi(c_1, \ldots, c_n) = (c_1 : \cdots : c_n : 1)$.

DEFINITION 16 (PROJECTIVE CLOSURE). Let V be an algebraic subset of $A^n(\mathbb{C})$. The smallest projective set $\mathbf{h}(V)$ in $P^n(\mathbb{C})$ containing V (i.e., containing $\phi(V)$) is called the *projective closure* of V.

On the other hand, if V is a projective set in $P^n(\mathbb{C})$, we set $\mathbf{a}(V) = \{ (a_1, \ldots, a_n) \mid (a_1 : \cdots : a_n : 1) \in V \}$.

DEFINITION 17 (AFFINE RESTRICTION). If V is a projective set in $P^n(\mathbb{C})$ then $\mathbf{a}(V)$ is called the *affine restriction* of V.

PROPOSITION 18. *If $V = \mathfrak{V}(\mathfrak{a})$ is a projective set then $\mathbf{a}(V) = \mathfrak{v}(a(\mathfrak{a}))$, i.e., the affine restriction of V is the zeroset of the dehomogenization of $\mathfrak{a}$. In particular $\mathbf{a}(V)$ is an algebraic set.*

Proof. This follows directly from the definitions. □

It is not equally easy to describe the projective closure with equations. If the projective set V is described with equations, i.e., as a zeroset of an ideal given by its generators, we need the Nullstellensatz to be able to describe the projective closure of V. This is the reason for restricting ourselves to algebraic sets in $A^n(\mathbb{C})$ here.

PROPOSITION 19. *Let $\mathfrak{a}$ be an ideal in $\mathbb{C}[x_1, \ldots, x_n]$. Then $\mathbf{h}(\mathfrak{v}(\mathfrak{a})) = \mathfrak{V}(h(\mathfrak{a}))$. In words, the projective closure of $\mathfrak{v}(\mathfrak{a})$ is the zeroset of the homogenization of $\mathfrak{a}$.*

Proof. We shall show that $\mathfrak{V}(h(\mathfrak{a}))$ is the smallest projective set containing $\mathfrak{v}(\mathfrak{a})$. We first show that $\mathfrak{V}(h(\mathfrak{a}))$ contains $\mathfrak{v}(\mathfrak{a})$. Suppose that $(a_1, \ldots, a_n) \in \mathfrak{v}(\mathfrak{a})$. We must show that $(a_1 : \cdots : a_n : 1) \in \mathfrak{V}(h(\mathfrak{a}))$. Let $g \in h(\mathfrak{a})$. Since $\mathfrak{a} = a(h(\mathfrak{a}))$, we have $a(g) \in \mathfrak{a}$. But $g = x_{n+1}^m h(a(g))$ for some m. Hence $g(a_1, \ldots, a_n, 1) = a(g)(a_1, \ldots, a_n) = 0$, so $\mathfrak{v}(\mathfrak{a})$ is contained in $\mathfrak{V}(h(\mathfrak{a}))$. For the second part of the proof suppose $W = \mathfrak{V}(\langle f_1, \ldots, f_s \rangle)$ is a projective set containing $\mathfrak{v}(\mathfrak{a})$. We shall show that W contains $\mathfrak{V}(h(\mathfrak{a}))$. Since $f_i(a_1, \ldots, a_n, 1) = 0$ for every $(a_1, \ldots, a_n) \in \mathfrak{v}(\mathfrak{a})$, we get $a(f_i)(x_1, \ldots, x_n) \in \mathfrak{i}(\mathfrak{v}(\mathfrak{a})) = \sqrt{\mathfrak{a}}$, so $(a(f_i))^m \in \mathfrak{a}$ for some m. This implies that $h((a(f_i))^m) = (h(a(f_i)))^m \in h(\mathfrak{a})$. Thus $(h(a(f_i)))^m$ is zero on $\mathfrak{V}(h(\mathfrak{a}))$, hence $h(a(f_i))$ is zero on $\mathfrak{V}(h(\mathfrak{a}))$. But $h(a(f_i)) = x_{n+1}^{-k} f_i$ for some $k \geq 0$, and hence f_i is zero on $\mathfrak{V}(h(\mathfrak{a}))$, so $\mathfrak{V}(h(\mathfrak{a})) \subseteq \mathfrak{V}(\langle f_1, \ldots, f_s \rangle)$. □

COROLLARY 20. *Let $\mathfrak{a}$ be an ideal in $\mathbb{C}[x_1, \ldots, x_n]$. If $\{g_1, \ldots, g_t\}$ is a Gröbner basis of $\mathfrak{a}$ then $\mathfrak{V}(h(g_1), \ldots, h(g_t))$ is the projective closure of $\mathfrak{v}(\mathfrak{a})$.*

Proof. This follows from Proposition 33 in Section 8.3. □

REMARK. One could now rather directly use the results of the previous chapter to draw the following geometric consequences. Let V be an algebraic set in $A^n(\mathbb{C})$. If V is irreducible then the projective closure is irreducible. If $V = \cup V_i$ is the irredundant decomposition in irreducible components, then $\mathbf{h}(V) = \cup \mathbf{h}(V_i)$ is the decomposition in irreducible components, and no $\mathbf{h}(V_i)$ is contained in the "hyperplane at infinity" defined by $x_{n+1} = 0$. We have that $\mathbf{a}(\mathbf{h}(V)) = V$. If W is an irreducible projective set which is not contained in the hyperplane at infinity, then $\mathbf{a}(W)$ is irreducible. If $W = \cup W_i$ is the irredundant decomposition of W in irreducible components, then $\mathbf{a}(W) = \cup_{i \in I} \mathbf{a}(W_i)$ is the decomposition in irreducible components, where the union is taken over the set I for which W_i is not contained in the hyperplane at infinity for $i \in I$.

Exercise

1. Determine the projective closure in $P^2(\mathbb{C})$ of the algebraic subset of $A^2(\mathbb{C})$ given by $\mathfrak{v}(\langle x_1 x_2 - 1, x_1^2 - x_2 \rangle)$.

Chapter 10

The Associated Graded Ring

We will in this section consider an algebraic construction used to study an algebraic set V in k^n near a point $P \in V$. If $P = (a_1, \ldots, a_n)$, we can make a translation $X_1 = x_1 - a_1, \ldots, X_n = x_n - a_n$, and hence we may assume (as we do from now on) that $P = (0, \ldots, 0)$. If $f(x,y) = 0$ is a curve passing through the origin, we have a well-defined tangent at the origin if $(\frac{\partial f}{\partial x}(0,0), \frac{\partial f}{\partial y}(0,0)) \neq (0,0)$. In that case we say that $(0,0)$ is a regular point on the curve. If $f(0,0) = \frac{\partial f}{\partial x}(0,0) = \frac{\partial f}{\partial y}(0,0) = 0$, we say that $(0,0)$ is a singular point on the curve. If $f(x,y) = f_0 + f_1(x,y) + f_2(x,y) + \cdots$, where $f_i(x,y)$ is the homogeneous component of degree i of f, then $f(0,0) = 0$ means that $f_0 = 0$ and $(0,0)$ is a singular point if and only if $f_0 = f_1(x,y) = 0$. If $f(x,y) = f_m(x,y) + f_{m+1}(x,y) + \cdots$, where $f_m(x,y) \neq 0$, then $f_m(x,y) = 0$ approximates the behaviour of $f(x,y) = 0$ near the origin as we are used to from the theory of Taylor expansions. We call the ideal $\langle f_m(x,y) \rangle$ the cone of the ideal $\langle f(x,y) \rangle$. We now turn to the general situation.

DEFINITION 1 (CONE OF AN IDEAL). If $f(x_1, \ldots, x_n) = f_m(x_1, \ldots, x_n) + f_{m+1}(x_1, \ldots, x_n) + \cdots$ with $f_m(x_1, \ldots, x_n) \neq 0$, we set $m(f) = f_m$. Let $\mathfrak{a} \subseteq k[x_1, \ldots, x_n]$ with $(0, \ldots, 0) \in \mathfrak{v}(\mathfrak{a})$. The *cone* of $\mathfrak{a}$ is

$$m(\mathfrak{a}) = \langle \{ m(f) \mid f \in \mathfrak{a} \} \rangle .$$

LEMMA 2. *We have* $m(fg) = m(f)m(g)$.

Proof. This is trivial. □

If $\mathfrak{a}$ is principal, it is easy to determine $m(\mathfrak{a})$.

PROPOSITION 3. *If $\mathfrak{a} = \langle f \rangle$ then $m(\mathfrak{a}) = \langle m(f) \rangle$.*

Proof. If $g \in \langle f \rangle$ then $g = fg_1$ for some g_1 and $m(g) = m(fg_1) = m(f)m(g_1) \in m(\langle f \rangle)$. Hence $m(\langle f \rangle) \subseteq \langle m(f) \rangle$. The other inclusion is trivial. □

If $\mathfrak{a} = \langle f_1, \ldots, f_k \rangle$ is not principal then $\langle m(f_1), \ldots, m(f_k) \rangle \subseteq m(\mathfrak{a})$, but there might be inequality.

EXAMPLE 4. If $\mathfrak{a} = \langle x + y^2, xy + x^3 \rangle$ then $\langle m(x + y^2), m(xy + x^3) \rangle = \langle x, xy \rangle = \langle x \rangle$, but $y^3 - x^3 = y(x + y^2) - (xy + x^3) \in \mathfrak{a}$, so $y^3 - x^3 = m(y^3 - x^3) \in m(\mathfrak{a})$ and $y^3 - x^3 \notin \langle x \rangle$. □

THEOREM 5. *Let $\mathfrak{a}$ be an ideal in $A = k[x_1, \ldots, x_n]$. Let $\prec$ be Degrevlex on A and extend $\prec$ on $k[x_1, \ldots, x_{n+1}]$ to the product ordering $\prec_1$ where the first block consists of x_{n+1} (so that $m_1 x_{n+1}^d \succ_1 m_2 x_{n+1}^e$ if $d > e$ or if $d = e$ and $m_1 \succ m_2$, where the m_i's are monomials in A). Let $\{g_1, \ldots, g_t\}$ be a Gröbner basis of $h(\mathfrak{a})$ in $\succ_1$. Then $m(\mathfrak{a}) = \langle m(a(g_1)), \ldots, m(a(g_t)) \rangle$.*

Proof. Since $g_i \in h(\mathfrak{a})$ we have $a(g_i) \in a(h(\mathfrak{a})) = \mathfrak{a}$. Thus we have the inclusion $\langle m(a(g_1)), \ldots, m(a(g_t)) \rangle \subseteq m(\mathfrak{a})$. To show the other inclusion, let $f \in \mathfrak{a}$. We must show that $m(f) \in \langle m(a(g_1)), \ldots, m(a(g_t)) \rangle$. Let $f = m(f) + \cdots$ be the decomposition of f in homogeneous components. Then $h(f) = m(f)x_{n+1}^d + \cdots$ for some d, and because of the ordering we have chosen, we have $\operatorname{lm}(h(f)) = x_{n+1}^d \operatorname{lm}(m(f))$. Since $\{g_1, \ldots, g_t\}$ is a Gröbner basis of $h(\mathfrak{a})$, we have that $\operatorname{lm}(g_i)$ divides $\operatorname{lm}(h(f))$ for some i. But $\operatorname{lm}(g_i) = x_{n+1}^e \operatorname{lm}(m(a(g_i)))$ for some e. Hence $\operatorname{lm}(m(f)) = cn \operatorname{lm}(m(a(g_i)))$ for some monomial n and some $c \in k$. Now consider $h = f - cna(g_i) \in \mathfrak{a}$. If $\deg(m(h)) > \deg(m(f))$, we have $m(f) = m(cna(g_i)) = cnm(a(g_i)) \in \langle m(a(g_1)), \ldots, m(a(g_t)) \rangle$. If $\deg(m(h)) = \deg m(f))$, we get $\operatorname{lm}(m(h)) \prec \operatorname{lm}(m(f))$, and we can repeat this construction on h. Since the leading terms gets smaller, we are through after a finite number of steps. □

REMARK. The theorem gives an algorithm to determine $m(\mathfrak{a})$. Unfortunately the method is not very efficient. To calculate $h(\mathfrak{a})$ we need to calculate a Gröbner basis of $\mathfrak{a}$ and then homogenize each element in this basis; see Proposition 33 in Section 8.3. Then we need to change the ordering and make another calculation of the Gröbner basis of $h(\mathfrak{a})$ and finally dehomogenize these basis elements. There are more efficient algorithms which will be mentioned later.

We will now look at a more general algebraic construction which in a special case gives the ring $k[x_1,\ldots,x_n]/m(\mathfrak{a})$.

Let $\mathfrak{a}$ be an ideal in a ring R. We denote the subset $\{ f+\mathfrak{a}^{k+1} \mid f \in \mathfrak{a}^k \}$ of $R/\mathfrak{a}^{k+1}$ by $\mathfrak{a}^k/\mathfrak{a}^{k+1}$. Elements in $\mathfrak{a}^k/\mathfrak{a}^{k+1}$ are added as in $R/\mathfrak{a}^{k+1}$. Consider the set $G_\mathfrak{a}(R) = \oplus_{i\geq 0}\mathfrak{a}^k/\mathfrak{a}^{k+1}$, i.e., the set of infinite sequences $(a_0,a_1,\ldots)$ with $a_k \in \mathfrak{a}^k/\mathfrak{a}^{k+1}$ and with only finitely many $a_i \neq 0$. The elements in $G_\mathfrak{a}(R)$ are added coordinatewise, and we will now define a multiplication, making $G_\mathfrak{a}(R)$ into a ring. For $f+\mathfrak{a}^{k+1} \in \mathfrak{a}^k/\mathfrak{a}^{k+1}$ and $g+\mathfrak{a}^{l+1} \in \mathfrak{a}^l/\mathfrak{a}^{l+1}$ we define $(f+\mathfrak{a}^{k+1})(g+\mathfrak{a}^{l+1}) = fg+\mathfrak{a}^{k+l+1}$. Then we extend the multiplication to all of $G_\mathfrak{a}(R)$ via the distributive law. It is easy to see that the multiplication is well defined and that $G_\mathfrak{a}(R)$ becomes a ring. $G_\mathfrak{a}(R)$ is called the *associated graded ring of R with respect to $\mathfrak{a}$.*

PROPOSITION 6. *Let $R = k[x_1,\ldots,x_n]/\mathfrak{a}$ for some $\mathfrak{a} \subseteq \langle x_1,\ldots,x_n\rangle = \mathfrak{m}$. Then $G_{\overline{\mathfrak{m}}}(R) \simeq k[x_1,\ldots,x_n]/m(\mathfrak{a})$, where $\overline{\mathfrak{m}}$ denotes the image of $\mathfrak{m}$ in R.*

Proof. We define a map $\phi : k[x_1,\ldots,x_n] \to G_{\overline{\mathfrak{m}}}(R)$ as follows. If f is homogeneous of degree k, then $\bar{f} \in \overline{\mathfrak{m}}^k$ and we let $\phi(f) = \bar{f}+\overline{\mathfrak{m}}^{k+1}$. If $f = f_0+f_1+\cdots$ is the decomposition of f in homogeneous components, we set $\phi(f) = \phi(f_0)+\phi(f_1)+\cdots = (\phi(f_0),\phi(f_1),\ldots)$. If $f \in \ker(\phi)$, then all homogeneous components of f belong to $\ker(\phi)$. If f is homogeneous of degree k then $\phi(f) = \bar{f}+\overline{\mathfrak{m}}^{k+1}$, so $\phi(f) = 0$ if and only if $f+g \in \mathfrak{a}$ for some $g \in \overline{\mathfrak{m}}^{k+1}$, thus $\phi(f) = 0$ if and only if $f = m(h)$ for some $h \in \mathfrak{a}$. Since $\ker(\phi) = m(\mathfrak{a})$, we have $G_{\overline{\mathfrak{m}}}(R) \simeq k[x_1,\ldots,x_n]/m(\mathfrak{a})$ by the homomorphism theorem, Theorem 61 in Section 1.7, the map ϕ being surjective. □

DEFINITION 7 (TANGENT CONE). Let V be an algebraic set in $A^n(k)$ such that $(0,\ldots,0) \in V$. The *tangent cone* of V at the origin is the algebraic set $C(V) = \mathfrak{v}(m(\mathfrak{i}(V)))$.

REMARK. Since $m(\mathfrak{i}(V))$ is a homogeneous ideal, $C(V)$ consists of lines through the origin. Such a set is called a cone at the origin. In the same way as a tangent to a curve is the limit of secant lines, one can show that $C(V)$ consists of all limits of secant lines in V through the origin.

If V is given by equations, one likes to be able to describe $C(V)$ with equations. This is possible if $V \subseteq A^n(\mathbb{C})$ (we need the Nullstellensatz).

THEOREM 8. *Let $\mathfrak{a} \subseteq \mathbb{C}[x_1,\ldots,x_n]$ with $(0,\ldots,0) \in \mathfrak{v}(\mathfrak{a}) = V$. Then $C(V) = \mathfrak{v}(m(\mathfrak{a}))$.*

Proof. We have $C(V) = \mathfrak{v}(m(\sqrt{\mathfrak{a}})) \subseteq \mathfrak{v}(m(\mathfrak{a}))$ since $m(\mathfrak{a}) \subseteq m(\sqrt{\mathfrak{a}})$. Suppose that $f \in \sqrt{\mathfrak{a}}$. We must show that $m(f)$ is zero on $\mathfrak{v}(m(\mathfrak{a}))$. We have $f^k \in \mathfrak{a}$ for some k, so $m(f^k) = (m(f))^k$ is zero on $\mathfrak{v}(m(\mathfrak{a}))$, hence $m(f)$ is zero on $\mathfrak{v}(m(\mathfrak{a}))$. □

Exercises

1. Determine $m(\mathfrak{a})$ if $\mathfrak{a} = \langle x + y^2, xy + x^3 \rangle$.

2. Show that the multiplication in $G_{\mathfrak{a}}(R)$ is well defined and that $G_{\mathfrak{a}}(R)$ is a ring.

3. Determine the tangent cone of $V = \mathfrak{v}(\langle x^2 - y^3 \rangle) \subseteq A^2(\mathbb{C})$ at the origin. Draw a picture of (the real part of) the curve V and its tangent cone.

4. $V = \mathfrak{v}(\langle xy(x + y - 1) \rangle)$ in $A^2(\mathbb{C})$ is the union of three lines. Determine the tangent cone of V

   (a) at the origin

   (b) at $(0, 1)$

   (c) at $(2, 0)$

5. Determine the tangent cone of $V = \mathfrak{v}(\langle x^2 + y^2 - z^4 \rangle)$ at the origin. Draw a picture of (the real part of) the surface V and its tangent cone.

6. Determine the tangent cone of $V = \mathfrak{v}(\langle x + y^2, xy + x^3 \rangle)$

   (a) at the origin

   (b) at $(-1, -1)$

Chapter 11

Hilbert Series

If $R = k[x_1, \dots, x_n]/\mathfrak{a}$ is a homogeneous algebra we know that $R = \oplus R_i$, where R_i is the vector space of homogeneous elements of degree i. The dimensions $\dim_k R_i$ have geometrical interpretations. It is practical to save information about all these dimensions in a generating function. In order to be able to define this function and study it, we need some preliminary work on formal power series.

11.1 Formal Power Series

Let R be a ring. A *formal power series* with coefficients in R is a formal expression $\sum_{n\geq 0} a_n X^n$ where $a_n \in R$. Note that we allow infinitely many a_n to be nonzero. We define the sum and product of two formal power series just as for polynomials, so

$$\sum_{n\geq 0} a_n X^n + \sum_{n\geq 0} b_n X^n = \sum_{n\geq 0} c_n X^n$$

where $c_n = a_n + b_n$ and

$$\sum_{n\geq 0} a_n X^n \cdot \sum_{n\geq 0} b_n X^n = \sum_{n\geq 0} d_n X^n$$

where $d_n = a_0 b_n + a_1 b_{n-1} + \dots + a_n b_0$. The set of such formal series becomes a ring denoted $R[[X]]$.

In the ring $R[[X]]$ we have $(1 - X)(1 + X + X^2 + \cdots) = 1$ (which is easily checked). Thus $1 - X$ has a multiplicative inverse. We can say exactly which elements in $R[[X]]$ have multiplicative inverses.

PROPOSITION 1. *The formal series $F(X) = \sum_{n\geq 0} a_n X^n \in R[[X]]$ has an inverse in $R[[X]]$ if and only if a_0 has an inverse in R.*

Proof. If $F(X)G(X) = 1$ we get, if $G(X) = \sum_{n\geq 0} b_n X^n$, that $a_0 b_0 = 1$, so a_0 has an inverse in R. Suppose for the converse that a_0 has an inverse in R. We construct the coefficients for the inverse $G(X) = \sum_{n\geq 0} b_n X^n$ to $F(X)$ inductively. First we get $a_0 b_0 = 1$, so $b_0 = a_0^{-1}$. Suppose we have constructed $b_0, \ldots, b_{n-1}$. Next we get $b_n a_0 + b_{n-1}a_1 + \cdots + b_0 a_n = 0$, so $b_n = -a_0^{-1}(b_{n-1}a_1 + \cdots + b_0 a_n)$. □

Exercises

1. Show that $R[[X]]$ is a ring with unit element $\sum b_n X^n$, where $b_0 = 1$ and $b_n = 0$ if $n > 0$.

2. If $F(X) = \sum_{n\geq 0} a_n X^n$ we define the formal derivative of $F(X)$ to be the series $F'(X) = \sum_{n\geq 0} n a_n X^{n-1}$.

   (a) Show that $(FG)' = F'G + G'F$.

   (b) Show that $(F^n)' = nF^{n-1}F'$.

   (c) Show that $G' = 0$ if and only if G is a constant.

   (d) Show that $(G^{-1})' = -G'G^{-2}$.

   (e) Show that $(G^{-n})' = -nG'G^{-n-1}$.

   (f) Show that $(1-X)^{-n} = \sum_{m\geq 0} \binom{n+m-1}{m} X^m$.

3. This exercise gives an alternative way to determine the series $(1-X)^{-n}$.

   (a) Let n be a positive integer. Show by induction on m that
$$\binom{n-1}{0} + \binom{n}{1} + \cdots + \binom{n+m-1}{m} = \binom{n+m}{m}$$

   (b) Let $F(X) = 1 + X + X^2 + \cdots$. Show that $(F(X))^n = \sum_{m\geq 0} \binom{n+m-1}{m} X^m$.

11.2 Hilbert Series

Let $R = k[x_1, \ldots, x_n]/\mathfrak{a}$ be a homogeneous algebra. We have seen that $R = \oplus_{i \geq 0} R_i$, where R_i is the k-vector space of homogeneous elements of degree i in R. The dimension of R_i is finite, since there are finitely many monomials of degree i, and $h(i) = \dim_k R_i$ is called the *Hilbert function* of R. As usual it is convenient to gather the information about the $h(i)$'s (which in general are infinitely many) in a generating function, a formal power series in $\mathbb{Z}[[X]]$.

DEFINITION 2 (HILBERT SERIES). If $R = \oplus_{i \geq 0} R_i$ is a homogeneous algebra, then $H_R(X) = \sum_{i \geq 0} h(i) X^i$ is called the *Hilbert series* of R.

EXAMPLE 3. If $R = k[x, y]$ then R_i is generated by $x^i, x^{i-1}y, \ldots, y^i$. These monomials are linearly independent, so $\dim_k R_i = i + 1$. Thus $H_R(X) = 1 + 2X + 3X^2 + \cdots = (1 - X)^{-2}$. □

EXAMPLE 4. If $R = k[x, y]/\langle x^2, xy \rangle$ then (images of) the following monomials constitute a k-basis for R: $1, x, y, y^2, y^3, y^4, \ldots$. Thus we have $h(0) = 1, h(1) = 2$, and $h(i) = 1$ if $i > 1$. Thus $H_R(X) = 1 + 2X + X^2 + X^3 + \cdots = (1 + X - X^2)/(1 - X)$. □

We will in the following several times use Theorem 74 in Section 1.9, which states that if $f : V \longrightarrow W$ is a surjective homomorphism of vector spaces, then $\dim W = \dim V + \dim \ker(f)$.

We have seen, see Theorem 12 in Section 8.1, that in any homogeneous algebra, there is either a homogeneous *socle element*, i.e., an element annihilating every homogeneous element of positive degree, or a homogeneous nonzerodivisor. We will use this to determine what the Hilbert series for homogeneous algebras look like.

LEMMA 5. *Suppose that $f \in k[x_1, \ldots, x_n]$ is homogeneous of degree k and that the image of f in the homogeneous algebra $R = k[x_1, \ldots, x_n]/\mathfrak{a}$ is a nonzerodivisor (i.e., that $\mathfrak{a} = \mathfrak{a} : \langle f \rangle$). Then $H_{R/\langle \bar{f} \rangle}(X) = (1 - X^k) H_R(X)$.*

Proof. The map $R_i \to R_{i+k}$ defined by $\bar{f}_i \mapsto \bar{f}\bar{f}_i = \overline{f f_i}$ is an injective homomorphism of vector spaces since $\bar{f}$ is a nonzerodivisor, so $\dim_k \langle \bar{f} \rangle_{i+k} = \dim_k R_i$. The map $R_{i+k} \to R_{i+k}/\langle \bar{f} \rangle_{i+k}$ is a surjective homomorphism of vector spaces, so

$$h_R(i+k) = \dim_k R_{i+k} = \dim_k \langle \bar{f} \rangle_{i+k} + \dim_k R_{i+k}/\langle \bar{f} \rangle_{i+k} =$$

$$\dim_k R_i + \dim_k S_{i+k} = h_R(i) + h_S(i+k).$$

where $S = R/\langle f\rangle$. In $\mathbb{Z}[[X]]$ we get

$$h_R(i+k)X^{i+k} = h_R(i)X^{i+k} + h_S(i+k)X^{i+k},$$

so

$$\sum_{i\geq 0} h_R(i+k)X^{i+k} = \sum_{i\geq 0} h_R(i)X^{i+k} + \sum_{i\geq 0} h_S(i+k)X^{i+k}.$$

If $j < k$ we have $R_j \simeq S_j$, so $h_R(j) = h_S(j)$. Hence we get

$$\sum_{i\geq 0} h_R(i)X^i = X^k \sum_{i\geq 0} h_R(i)X^i + \sum_{i\geq 0} h_S(i)X^i,$$

i.e.,

$$H_R(X) = X^k H_R(X) + H_S(X).$$

□

LEMMA 6. *Let $R = k[x_1,\ldots,x_n]/\mathfrak{a}$ be a homogeneous algebra and suppose that f is a homogeneous polynomial such that $\bar{f}$ is a socle element in R, i.e., such that $\bar{f} \in \langle 0\rangle : \langle \overline{x_1}\ldots,\overline{x_n}\rangle$ or equivalently $f \in \mathfrak{a} : \langle x_1,\ldots,x_n\rangle$. Then $H_{R/\langle \bar{f}\rangle}(X) = H_R(X) - X^{\deg f}$.*

Proof. We have $\dim_k \langle \bar{f}\rangle_i = 1$ if $i = \deg(f)$ and $\dim_k \langle \bar{f}\rangle = 0$ if $i \neq \deg(f)$ since $\bar{f}\bar{g} = \bar{0}$ for every homogeneous g of positive degree. Hence $\dim_k(R/\langle \bar{f}\rangle_i = \dim_k R_i - \dim_k \langle \bar{f}\rangle_i$ equals $\dim_k R_i - 1$ if $i = \deg(f)$ and equals $\dim_k R_i$ otherwise. Thus

$$H_{R/\langle \bar{f}\rangle}(X) = \sum \dim_k(R/\langle \bar{f}\rangle)_i X^i = \sum \dim_k R_i X^i - X^{\deg(f)} =$$

$$H_R(X) - X^{\deg(f)}.$$

□

We are ready to prove our main theorem for Hilbert series of homogeneous algebras.

THEOREM 7. *Let $R = k[x_1,\ldots,x_n]/\mathfrak{a}$ be a homogeneous algebra. Then $H_R(X) = p(X)/(1-X)^d$ for some polynomial $p(X) \in \mathbb{Z}[X]$ with $p(0) = 1$ and some $d \leq n$.*

Proof. Since the set of monomials outside $l(\mathfrak{a})$ constitute a vector space basis for R, see Proposition 15 in Section 3.6, we see that $H_R(X) = H_S(X)$, where $S = k[x_1,\ldots,x_n]/l(\mathfrak{a})$. Then we note that we can assume that k is an infinite field, which is the result of the following reasoning. Let Y be a variable and let $K = k(Y)$ be the fraction field of $k[Y]$. It is clear that K is infinite. Let $\mathfrak{a} = \langle f_1(x_1,\ldots,x_n),\ldots,f_k(x_1,\ldots,x_n)\rangle \subseteq k[x_1,\ldots,x_n]$ and let $\mathfrak{a}'$ be generated by the same elements but now as an ideal in $K[x_1,\ldots,x_n]$. Calculating a Gröbner basis for $\mathfrak{a}'$ is exactly the same as for $\mathfrak{a}$, hence the two ideals have "the same" initial ideals, and we get $H_R(X) = H_{K[x_1,\ldots,x_n]/\mathfrak{a}'}(X)$. We assume from now on that k is an infinite field. Then we can use the result in Theorem 12 of Section 8.1 that R either contains a nonzerodivisor of degree 1 or a homogeneous socle element. Since $0 : \langle \overline{x_1},\ldots,\overline{x_n}\rangle = \langle \overline{s_1},\ldots,\overline{s_t}\rangle$ is a finitely generated homogeneous ideal in R, we see that there exists a homogeneous socle element $\bar{s}$ of maximal degree, namely of degree $\max\{\deg \overline{s_i}\}$. (We have $\bar{f}\overline{s_i} = 0$ for each homogeneous element f of positive degree.) Then $H_{R/\langle\bar{s}\rangle}(X) = H_R(X) - X^{\deg s}$ by Lemma 6. Let $\bar{t}$ be a homogeneous socle element of maximal degree in $R/\langle\bar{s}\rangle$. Since R and $R/\langle\bar{s}\rangle$ coincide in degrees $> \deg s$, we have $\deg t \le \deg s$. We continue like this to get an ideal $\langle s,t,\ldots\rangle$ such that $k[x_1,\ldots,x_n]/\mathfrak{a} + \langle s,t,\ldots\rangle$ has no socle. (The process is finite since the degrees never increase, and $\dim_k \oplus_{i\le \deg s} R_i$ is finite.) Hence we eventually reach a ring $R_1 = R/\mathfrak{s}$ without socle and $H_{R/\mathfrak{s}}(X) = H_R(X) - q(X)$, where $q(X) \in \mathbb{Z}[X]$ has degree $\deg s$. Then R_1 contains a nonzerodivisor of degree 1, which after a linear change of coordinates, we can assume to be one of the $\overline{x_i}$'s, and $H_{R_1/\langle\overline{x_i}\rangle}(X) = (1-X)H_{R_1}(X)$. We now use induction over the number of variables n. If $n = 0$ we have $R = k$ so $H_R(X) = 1$. Suppose that the statement is proved for $n-1$ variables, so $H_{R_1/\langle\overline{x_i}\rangle}(X) = p_1(X)/(1-X)^e$ with $p_1(X) \in \mathbb{Z}[X]$ and $e \le n-1$. Then $(1-X)(H_R(X) - q(X)) = p_1(X)/(1-X)^e$, so $H_R(X) = (p_1(X) + (1-X)^{e+1}q(X))/(1-X)^{e+1}$. □

For large i, $h_R(i)$ is a polynomial in i, the *Hilbert polynomial* of R.

COROLLARY 8. *Let $R = k[x_1,\ldots,x_n]/\mathfrak{a}$ be a homogeneous algebra with Hilbert series $p(X)/(1-X)^d$, where $p(1) \ne 0$. Then the Hilbert function of R, $h_R(i) = \dim_k R_i$ is a polynomial of degree $d-1$ in i with highest coefficient $p(1)/(d-1)!$ for large i. (If $d = 0$, $h_R(i) = 0$ for large i.)*

Proof. Suppose that $p(X) = a_0 + a_1X + \cdots + a_kX^k$. Then $H_R(X) = 1/(1-X)^d + a_1X/(1-X)^d + \cdots + a_kX^k/(1-X)^d$. The coefficient of X^i in $a_jX^j/(1-X)^d$ equals a_j times the coefficient of X^{i-j} in $1/(1-X)^d$ which is $a_j\binom{i-j+d-1}{d-1}$ which, if $i \ge j$, is a polynomial of degree $d-1$ in i with highest coefficient $a_j/(d-1)!$. Hence the coefficient $h_i(R)$ of X^i in $H_R(X)$ equals a polynomial in i of degree $d-1$ with highest coefficient $(1 + a_1 + \cdots + a_k)/(d-1)! = p(1)/(d-1)!$ if $i \ge k$. □

EXAMPLE 9. Let $R = k[x,y,z]/\langle x^2, xyz\rangle$. If $i \ge 3$ the nonzero monomials of degree i in R are $x^jy^{i-j}, 0 \le j \le 1, x^jz^{i-j}, 0 \le j \le 1$, and $y^jz^{i-j}, 1 \le$

$j \leq i - 1$. They are $i + 3$ in number, hence the Hilbert polynomial of R is $i + 3$. □

Exercises

1. Show that the number of monomials of degree i in $k[x_1, \ldots, x_n]$ is $\binom{n+i-1}{i}$.

2. Show that $H_{k[x_1,\ldots,x_n]}(X) = (1 - X)^{-n}$.

3. Determine the Hilbert series $H_R(X)$ and the Hilbert polynomials $h_R(i)$ of

   (a) $k[x, y, z]/\langle xy, xz, yz\rangle$

   (b) $k[x, y, z]/\langle x^2, y^2, z^2\rangle$

   (c) $k[x, y, z]/\langle x^2, xy, xz\rangle$

11.3 Geometric Meaning of Hilbert Series

Let $\mathfrak{b}$ be a homogeneos ideal in $k[x_1, \ldots, x_n, x_{n+1}]$. In this section we will show that the Hilbert series of $k[x_1, \ldots, x_{n+1}]/\mathfrak{b}$ contains geometric information on $\mathfrak{V}(\mathfrak{b})$ (or on $\mathfrak{v}(\mathfrak{a})$ if $\mathfrak{b} = h(\mathfrak{a})$). We start with a purely algebraic definition.

DEFINITION 10 (DIMENSION OF AN ALGEBRA). Let R be a homogeneous algebra with $H_R(X) = p(X)/(1 - X)^d$, where $p(1) \neq 0$. Then we define the *dimension* of R to be d.

EXAMPLE 11. Using Corollary 8 and Example 9, we see that the dimension of $k[x, y, z]/\langle x^2, xyz\rangle$ is 2, since $\deg i + 3 = 1$. □

The following lemma is an immediate consequence of the definition.

LEMMA 12. *If $\bar{f}$ is a homogeneous nonzerodivisor in the homogeneous algebra R, then* $\dim R/\langle \bar{f}\rangle = \dim R - 1$.

Proof. If $H_R(X) = p(X)/(1 - X)^d$ with $p(1) \neq 0$, then $H_{R/\langle \bar{f}\rangle}(X) = (1 - X^k)p(X)/(1 - X)^d = (1 + X + \cdots + X^{k-1})p(X)/(1 - X)^{d-1}$, and $kp(1) \neq 0$, where $k = \deg \bar{f}$. □

The Hilbert series has geometric meaning.

DEFINITION 13 (DIMENSION OF AN ALGEBRAIC SET). Let $\mathfrak{a}$ be an ideal in $\mathbb{C}[x_1, \ldots, x_n]$ and let $V = \mathfrak{v}(\mathfrak{a})$. The *dimension* of V is defined to be $\dim V = \dim(\mathbb{C}[x_1, \ldots, x_{n+1}]/h(\mathfrak{a})) - 1$.

In order to show that this is a good definition, we must prove that the dimension does not depend on the ideal $\mathfrak{a}$ used to define V. It is enough to prove that $\dim \mathbb{C}[x_1, \ldots, x_n]/\mathfrak{a} = \dim \mathbb{C}[x_1, \ldots, x_n]/\sqrt{\mathfrak{a}}$.

DEFINITION 14 (DIMENSION OF A PROJECTIVE SET). Let $\mathfrak{a}$ be a homogeneous ideal in $\mathbb{C}[x_1, \ldots, x_{n+1}]$ and let $V = \mathfrak{V}(\mathfrak{a})$. The *projective dimension* of V is defined to be $\dim V = \dim(\mathbb{C}[x_1, \ldots, x_{n+1}]/\mathfrak{a}) - 1$.

We must also in this case show that $\dim V$ is well defined. We need some lemmas, but first we give an example.

EXAMPLE 15. Let $\mathfrak{a} = \langle x \rangle \cap \langle y \rangle \cap \langle x - y \rangle = \langle x^2y - xy^2 \rangle$. As an affine algebraic set in $A^2(\mathbb{C})$, $\mathfrak{v}(\mathfrak{a})$ is the union of three lines through the origin. As a projective set in $P^1(\mathbb{C})$, $\mathfrak{V}(\mathfrak{a})$ consists of the three points $(0:1)$, $(1:0)$, and $(1:1)$. We have $h(\mathfrak{a}) = \langle x^2y - xy^2 \rangle \subseteq \mathbb{C}[x, y, z]$, so

$$H_{\mathbb{C}[x,y,z]/h(\mathfrak{a})}(X) = (1 - X^3)/(1 - X)^3 = (1 + X + X^2)/(1 - X)^2.$$

The dimension of the affine algebraic subset $\mathfrak{v}(\mathfrak{a})$ of $A^2(\mathbb{C})$ is thus one, as it should be. Since

$$H_{\mathbb{C}[x,y]/\langle x^2y - xy^2 \rangle}(X) = (1 - X^3)/(1 - X) = (1 + X + X^2)/(1 - X),$$

the dimension of the projective algebraic subset $\mathfrak{V}(\mathfrak{a})$ of $P^1(\mathbb{C})$ is zero. □

LEMMA 16. *Let R and S be homogeneous algebras. If $h_R(i) \leq h_S(i)$ for all large i, then $\dim R \leq \dim S$.*

Proof. Both $h_R(i)$ and $h_S(i)$ are polynomials $p(i)$ and $q(i)$, respectively, for large i. If $h_R(i) \leq h_S(i)$ for all large i, we must have $\deg p \leq \deg q$, hence $\dim R \leq \dim S$, by Corollary 8. □

LEMMA 17. *Let $\mathfrak{a} \subseteq \mathfrak{b}$ be homogeneous ideals in $A = k[x_1, \ldots, x_n]$. Then $\dim k[x_1, \ldots, x_n]/\mathfrak{b} \leq \dim k[x_1, \ldots, x_n]/\mathfrak{a}$.*

Proof. We have $\mathfrak{a}_i \subseteq \mathfrak{b}_i$ so $\dim_k(\mathfrak{a}_i) \leq \dim_k(\mathfrak{b}_i)$. Hence

$$h_{A/\mathfrak{b}}(i) = \dim_k(A/\mathfrak{b})_i = \dim_k A_i - \dim_k \mathfrak{b}_i \leq$$

$$\dim_k A_i - \dim_k \mathfrak{a}_i = \dim_k(A/\mathfrak{a})_i = h_{A/\mathfrak{a}}(i).$$

Thus the conclusion follows from Lemma 17. □

LEMMA 18. *If $\mathfrak{a}$ is a monomial ideal in $A = k[x_1, \ldots, x_n]$ then* $\dim A/\mathfrak{a} = \dim A/\sqrt{\mathfrak{a}}$.

Proof. Let e be $\max \{ k \mid x_{i_1} \cdots x_{i_k} \notin \sqrt{\mathfrak{a}} \}$ for some $1 \leq i_1 < \cdots < i_k \leq n$. We will show that $\dim A/\mathfrak{a} = \dim A/\sqrt{\mathfrak{a}} = e$. There is a sequence $1 \leq i_1 < \cdots < i_e \leq n$ such that $x_{i_1} \cdots x_{i_e} \notin \sqrt{\mathfrak{a}}$. Hence $B = k[x_{i_1}, \cdots, x_{i_e}]$ is a subring of $A/\sqrt{\mathfrak{a}}$. Since $\binom{i+e-1}{e-1} = h_i(B) \leq h_i(A/\sqrt{\mathfrak{a}})$ and $\binom{i+e-1}{e-1}$ is a polynomial of degree $e - 1$, we get that $\dim A/\sqrt{\mathfrak{a}} \geq e$ by Lemma 16. Lemma 17 gives $\dim A/\sqrt{\mathfrak{a}} \leq \dim A/\mathfrak{a}$. For every sequence $1 \leq j_1 < \cdots < j_{e+1} \leq n$ we have $x_{j_1} \cdots x_{j_{e+1}} \in \sqrt{\mathfrak{a}}$, so $(x_{j_1} \cdots x_{j_{e+1}})^N \in \mathfrak{a}$ for some N. Since there are only finitely many sequences of length $e + 1$, we can choose N large enough so that $(x_{j_1} \cdots x_{j_{e+1}})^N \in \mathfrak{a}$ for every sequence $1 \leq j_1 < \cdots < j_{e+1} \leq n$. Let $\mathfrak{b}$ be the ideal generated by all $(x_{j_1} \cdots x_{j_{e+1}})^N$. If we can show that $\dim A/\mathfrak{b} = e$, the conclusion will follow since then $e \leq \dim A/\sqrt{\mathfrak{a}} \leq \dim A/\mathfrak{a} \leq \dim A/\mathfrak{b} = e$. That $\dim A/\mathfrak{b} = e$ is left until Exercise 1 below. □

PROPOSITION 19. *If $\mathfrak{a}$ is a homogeneous ideal in $k[x_1, \ldots, x_n]$ then we have* $\dim k[x_1, \ldots, x_n]/\mathfrak{a} = \dim k[x_1, \ldots, x_n]/\sqrt{\mathfrak{a}}$.

Proof. We know that $\dim k[x_1, \ldots, x_n]/\mathfrak{b} = \dim k[x_1, \ldots, x_n]/\sqrt{\mathfrak{b}}$ for any monomial ideal $\mathfrak{b}$. We claim that $l(\sqrt{\mathfrak{a}}) \subseteq \sqrt{l(\mathfrak{a})}$. If $m \in l(\sqrt{\mathfrak{a}})$ then $m = \mathrm{lm}(f)$ for some $f \in \sqrt{\mathfrak{a}}$. Then $f^k \in \mathfrak{a}$ for some k and $\mathrm{lm}(f^k) = (\mathrm{lm}(f))^k \in l(\mathfrak{a})$, so $m \in \sqrt{l(\mathfrak{a})}$. Since $l(\mathfrak{a}) \subseteq l(\sqrt{\mathfrak{a}}) \subseteq \sqrt{l(\mathfrak{a})}$ we get $\dim k[x_1, \ldots, x_n]/l(\mathfrak{a}) \geq \dim k[x_1, \ldots, x_n]/l(\sqrt{\mathfrak{a}}) \geq \dim k[x_1, \ldots, x_n]/\sqrt{l(\mathfrak{a})}$. Lemma 18 gives equality, thus

$$\dim k[x_1, \ldots, x_n]/\mathfrak{a} = \dim k[x_1, \ldots, x_n]/l(\mathfrak{a}) = \dim k[x_1, \ldots, x_n]/l(\sqrt{\mathfrak{a}}) =$$

$$\dim k[x_1, \ldots, x_n]/\sqrt{\mathfrak{a}}.$$

□

COROLLARY 20. *If V is an affine algebraic set in $A^n(\mathbb{C})$ or a projective set in $P^n(\mathbb{C})$ then $\dim V$ is well defined, i.e. the definition does not depend on the ideal used to define V.*

Proof. Let V be an algebraic set. We must show that $\dim V$ does not depend on the ideal used to define V. If $\mathfrak{v}(\mathfrak{a}) = \mathfrak{v}(\mathfrak{b})$, then $\sqrt{\mathfrak{a}} = \sqrt{\mathfrak{b}}$, so it suffices to show that $\dim \mathbb{C}[x_1, \ldots, x_n]/\mathfrak{a} = \dim \mathbb{C}[x_1, \ldots, x_n]/\sqrt{\mathfrak{a}}$. In Degrevlex $l(h(\mathfrak{a}))$ and $l(\mathfrak{a})$ have the same generators, see Proposition 33 in Section 8.3, hence x_{n+1} is a nonzerodivisor in $R = \mathbb{C}[x_1, \ldots, x_{n+1}]/l(h(\mathfrak{a}))$

and $R/\langle x_{n+1}\rangle \simeq \mathbb{C}[x_1,\ldots,x_n]/l(\mathfrak{a})$, so $\dim R/\langle x_{n+1}\rangle = \dim R - 1$ by Lemma 12. Hence

$$\dim \mathbb{C}[x_1,\ldots,x_{n+1}]/h(\mathfrak{a}) = \dim \mathbb{C}[x_1,\ldots,x_{n+1}]/l(h(\mathfrak{a})) =$$

$$1+\dim \mathbb{C}[x_1,\ldots,x_n]/l(\mathfrak{a}) = 1+\dim \mathbb{C}[x_1,\ldots,x_n]/l(\sqrt{\mathfrak{a}}) =$$

$$\dim \mathbb{C}[x_1,\ldots,x_{n+1}]/l(h(\sqrt{\mathfrak{a}})) = \dim \mathbb{C}[x_1,\ldots,x_{n+1}]/h(\sqrt{\mathfrak{a}}).$$

The proof that the dimension of a projective set is well defined is similar (and easier). □

COROLLARY 21. *If* $\mathfrak{a}$ *is an ideal in* $\mathbb{C}[x_1,\ldots,x_n]$*, then* $\mathfrak{v}(\mathfrak{a})$ *is finite if and only if* $\dim \mathfrak{v}(\mathfrak{a}) = 0$.

Proof. In Degrevlex we have

$$H_{\mathbb{C}[x_1,\ldots,x_{n+1}]/h(\mathfrak{a})}(X) = H_{\mathbb{C}[x_1,\ldots,x_{n+1}]/l(h(\mathfrak{a}))}(X) =$$

$$H_{\mathbb{C}[x_1,\ldots,x_{n+1}]/l(\mathfrak{a})}(X) = H_{\mathbb{C}[x_1,\ldots,x_n]/l(\mathfrak{a})}(X)/(1-X).$$

We know that $\mathfrak{v}(\mathfrak{a})$ is finite if and only if for each i we have $x_i^{n_i} \in l(\mathfrak{a})$ for some n_i; see Proposition 33 in Section 8.3. Thus $H_{\mathbb{C}[x_1,\ldots,x_n]/l(\mathfrak{a})}(X)$ is a polynomial if and only if $\mathfrak{v}(\mathfrak{a})$ is finite. But $H_{\mathbb{C}[x_1,\ldots,x_n]/l(\mathfrak{a})}(X)$ is a polynomial if and only if $\dim \mathbb{C}[x_1,\ldots,x_n]/l(\mathfrak{a}) = 0$. □

COROLLARY 22. *Let* $\mathfrak{a} = \langle f_1,\ldots,f_k\rangle$ *and suppose* $H_{\mathbb{C}[x_1,\ldots,x_{n+1}]/h(\mathfrak{a})}(X) = p(X)/(1-X)$*. Then the number solutions to* $f_1 = \cdots = f_k = 0$ *(counted with multiplicity) is* $p(1)$.

Proof. The number of solutions is $\dim_{\mathbb{C}} \mathbb{C}[x_1,\ldots,x_n]/l(\mathfrak{a})$; see Theorem 7 in Section 6.2.2. If $H_{\mathbb{C}[x_1,\ldots,x_n]/l(\mathfrak{a})}(X) = 1 + a_1X + \cdots + a_mX^m = p(X)$, then $\dim_{\mathbb{C}} \mathbb{C}[x_1,\ldots,x_n]/l(\mathfrak{a}) = 1 + a_1 + \cdots + a_m = p(1)$. □

REMARK. We have proved that the dimension of a monomial ring R equals the maximum number d such that $k[x_{i_1},\ldots,x_{i_d}]$ is a subring of R for some $1 \le i_1 < \cdots i_d \le n$. This fact has a far-reaching generalization in the Noether normalization lemma, which states in particular that for any $\mathfrak{a}$, $\dim \mathbb{C}[x_1,\ldots,x_n]/\mathfrak{a}$ equals the maximal number of homogeneous elements $l_1,\ldots,l_d$ of degree 1, such that the subring $k[l_1,\ldots,l_d]$ is a polynomial ring.

The possibility of using Hilbert series to calculate of the dimension of an algebraic set, or to determine the number of solutions for a system of equations of course depends on our ability to determine the Hilbert series. We have not so far mentioned the problem of how to calculate the Hilbert series algorithmically. We will later on see that there is an efficient algorithm.

REMARK. Our definition of the dimension of an algebraic or a projective set is not very intuitive. One can show that the following is an equivalent definition. If V is irreducible then $\dim V = \max\{\, k \mid V_0 \subset V_1 \subset \cdots \subset V_k = V \,\}$ where V_i are irreducible nonempty algebraic sets and the inclusions are strict. If V is not irreducible, $\dim V$ is the maximum of the dimensions of the irreducible components of V. An irreducible algebraic set V is defined by a prime ideal, $V = \mathfrak{v}(\mathfrak{p})$. There is a corresponding equivalent formulation of the dimension of a ring, namely $\dim R = \max\{\, k \mid \mathfrak{p}_0 \subset \cdots \subset \mathfrak{p}_k \,\}$, where $\mathfrak{p}_0 \subset \cdots \subset \mathfrak{p}_k$ is a strictly increasing sequence of proper prime ideals in R. In case $R = k[x_1, \ldots, x_n]/\mathfrak{p}$, one can show that all maximal chains of prime ideals have the same length. There is also a concept of "local dimension at a point P on V". The local dimension is the maximal dimension of the components of V which pass through P. Algebraically, one can calculate the local dimension as the dimension of the associated graded ring with respect to the maximal ideal which corresponds to P. There are corresponding algebraic versions for the projective dimension of a projective set.

Exercises

1. Let N be a positive number and let $R = k[x_1, \ldots, x_n]/\mathfrak{b}$, where $\mathfrak{b}$ is generated by all $(x_{j_1} \cdots x_{j_{e+1}})^N$, $1 \le j_1 < \cdots < j_{e+1} \le n$. Let $i > n^N$.

   (a) Show that if $x_1^{m_1} \cdots x_n^{m_n}$ is a monomial of degree i then $m_i \ge N$ for at least one i.

   (b) Show that the number of monomials $x_1^{m_1} \cdots x_n^{m_n}$ of degree i with $m_1 \ge N$, $m_2, \ldots, m_n < N$, is constant (i.e., independent of i).

   (c) Show that the number of monomials $x_1^{m_1} \cdots x_n^{m_n}$ of degree i with exactly one $m_i \ge N$, is constant.

   (d) Show that the number of monomials $x_1^{m_1} \cdots x_n^{m_n}$ of degree i with $m_1, \ldots, m_k \ge N$, $m_{k+1}, \ldots, m_n < N$, is a polynomial of degree $k - 1$.

   (e) Show that the number of monomials $x_1^{m_1} \cdots x_n^{m_n}$ of degree i with exactly k of the m_i's $\ge N$, is a polynomial of degree $k - 1$.

   (f) Show that a monomial $x_1^{m_1} \cdots x_n^{m_n}$ is nonzero in R if and only if at most e of the m_i's are $\ge N$.

   (g) Show that $\dim R = e$.

2. Let $l_1, \ldots, l_k$ be linearly independent homogeneous elements of degree one in $\mathbb{C}[x_1, \ldots, x_n]$, and let $\mathfrak{a} = \langle l_1, \ldots, l_k \rangle$. Show that $\dim \mathfrak{v}(\mathfrak{a}) = n - k$.

3. Show that $\dim A^n(\mathbb{C}) = n$ and $\dim P^n(\mathbb{C}) = n$.

4. Show that if V is a hypersurface in $A^n(\mathbb{C})$ or in $P^n(\mathbb{C})$, i.e., $V = \mathfrak{v}(\langle f \rangle)$ for some $f \in \mathbb{C}[x_1, \ldots, x_n]$, or $V = \mathfrak{V}(\langle f \rangle)$ for some homogeneous $f \in \mathbb{C}[x_1, \ldots, x_{n+1}]$ respectively, then $\dim V = n - 1$.

5. Let $f_1, \ldots, f_k$ be homogeneous elements in $A = \mathbb{C}[x_1, \ldots, x_n]$. The sequence $(f_1, \ldots, f_k)$ is said to be an *A-sequence* or a *regular sequence* if $\overline{f_i}$ is a nonzerodivisor in $A/\langle f_1, \ldots, f_{i-1} \rangle$ for $i = 1, \ldots, n$. Show that if $\mathfrak{a} = \langle f_1, \ldots, f_k \rangle$ where $(f_1, \ldots, f_k)$ is an A-sequence, then $\dim \mathfrak{v}(\mathfrak{a}) = n - k$. The algebraic set $\mathfrak{v}(\mathfrak{a})$ is called a *complete intersection* if $\mathfrak{a}$ is generated by an A-sequence.

6. Show that if $\mathfrak{v}(\langle f_1, \ldots, f_k \rangle)$ is a complete intersection then $k \leq n$.

7. Let $f_1, \ldots, f_k$ be homogeneous in $\mathbb{C}[x_1, \ldots, x_n]$ with $\deg f_i = d_i$. Show that $H_{\mathbb{C}[x_1,\ldots,x_n]/\langle f_1,\ldots,f_k \rangle}(X) \geq (\prod_{i=1}^k (1 - X^{d_i}))/(1 - X)^n$ with equality if and only if $(f_1, \ldots, f_k)$ is an A-sequence. (Here $\sum a_i X^i \geq \sum b_i X^i$ means $a_i \geq b_i$ for all i.) Show that a permutation of an A-sequence is an A-sequence.

Chapter 12

Variations of Gröbner Bases

We will in this chapter mention how the concept of Gröbner bases can be extended to more general situations and look at some variations of the concept. The nature of this chapter will be more a survey, and some sections will consist mainly of pointers to the litterature.

12.1 Gröbner Bases for Modules

Let $A = k[x_1, \ldots, x_n]$ and let $F = A^n = \{ (a_1, \ldots, a_n) \mid a_i \in A \}$. F is called a *free* A-module, and is equipped with the basis $e_1 = (1, 0, \ldots, 0), e_2 = (0, 1, \ldots, 0), \ldots, e_n = (0, 0, \ldots, 1)$. We will extend the notion of Gröbner bases from ideals in A to submodules of F. This can be done without any real difficulty. If S is a subset of F, we denote the submodule *generated* by S by $\langle S \rangle$, i.e., $\langle S \rangle = \{ \sum a_i s_i \mid a_i \in A, s_i \in S \}$. A *monomial* in F is an element me_i, where m is a usual monomial in A, and a *monomial submodule* of F is a submodule generated by monomials. A *monomial ordering* of F is a total ordering of the monomials of F such that if $m_1e_i \prec m_2e_j$ then $mm_1e_i \prec mm_2e_j$ for every monomial m in A, and such that $m_1e_i \prec mm_1e_i$ for every monomial $m \neq 1$ in A. Given a monomial ordering $\prec$ on A, we can use $m_1e_i \prec m_2e_j$ if $i < j$ or if $i = j$ and $m_1 \prec m_2$. We can define the leading monomial, $\text{lm}(f)$, for each $f \in F$ since f is a linear combination of monomials and, if H is a submodule of F, then $l(H)$ is the submodule $\langle \text{lm}(h) | h \in H \rangle$. For each submodule H, $l(H)$ is a monomial submodule and a generalization of Dickson's Lemma shows that monomial submodules are finitely generated. If H is a submodule of F, then $G = \{g_1, \ldots, g_s\}$ is a

Gröbner basis of H if $g_i \in H$ and $l(H) = \langle \mathrm{lm}(g_i) \rangle$.

An *A-module* is a nonempty set whose elements can be added and multiplied with elements in A so that the rules for a vector space are satisfied (with the field k exchanged for the ring A). If H is a submodule of F, then the set of cosets $F/H = \{ f + H \mid f \in F \}$ constitute an A-module with addition $(f_1 + H) + (f_2 + H) = (f_1 + f_2) + H$ and multiplication $a(f+H) = af + H, a \in A$. The zero element in F/H is $H = 0 + H$. We can define the reduction of an element with respect to a sequence of elements, in a similar way to ideals. A Gröbner basis of H gives rise to normal forms for elements in F/H. If $f_1, f_2 \in F$ and $\mathrm{lm}(f_1) = m_1 e_i, \mathrm{lm}(f_2) = m_2 e_j$ we define the S-"polynomial" of f_1 and f_2 as

$$S(f_1, f_2) = m_2/\gcd(m_1, m_2) f_1 - m_1/\gcd(m_1, m_2) f_2$$

if $i = j$ and $S(f_1, f_2) = 0$ otherwise. Then the criterion is still valid for a set $G = \{g_1, \ldots, g_s\}$ to be a Gröbner basis for the submodule they generate if and only $S(g_i, g_j)$ reduces to 0 with respect to G for every (i, j). As a consequence we get that submodules of F are finitely generated and that Gröbner bases can be constructed with Buchberger's algorithm.

EXAMPLE 1. As an example of the use of Gröbner bases for modules we mention an old theorem by Hilbert, the so-called Syzygy Theorem. We will describe this theorem by means of an easy example. Consider the sequence $\mathbf{s} = (x, y, z)$ of elements in $A = k[x, y, z]$. We recall that a syzygy of $\mathbf{s}$ is an element $(f_1, f_2, f_3) \in A^3$ such that $f_1 x + f_2 y + f_3 z = 0$. We have seen that the set of syzygies of $\mathbf{s}$ constitutes a submodule of A^3. In this case it is generated by the three elements $(y, -x, 0), (z, 0, -x)$, and $(0, z, -y)$, cf. Section 3.9. The syzygies of $\mathbf{s}$ describe the relations among the elements of $\mathbf{s}$. We note that we can describe the syzygies as the kernel of the map $\Phi_1 : A^3 \longrightarrow A$ defined by $\Phi_1((r_1, r_2, r_3)) = r_1 x + r_2 y + r_3 z$. We denote the sequence $((y, -x, 0), (z, 0, -x), (0, z, -y))$ of the generators for the syzygy module by $\mathrm{syz}_1(\mathbf{s})$. Now we turn to the relations of the relations, i.e.,

$$\{ (r_1, r_2, r_3) \mid r_1(y, -x, 0) + r_2(z, 0, -x) + r_3(0, -z, y) = (0, 0, 0), r_i \in A \}.$$

It is not hard to see that this is a submodule of A^3, generated by the single element $(z, -y, x)$. This submodule consists of the *second syzygies* of $\mathbf{s}$. As for the (first) syzygies, we can describe the submodule of second syzygies as the kernel of a map, namely $\Phi_2 : A^3 \longrightarrow A^3$ defined by $\Phi_2((r_1, r_2, r_3)) = r_1(y, -x, 0) + r_2(z, 0, -x) + r_3(0, -z, y)$. We denote the sequence $((z, -y, x))$ (consisting of one element) by $\mathrm{syz}_2(\mathbf{s})$. A third syzygy would be an element r in A such that $r(z, -y, x) = (0, 0, 0)$. This gives $r = 0$. Hence $\ker(\Phi_3) = 0$, if $\Phi_3 : A \longrightarrow A^3, \Phi_3(r) = r(z, -y, x)$. The third module of syzygies is the zero submodule (i.e., the zero ideal) of A. □

The construction of the first syzygy of a sequence of elements is the submodule analogue of calculating $\langle 0 \rangle : \mathfrak{a}$ for an ideal $\mathfrak{a}$, and similarly for higher syzygies. This can be done with Gröbner bases for modules. The Syzygy Theorem states the following. Let $\mathbf{f} = (f_1, \ldots, f_s)$ be a sequence of homogeneous elements in $A = k[x_1, \ldots, x_n]$. Let $\mathrm{syz}_1(\mathbf{f}) = (g_1, \ldots, g_t)$ be a sequence of homogeneous elements that minimally generate $\ker(\Phi_1)$, where $\Phi_1 : A^s \longrightarrow A$ is defined by $\Phi_1((r_1, \ldots, r_s)) = r_1 f_1 + \cdots + r_s f_s$. Let $\mathrm{syz}_2(\mathbf{f})$ be a sequence minimally generating $\ker(\Phi_2)$, where $\Phi_2 : A^t \longrightarrow A^s$ is defined by $\Phi_2((r_1, \ldots, r_t)) = r_1 g_1 + \cdots + r_t g_t$ and so on. Then $\ker(\Phi_n)$ is the zero submodule. It can be shown that different choices of minimal generators for the kernels give isomorphic syzygy modules.

By using a suitable ordering, it is even possible to give a proof of this theorem using Gröbner bases. This was done in an unpublished Diplomarbeit by F.-O. Schreyer. We refer the interested reader to the book of D. Eisenbud.

Further Reading

D. Eisenbud, *Commutative algebra with a view toward algebraic geometry*, Springer 1995.

F.-O. Schreyer, *Die Berechnung von Syzygien mit dem verallgemeintern Weierstrass'schen Divisionssatz*, Diplomarbeit, Hamburg 1980.

Exercise

1. Let $A = k[x, y, z]$ and $\mathbf{s} = (x^2, xy, xz, y^2, yz, z^2)$. Show that the submodule of (first) syzygies is a submodule of A^6 generated by 8 elements. Show that the submodule of second syzygies is a submodule of A^8 generated by 3 elements. Show that the third module of syzygies is the zero submodule of A^3.

12.2 More General Orderings within More General Rings

We cannot expect a direct generalization of the concept of Gröbner bases to ideals in a power series ring $k[[x_1, \ldots, x_n]]$, since in general a power series does not have a leading term. There are also other rings which are of importance in commutative algebra and algebraic geometry. We will mention an important example. If V is an algebraic subset of $\mathbb{C}^n$ containing the origin, and one wants to study the behaviour of V in a neighbourhood of the origin, the natural ring to study is not $\mathbb{C}[x_1, \ldots, x_n]$ but its "localization at the origin", denoted $\mathbb{C}[x_1, \ldots, x_n]_{(x_1,\ldots,x_n)}$. This ring, which lies between $\mathbb{C}[x_1, \ldots, x_n]$ and the fraction field of $\mathbb{C}[x_1, \ldots, x_n]$, consists of $\{ f/g \mid f, g \in \mathbb{C}[x_1, \ldots, x_n], g(0, \ldots, 0) \neq 0 \}$. This is a *local ring*, i.e., there is a unique maximal ideal, namely $\langle x_1, \ldots, x_n \rangle$. Every ideal in $\mathbb{C}[x_1, \ldots, x_n]_{(x_1,\ldots,x_n)}$ can be generated by polynomials in $\mathbb{C}[x_1, \ldots, x_n]$, since a generator f/g can be replaced by f, g being invertible. If one wants to study ideals in $\mathbb{C}[x_1, \ldots, x_n]_{(x_1,\ldots,x_n)}$ by means of Gröbner bases, it turns out that one has to consider new kinds of orderings; we must demand that $x_i \prec 1$ for every i. This implies that one has to be more careful to get a terminating algorithm; it is not possible just to copy the Buchberger algorithm. One important concept to introduce is the *ecart* of a polynomial f, which is the difference between the highest and the lowest degree of a monomial occurring in f. When one has a choice when making a reduction, one has to minimize the ecart. The first to give a *standard basis* algorithm for these kinds of rings was Mora. He was interested in getting an algorithm to compute the tangent cone of an algebraic set. If $\langle f_1, \ldots, f_r \rangle \subseteq \mathfrak{m} = \langle x_1, \ldots, x_n \rangle \subseteq \mathbb{C}[x_1, \ldots, x_n]$ then the associated graded ring of $C[x_1, \ldots, x_n]/\langle f_1, \ldots, f_r \rangle$ with respect to $\overline{\mathfrak{m}}$ coincides with the associated graded ring of $C[x_1, \ldots, x_n]_{(x_1,\ldots,x_n)}/\langle f_1, \ldots, f_r \rangle$ with respect to $\overline{\mathfrak{m}}$. His results have subsequently been simplified and generalized by Gräbe and the Singular team. There is now a theory of standard bases for general orderings where some of the variables are $\succ 1$ and some are $\prec 1$. Standard bases have the same applications as Gröbner bases and there are also algorithms to make ideal operations, calculating syzygies and so on in this more general setting.

Further Reading

H.-G. Gräbe, *The tangent cone algorithm and homogenization*, J. Pure Appl. Algebra **97** (1994), 303–312.

H. Grassmann, G.M. Greuel, B. Martin, W. Neumann, G. Pfister, W. Pohl, H. Schönemann, and T. Siebert, *Standard bases, syzygies and their implementation in SINGULAR*, Beiträge zur angewandten Analysis und Informatik, pp. 69–96, Shaker Aachen 1994.

T. Mora, *An algorithm to compute the equations of tangent cones*, Proc. EUROCAM 82, Springer Lecture Notes in Computer Science (1982).

12.3 Gröbner Bases for Noncommutative Rings

In this section we will study ideals in the *noncommutative polynomial ring* over a field k (or the *free associative algebra* over k). The noncommutative polynomial ring $k < x_1, \ldots, x_n >$ consists of linear combinations of *monomials* (or *words*) $x_{i_1} \cdots x_{i_k}$, where the variables do not commute. There are for example nine monomials of degree 2 in $k < x, y, z >$, namely

$$xx = x^2, xy, xz, yx, yy = y^2, yz, zx, zy, zz = z^2.$$

An *ideal* in $B = k < x_1, \ldots, x_n >$ (or a *two-sided ideal*) is a nonempty subset $\mathfrak{a}$ of B such that

- $a, b \in \mathfrak{a} \Rightarrow a + b \in \mathfrak{a}$
- $a \in \mathfrak{a} \Rightarrow ras \in \mathfrak{a}$ for all $r, s \in B$

The foundations for noncommutative Gröbner bases were laid down by Bergman.

A *monomial ordering* in B is a total ordering of the monomials in B compatible with multiplication, i.e.,

- For any monomials m, n we have $m \prec n$ or $n \prec m$ or $m = n$
- $1 \prec m$ for every monomial $m \neq 1$
- $m_1 \prec m_2$ implies $mm_1n \prec mm_2n$ for any monomials m, n

There are lots of monomial orderings, but we will in this section stick to one choice, namely Deglex. Here $x_{i_1} \cdots x_{i_k} \succ x_{j_1} \cdots x_{j_m}$ if $k > m$ or if $k = m$ and $i_1 = j_1, \ldots, i_{s-1} = j_{s-1}$ and $i_s < j_s$ for some s. Thus we have made the choice that $x_1 \succ \cdots \succ x_n$. In $k < x_1, x_2, x_3 >$ the smallest monomials are

$$1 \prec x_3 \prec x_2 \prec x_1 \prec x_3^2 \prec x_3x_2 \prec x_3x_1 \prec x_2x_3 \prec$$
$$x_2^2 \prec x_2x_1 \prec x_1x_3 \prec x_1x_2 \prec x_1^2 \prec \cdots$$

Since any element $f \in B$ is a linear combination of monomials, we can define $\mathrm{lm}(f)$, and for an ideal $\mathfrak{a}$ we define $l(\mathfrak{a})$ to be $\langle \mathrm{lm}(a) | a \in \mathfrak{a} \rangle$. A Gröbner basis for an ideal $\mathfrak{a}$ is a set $G = \{g_i\} \subseteq \mathfrak{a}$ such that $l(\mathfrak{a}) = \langle \mathrm{lm}(g_i) \rangle$. The reduction process is not much different from the commutative case. Let $f \in B$ be monic and suppose $\mathrm{lm}\, f = m$ and $f = m + f'$. A monomial n has m as a subword if $n = n_1mn_2$ for some monomials n_1, n_2. The reduction of n with respect to f is then $n_1(-f')n_2$. With this piece of information one can easily recursively define the reduction of an element with respect to a sequence of elements as in the commutative case. There are, however, several problems when extending the theory of Gröbner bases to ideals in the noncommutative ring B. Perhaps the most serious is that the "word problem" is in general not decidable for ideals in B, i.e., one can show there is in general no algorithm to decide whether or not an element f belongs to a given ideal $\mathfrak{a}$. If we stick to homogeneous ideals this difficulty disappears, so we will restrict ourselves to homogeneous ideals. (The degree of a monomial is just the length, i.e., the number of letters in it, and an element is homogeneous if it is a linear combination of monomials of the same degree.) Another difficulty is that Gröbner bases might very well be infinite. What we can always achieve in the homogeneous case is an algorithm which for a given integer k produces the part of a Gröbner basis consisting of the elements of degree at most k. (In many cases one can describe the whole Gröbner basis even if it is infinite; see the examples below.) There are also some minor problems which force one to be cautious when extending the Buchberger algorithm to the noncommutative setting. These difficulties are due to the fact it is not as easy in the noncommutative case as in the commutative to define the least common multiple of two monomials. Let m and n be two monomials such that neither is a subword of the other. We define an *essential* common multiple of m and n to be a monomial w with $\deg(w) < \deg(m) + \deg(n)$ such that $w = mw_1 = w_2n$ for some monomials $w_1, w_2 \neq 1$. Let $e(m, n)$ be the set of essential common multiples of m and n. Suppose we want to calculate the Gröbner basis of an ideal $\langle f_1, \ldots, f_r \rangle$ and suppose that $\operatorname{lm}(f_i) = m_i$. The difference from the commutative case is now that we have, for each ordered pair (i, j), to make one S-polynomial calculation for each element in $e(m_i, m_j)$ which is not a strict subword of another element in $e(m_i, m_j)$ (*including the pairs* (i, i)). With this in mind one can prove the criterion for a set to be a Gröbner basis if and only if all S-polynomials reduce to zero, and we can get an algorithm as in the commutative case. (Note that the algorithm calculates the Gröbner basis degree by degree, so if the Gröbner basis is infinite, it will not terminate.) It is high time for some examples.

EXAMPLE 2. Let us determine the Gröbner basis for the ideal $\langle x_1^2 - x_1x_2 \rangle \subseteq k < x_1, x_2 >$. We have $\operatorname{lm}(x_1^2 - x_1x_2) = x_1^2$ and $e(x_1^2, x_1^2) = x_1^3$. We need to calculate the S-polynomial

$$S(x_1^2 - x_1x_2, x_1^2 - x_1x_2) =$$

$$x_1(x_1^2 - x_1x_2) - (x_1^2 - x_1x_2)x_1 = -x_1^2x_2 + x_1x_2x_1.$$

Now $x_1^2x_2$ reduces to $x_1x_2^2$ so the S-polynomial reduces to $x_1x_2x_1 - x_1x_2^2$, which cannot be reduced further, so it has to be adjoined to the basis. We have $\operatorname{lm}(x_1x_2x_1 - x_1x_2^2) = x_1x_2x_1$ and

$$e(x_1^2, x_1x_2x_1) = \{x_1^2x_2x_1\},$$

$$e(x_1x_2x_1, x_1^2) = \{x_1x_2x_1^2\},$$

$$e(x_1x_2x_1, x_1x_2x_1) = \{x_1x_2x_1x_2x_1\},$$

so we have three S-polynomial calculations to perform. We have

$$x_1(x_1x_2x_1 - x_1x_2^2) - (x_1^2 - x_1x_2)x_2x_1 = -x_1^2x_2^2 + x_1x_2^2x_1$$

which reduces to $x_1x_2^2x_1 - x_1x_2^3$ and has to be added to the basis. And

$$x_1x_2(x_1^2 - x_1x_2) - (x_1x_2x_1 - x_1x_2^2)x_1 = -x_1x_2x_1x_2 + x_1x_2^2x_1$$

which reduces to 0,

$$x_1x_2(x_1x_2x_1 - x_1x_2^2) - (x_1x_2x_1 - x_1x_2^2)x_2x_1 = -x_1x_2x_1x_2^2 + x_1x_2^3x_1$$

which reduces to $x_1x_2^3x_1 - x_1x_2^4$. It is not hard to prove by induction that the Gröbner basis will be

$$G = \{x_1^2 - x_1x_2, x_1x_2x_1 - x_1x_2^2, x_1x_2^2x_1 - x_1x_2^3, x_1x_2^3x_1 - x_1x_2^4, \ldots\}.$$

□

EXAMPLE 3. We now study the same example with the roles of x_1 and x_2 exchanged, i.e., the same example with another ordering. We determine the Gröbner basis of $\langle x_1x_2 - x_2^2\rangle \subseteq k < x_1, x_2 >$. Now $\mathrm{lm}(x_1x_2 - x_2^2) = x_1x_2$ and $e(x_1x_2, x_1x_2) = \emptyset$, so $G = \{x_1x_2 - x_2^2\}$ is a Gröbner basis. □

EXAMPLE 4. We will calculate the Gröbner basis for

$$\langle x_1^2 - x_1x_2, x_1x_2x_1 - x_2x_1x_2\rangle \subseteq k{<}x_1, x_2{>}.$$

The leading monomials are x_1^2 and $x_1x_2x_1$, respectively. We have

$$e(x_1^2, x_1^2) = \{x_1^3\},$$

$$e(x_1^2, x_1x_2x_1) = \{x_1^2x_2x_1\},$$

$$e(x_1x_2x_1, x_1^2) = \{x_1x_2x_1^2\},$$

and

$$e(x_1x_2x_1, x_1x_2x_1) = \{x_1x_2x_1x_2x_1\},$$

so we have to make four S-polynomial calculations. We get

$$(x_1^2 - x_1x_2)x_1 - x_1(x_1^2 - x_1x_2) = x_1^2x_2 - x_1x_2x_1$$

which reduces to $x_1x_2^2 - x_2x_1x_2 = f$. The other three S-polynomials reduce to 0. The only new leading monomial is $x_1x_2^2$. We have

$$e(x_1^2, x_1x_2^2) = x_1^2x_2^2, e(x_1x_2^2, x_1x_2x_1) = \emptyset,$$

and

$$e(x_1x_2^2, x_1x_2^2) = \emptyset.$$

We get

$$(x_1^2 - x_1x_2)x_2^2 - x_1(x_1x_2^2 - x_2x_1x_2) = x_1x_2x_1x_2 - x_1x_2^3$$

which reduces to 0. Hence the Gröbner basis is

$$G = \left\{x_1^2 - x_1x_2, x_1x_2x_1 - x_2x_1x_2, x_1x_2^2 - x_2x_1x_2\right\}.$$

□

Exercises

1. We have that $k[x_1, \ldots, x_n] \simeq k{<}x_1, \ldots, x_n{>} \,/\, \langle x_ix_j - x_jx_i\rangle$. Determine the Gröbner basis of $\langle x_ix_j - x_jx_i\rangle$ in $k{<}x_1, \ldots, x_n{>}$.

2. Since any commutative ring $k[x_1, \ldots, x_n]/\langle f_1, \ldots, f_r\rangle$ can be written $k{<}x_1, \ldots, x_n{>} \,/\, \langle x_ix_j - x_jx_i, f_1, \ldots, f_r\rangle$ one could hope that the commutative Gröbner basis of $\langle f_1, \ldots, f_r\rangle$ would help to give the noncommutative Gröbner basis of $\langle f_1, \ldots, f_r\rangle$. This is not so. Determine the Gröbner basis of

$$\langle x_1x_2 - x_2x_1, x_1x_3 - x_3x_1, x_2x_3 - x_3x_2, x_3x_2x_1\rangle \subseteq k{<}x_1, x_2, x_3{>}\,.$$

   In particular, show that it is infinite.

3. As for commutative homogeneous algebras, we can define the Hilbert series for a noncommutative algebra C as $H_C(X) = \sum \dim_k C_iX^i$. Consider the ideal

$$\left\langle x_1x_2 - x_1x_3, x_2x_3 - x_3x_2 - x_3^2/q, x_2x_4\right\rangle \subseteq k{<}x_1, x_2, x_3, x_4{>},$$

   where q is a positive integer. Show that the the Hilbert series is different for every choice of q, by calculating the Gröbner basis for the ideal.

Further Reading

G. Bergman, *The diamond lemma in ring theory*, Adv. Math. **29** (1978) 178–218.

12.4 Differential Gröbner Bases

Differential algebra is used to study systems of algebraic differential equations. We will give some basic definitions.

A map δ on a ring is called a *differential operator* if $\delta(a+b) = \delta a + \delta b$ and $\delta(ab) = (\delta a)b + a(\delta b)$. A differential ring is a ring with a finite set $\Delta = \{\delta_1, \ldots, \delta_m\}$ of commuting differential operators. A differential field is a differential ring which is a field. An ideal $\mathfrak{a}$ in a differential ring is called a *differential ideal* if $r \in \mathfrak{a}$ implies $\delta_i r \in \mathfrak{a}$ for all i. Let k be a differential field with a set $\Delta = \{\delta_1, \ldots, \delta_m\}$ of commuting differential operators. Let $\Theta = \{\delta_1^{i_1} \cdots \delta_m^{i_m}\}$ be the set of monomials in the differential operators. Consider the ring $k\{x_1, \ldots, x_n\} = k[x_{j,\delta} | \delta \in \Theta]$, a polynomial ring in infinitely many variables. We make $k\{x_1, \ldots, x_n\}$ into a differential ring by defining $\delta_i(x_{j,\delta}) = x_{j,\delta_i\delta}$. A monomial in $k\{x_1, \ldots, x_n\}$ is a monomial in the $x_{j,\delta}$'s. An admissible ordering of the monomials is a usual ordering such that furthermore $\text{lm}(\delta(M)) \succ M$ if $M \neq 1$, and if $M \succ M'$ then $\text{lm}(\delta(M)) \succ \text{lm}(\delta(M'))$. It is now possible to define reduction and differential Gröbner bases for differential ideals. Differential Gröbner bases are in general not finite. For the foundation of differential algebra, refer to the books below; for the theory of differential Gröbner bases to the articles.

Further Reading

G. Carrá Ferro, *Gröbner bases and differential algebra*, Proc. AAECC-6, Springer Lecture Notes in Computer Science **357** (1989), 152–163.

I. Kaplansky, *An introduction to differential algebra*, Hermann, 1957.

E. R. Kolchin, *Differential algebra and algebraic groups*, Academic Press, 1973.

F. Ollivier, *Standard bases of differential ideals*, Proc. AAECC-8, Springer Lecture Notes in Computer Science **508** (1990), 304–321.

12.5 SAGBI Bases

Let $f_1, \ldots, f_r \in k[x_1, \ldots, x_n] = A$ and consider $B = k[f_1 \ldots, f_r]$, the subalgebra of A generated by $f_1, \ldots, f_r$. The concept of SAGBI basis for B, short for Subalgebra Analog to Gröbner Basis for Ideals, was introduced by Robbiano and Sweedler, and has lately been used as a theoretical approach to questions about subalgebras of polynomial rings. We use an ordinary monomial ordering on the monomials of A. A SAGBI basis $S = \{s_i\}$ for B is a set of elements in B such that $k[\text{lm}(s_i)]$ equals the subalgebra $k[l(B)]$ generated by the leading monomials of elements in B. There are parallel concepts to those of reduction, S-polynomial, and the Buchberger algorithm for SAGBI bases. Some drawbacks of the theory are quite similar to those of noncommutative Gröbner bases. One can easily give examples of subalgebras with no finite SAGBI basis, of algebras with a finite SAGBI basis in one monomial ordering and no finite basis in another ordering, and of subalgebras with finite SAGBI bases in every ordering.

Further Reading

D. Kapur and K. Madlener, *A completion procedure for computing a canonical basis for a k-subalgebra*, in Computers and Mathematics, Cambridge 1989, pp. 1–11, Springer 1989.

L. Robbiano and M. Sweedler, *Subalgebra bases*, Proc. Salvador 1988, Lect. Notes Math. **1430**, 61–87, Springer 1990.

Chapter 13

Improvements to Buchberger's Algorithm

A lot of tuning may be done when implementing the Buchberger algorithm. We will only briefly mention some possibilities.

13.1 Choice of Ordering

For some problems one has a free choice of orderings. Even if some specific ordering is needed, such as Lex, one often still has the choice of ordering the variables. If there are n variables, this could be done in $n!$ ways. The ordering often has a dramatic effect on both timing and use of memory. There are theoretical reasons to believe that Degrevlex is most efficient on "most" problems, but in practice there are often some symmetries in problems one wants to treat, and sometimes a block ordering is more efficient. It seems extremely hard to find some general rules for determining the best ordering, and some experimenting is often necessary for large problems. It has even been proposed to change the ordering dynamically, when partial Gröbner bases have been found.

Further Reading

D. Bayer and M. Stillman, *A theorem on refining division orders by the reverse lexicographic order*, Duke Math. J. **55** (1987), 321–328.

S. R. Czapor, *A heuristic strategy for lexicographic Gröbner bases?*, Proc. ISSAC'91, 39–48, ACM Press 1991.

P. Gritzmann and B. Sturmfels, *Minkowski addition of polytopes: Computational complexity and applications to Gröbner bases*, SIAM J. Discrete Math **6** (1993), 246–269.

F. Hreinsdottir, *A case where choosing a product order makes the calculations of a Groebner basis much faster*, J. Symb. Comp. **18** (1994), 373–378.

13.2 Strategies

There is also a large freedom to choose the strategy for the ordering in which one calculates the S-polynomials. Sometimes one may use the "history" of the calculation, i.e., to use information about when new Gröbner basis elements were found. There are methods which for some problems speed up the calculation, but here too it is difficult to get a general rule.

Further Reading

A. Giovini, T. Mora, G. Niesi, L. Robbiano, and C. Traverso, *"One sugar cube, please" or selection strategies on the Buchberger algorithm*, Proc. ISSAC'91, 49–54, ACM Press 1991.

13.3 Unnecessary Reductions

A lot of time and memory are usually consumed by unnecessary S-polynomial calculations, i.e., calculating S-polynomials which, as it turns out, are reduced to 0. Some criteria do exist to detect, and thus avoid, these unnecessary calculations. It is easy to see that if $\gcd(\mathrm{lm}(f_1), \mathrm{lm}(f_2)) = 1$ then $S(f_1, f_2)$ will reduce to 0. This is one of Buchberger's criteria. The most commonly used criteria are listed in the article by Gebauer and Möller; they were subsequently improved by Backelin in his graph component algorithm.

Further Reading

J. Backelin and R. Fröberg, *How we proved that there are exactly 924 cyclic 7-roots*, Proc. ISSAC'91 pp. 103–111, ACM Press 1991.

B. Buchberger, *A criterion for detecting unnecessary reductions in the construction of Gröbner bases*, Lect. Notes in Comp. Science **72** pp. 3–21, Springer 1979.

R. Gebauer and H. M. Möller, *On an installation of Buchberger's algorithm*, J. Symb. Comp. **6** (1988), 275–286.

13.4 Homogenizing

For some problems it seems to be more efficient first to homogenize the generators of the ideal and then to dehomogenize the resulting Gröbner basis, cf. Section 8.3.

13.5 Factorizing

Sometimes it may be efficient to factorize a Gröbner basis element, in order to split the problem into smaller parts. If one wants to solve a system of equations and have a partial Gröbner basis $g_1, \ldots, g_N = g_N^{(1)} g_N^{(2)}$, one can split the problem in two parts, using $g_1, \ldots, g_{N-1}, g_N^{(1)}$ and $g_1, \ldots, g_{N-1}, g_N^{(2)}$ as inputs for the continuation.

Further Reading

S. R. Czapor, *Solving algebraic equations: Combining Buchberger's algorithm with multivariate factorization*, J. Symb. Comp. **7** (1989), 49–55.

13.6 Coefficients

When calculating a Gröbner basis of an ideal in $\mathbb{Q}[x_1, \ldots, x_n]$, most of the time is spent on calculations with big integers. Even if the final Gröbner basis have coefficients of moderate size, it seems unavoidable that the intermediate calculations often use very large integers. Thus it is essential that big integers are handled efficiently. If it is possible to use coefficients in $\mathbb{Z}_p$ for some prime p, this difficulty disappears, since the coefficients have fixed size. Nevertheless it turns out that a lot of time is spent on multiplication in $\mathbb{Z}_p$. We will describe a method to lower the cost for these multiplications. It is well known that, for each p, there is an $r \in \mathbb{Z}_p$ such that each nonzero element in $\mathbb{Z}_p$ is a power of r. Such an element r is called a generator for $\mathbb{Z}_p \setminus \{0\}$. (As an example, the reader can check that we can use $r = 3$ in $\mathbb{Z}_{17}$.) Let us denote ${}^r\log k = s$ if $k = r^s$. Then $k \cdot m = r^{{}^r\log k + {}^r\log m}$, and we can replace a multiplication with (a much cheaper) addition. We need to precalculate a generator r and two tables of size p, one with logarithms and one with antilogarithms. With this trick, one is often able to reduce the total time for a Gröbner basis calculation by a third.

13.7 Using the Hilbert Series

Suppose that for some reason we know the Hilbert series of $k[x_1, \ldots, x_n]/\mathfrak{a}$, where $\mathfrak{a}$ is a homogeneous ideal. Then we can use this information to speed up the calculation. If one calculates the Gröbner basis degree by degree (which by the way seems to be the most efficient method) and when all Gröbner basis element up to and including a certain degree d are determined, one calculates the Hilbert series of the factor ring modulo the ideal generated by this partial Gröbner basis. (In practice one uses the initial ideal, which gives the same result.) By comparing this Hilbert series with the one we know for the ideal, we can judge whether the calculation is ready (if they are equal), or if not, how many new Gröbner basis elements we will find in degree $d+1$. When next dealing with degree $d+1$, the calculation is stopped in that degree when the right number of new Gröbner basis elements are found. This often makes it possible to avoid most unnecessary S-polynomial calculations. Perhaps it seems strange to have information about the Hilbert series before a Gröbner basis has been calculated, but sometimes it is known from theoretical reason that the ring is a complete intersection. More generally, in fact, we need only a lower bound for the Hilbert series, and this can sometimes be achieved by theoretical reasoning; then we save time if this lower bound is actually achieved, so it is enough to make an intelligent guess of the Hilbert series, which can be shown to be a lower bound. Another instance where this method can be used is when we need the Gröbner basis in Lex. Then we can calculate the Hilbert series in a fast ordering, such as Degrevlex, before using it for the calculation in Lex.

Further reading

P. Gianni, T. Mora, L. Robbiano, and C. Traverso, *Hilbert functions and Buchberger algorithm*, preprint.

13.7.1 Calculation of Hilbert Series

We will now sketch an efficient algorithm to calculate the Hilbert series of a polynomial ring modulo a monomial ideal. If $\mathfrak{a}$ is any homogeneous ideal, we know that $H_{k[x_1,\ldots,x_n]/\mathfrak{a}}(X) = H_{k[x_1,\ldots,x_n]/(\mathfrak{a})}(X)$, so the algorithm can be used for any homogeneous algebra $k[x_1, \ldots, x_n]/\mathfrak{a}$ once we have calculated a Gröbner basis for its defining ideal $\mathfrak{a}$. The idea is to split the calculation into two simpler parts. Let $\mathfrak{a}$ be any homogeneous ideal, and let $R = k[x_1, \ldots, x_n]/l(\mathfrak{a}) = \oplus_{i \geq 0} R_i$ be the decomposition into homogeneous parts. Let m be any monomial. The multiplication by m gives for each i a linear map of vector spaces $R_i \to R_{i+d}$, where $d = \deg(m)$. The kernel of this map is $(\mathfrak{a} : \langle m \rangle / \mathfrak{a})_i$ and the image is $((\mathfrak{a} + \langle m \rangle)/\mathfrak{a})_{i+d}$. Hence $\dim_k(\mathfrak{a} : \langle m \rangle / \mathfrak{a})_i + \dim_k((\mathfrak{a} + \langle m \rangle)/\mathfrak{a})_{i+d} = \dim_k R_i$. Now $\dim_k(\mathfrak{a} : \langle m \rangle / \mathfrak{a})_i = \dim_k R_i - \dim_k(R/\mathfrak{a} : \langle m \rangle)_i$ and $\dim_k((\mathfrak{a} + \langle m \rangle)/\mathfrak{a}) = \dim_k R_{i+d} - \dim_k(R/\langle m \rangle)_{i+d}$, so we get $\dim_k R_{i+d} = \dim_k(R/\mathfrak{a} : \langle m \rangle)_i + \dim_k(R/\langle m \rangle)_{i+d}$. Multiplying this equality with X^{i+d} and summing over i, we get $H_R(X) = H_{R/\langle m \rangle}(X) + X^d H_{R/\mathfrak{a}:\langle m \rangle}(X)$. There are different possibilities to choose the monomial m, in general the choice of a variable is sufficiently good; the time to calculate the Hilbert series is then usually only a very small fraction of the time to calculate the initial ideal. With this choice, one of the parts, $R/\langle m \rangle$, is a ring with one less variable; the other part, $R/\mathfrak{a} : \langle m \rangle$, has a defining ideal with generators of smaller degree. If $R = k[x_1, \ldots, x_n]/\left\langle x_{i_1}^{k_1}, \ldots, x_{i_s}^{k_s} \right\rangle$ we see as above that $H_R(X) = (1 - X^{k_1}) \cdots (1 - X^{k_s})/(1 - X)^n$, so we can use such rings as base cases and stop the calculation when one of them is reached. There are also other ways to choose the base cases.

Further Reading

A. Bigatti, M. Caboara, and L. Robbiano, *On the computation of Hilbert-Poincaré series*, Proc. AAECC'2 (1991), 21–33.

A. Bigatti, P. Conti, L. Robbiano, and C. Traverso, *A "divide and conquer" algorithm for Hilbert-Poincaré series, multiplicity, and dimension of monomial ideals*, Proc. AAECC (1993), Lect. Notes in Comp. Science **673** pp. 76–88, Springer 1993.

13.8 Change of Ordering

As we have indicated, it is usually much more time-consuming to calculate a Gröbner basis in Lex than in Degrevlex. We have mentioned the possibility of using Hilbert series in order to facilitate the calculation of a Gröbner basis in Lex from a Gröbner basis in Degrevlex. For zero-dimensional problems, there is also a method using pure linear algebra.

Further Reading

J. C. Faugère, P. Gianni, D. Lazard, and T. Mora, *Efficient computation of zero-dimensional Gröbner bases by change of ordering*, J. Symb. Comp. **16** (1993), 329–344.

13.9 Tracing

A last method we will mention is to use a "trace" of a calculation in a finite characteristic. Suppose that we want to calculate the Gröbner basis of an ideal $\mathfrak{a}$ in $\mathbb{Q}[x_1, \ldots, x_n]$, and that we know the Hilbert series of $\mathbb{Q}[x_1, \ldots, x_n]/\mathfrak{a}$. Then we can make the calculation for "the same" ideal in $\mathbb{Z}_p[x_1, \ldots, x_n]/\mathfrak{a}$ for some prime p. Usually the growth of coefficients is the main problem, both with regard to time and space, when calculating with rational coefficients. This problem does not occur when calculating over a finite field, since the coefficients have a fixed size. Thus there is often a dramatic difference in time and space in favour of finite fields. If we are not unlucky with the choice of p, we will get as our answer the true Gröbner basis with the coefficients reduced mod p. If we save the information about which of the S-polynomials gave new Gröbner basis elements (i.e., did not reduce to 0), we can avoid *all* unnecessary S-polynomial calculations by calculating only these S-polynomials in $\mathbb{Q}[x_1, \ldots, x_n]$. Again, if we are not unlucky, we will get the true Gröbner basis. (This can be checked with the Hilbert series.) If we *are* unlucky, we could try another p.

Chapter 14

Software

All major computer algebra systems such as Axiom, Macsyma, Maple, Mathematica, and Reduce contain Gröbner basis algorithms. There has been a lot of work on these lately, and they have become much more efficient over the past few years. The most efficient programs are, however, more specialized, and we will mention some free software.

The most commonly used is certainly **Macaulay**, created by D. Bayer and M. Stillman. In addition to Gröbner bases for ideals and modules, many commands are built in, such as ideal operations, calculation of syzygies and Hilbert series. Furthermore, there are lots of scripts for more advanced concepts, produced by the creators and by D. Eisenbud, for commutative algebra and algebraic geometry. Macaulay uses only finite fields as coefficients and behaves best on homogeneous input. It is available via anonymous ftp on `zariski.harvard.edu`. M. Stillman and D. Grayson are developing a new version, **Macaulay2**, which is also able to use more general coefficients.

CoCoa has been developed by a group at the University of Genova. It has a nice interface, and like Macaulay lots of scripts for commutative algebra and algebraic geometry. It also offers lots of strategy choices and different orderings, and allows more general coefficients than Macaulay. CoCoa is available via anonymous ftp at `lancelot.dima.unige.it`.

Singular has been developed by a group at the University of Kaiserslautern. It too has lots of scripts for commutative algebra and algebraic geometry; of strategy choices and allows general coefficients. Singular is probably the most efficient program for some classes of problems from algebraic geometry when orderings are needed where some or all of the variables are smaller than 1 are needed. Singular is available via anonymous ftp at saturn.mathematik.uni.kl.de.

Cali, created by H.-G. Gräbe at the University of Leipzig, is based on Reduce and has characteristics similar to CoCoa and Singular. Cali is available through the Reduce library at redlib@rand.org.

Bergman, created by J. Backelin at the University of Stockholm, calculates Gröbner bases of homogeneous ideals in commutative and noncommutative polynomial rings with coefficients in $\mathbb{Q}$ or $\mathbb{Z}_p$. It has a limited choice of orderings but some choices of strategies. It has an extension, called Anick, made by V. Ufnarovski, University of Lund, S. Cojacaru and A. Podoplelov, University of Cisinau, for calculating higher syzygies in the noncommutative case. Bergman is probably the most efficient program for noncommutative Gröbner basis calculations. It is available via anonymous ftp at ftp.matematik.su.se.

Chapter 15

Hints to Some Exercises

Section 1.2.4

2. In $\mathbb{Z}_4[x]$ we have $[2]x([2]x + 1) = [2]x$.
3. In $\mathbb{Z}_4[x]$ we have $([2]x + 1)^2 = [1]$.

Section 1.2.5

1. Show that if $n = n_1 n_2$, then $(n_1 \cdot 1) \cdot (n_2 \cdot 1) = n \cdot 1$.

Section 1.3.2

4. If R is a field and $\mathfrak{a}$ is an ideal $\neq \langle 0 \rangle$, show that $1 \in \mathfrak{a}$. If R is a ring with $\langle 0 \rangle$ and R as the only ideals then, if $r \neq 0$, we get $\langle r \rangle = R$. Conclude that r has a multiplicative inverse.

Section 1.3.4

2. Show that $d(1) \leq d(a)$ for every $a \neq 0$, and that $d(a) \leq d(1)$ if a is invertible.
3. If $a = bu$ then $b = au^{-1}$.

Section 1.6

1. If $x^3 + 3x + 1$ were reducible in $\mathbb{Q}[x]$, it would be reducible in $\mathbb{Z}[x]$ and thus would have a factor $x - a, a \in \mathbb{Z}$.
2. Suppose $f(x) = (b_0 + \cdots + b_k x^k)(c_0 + \cdots + c_{n-k} x^{n-k})$. Since $a_0 = b_0 c_0$ we get that p divides exactly one of b_0, c_0. Suppose p divides b_0. Then p

can't divide b_i for all i, because then p would divide a_n. Suppose p divides $b_0, \ldots, b_{l-1}$ but not b_l. Consider the coefficient of x^l in the product.
4. If $q(x)$ is reducible, then $q(x+1)$ is reducible.

Section 1.7

6. Show that $f : \mathbb{R}[x]/\langle x^2+1\rangle \longrightarrow \mathbb{C}$ defined by $f(a+bx+\langle x^2+1\rangle) = a+bi$ is an isomorphism.
8. Compare with Exercise 6.
13. To show the isomorphism, consider the map $f : \mathbb{R}[x] \longrightarrow R$ defined by $f(a_0 + a_1x + \cdots + a_kx^k) = (a_0, a_1)$.
14. If we write $x_2^2 = (x_2^2 - x_1^3) + x_1^3$, we see that we write any element in $f \in k[x_1, x_2]$ as $f = g_0(x_1) + x_2g_1(x_1) + g(x_1, x_2)$ with $g \in \langle x_1^3 - x_2^2\rangle$. It is obvious that $\langle x_1^3 - x_2^2\rangle \subseteq \ker(\phi)$. Now $\phi(f) = g_0(x^2) + x^3g_1(x^2)$ and $g_0(x^2)$ contains only even powers of x, $x^3g_1(x^2)$ contains only odd powers of x.
15. If $\mathfrak{a}$ is an ideal in R containing $\ker(\phi)$, show that $\phi(\mathfrak{a}) = \{\,\phi(a) \mid a \in \mathfrak{a}\,\}$ is an ideal in S. If $\mathfrak{b}$ is an ideal in S, show that $\phi^{-1}(\mathfrak{b}) = \{\,a \in R \mid \phi(a) \in \mathfrak{b}\,\}$ is an ideal in R which contains $\ker(\phi)$. Show that $\phi(\phi^{-1}(\mathfrak{b})) = \mathfrak{b}$ and that $\phi^{-1}(\phi(\mathfrak{a})) = \mathfrak{a}$.
16c. Use $p = 2$ in Exercise 16b and Exercise 5 in Section 1.6.
16d. Use $p = 3$ in Exercise 16b and Exercise 5 in Section 1.6.

Section 1.8

2. Show that $\mathbb{R}[x_1, x_2]/\langle x_1^2+1, x_2\rangle \simeq \mathbb{R}[x_1]/\langle x_1^2+1\rangle \simeq \mathbb{C}$.
3. Use Exercise 14 in Section 1.7. Note that $k[x^2, x^3]$ is an integral domain.
6. Show that $R/\{(0,b)\} \simeq \mathbb{R}$.
7. If $\mathfrak{a}$ is an ideal in R containing $\ker(\phi)$, show that $\phi(\mathfrak{a}) = \{\,\phi(a) \mid a \in \mathfrak{a}\,\}$ is an ideal in S. If $\mathfrak{b}$ is an ideal in S, show that $\phi^{-1}(\mathfrak{b}) = \{\,a \in R \mid \phi(a) \in \mathfrak{b}\,\}$ is an ideal in R. Show that $\phi(\phi^{-1}(\mathfrak{b})) = \mathfrak{b}$ and that $\phi^{-1}(\phi(\mathfrak{a})) = \mathfrak{a}$.

Section 1.9

2. Let $v_1, \ldots, v_n$ be a basis for V. Show that the map $f : V \longrightarrow k^n$ given by $f(c_1v_1 + \cdots + c_nv_n) = (c_1, \ldots, c_n)$ is an isomorphism.

Chapter 2

6. The smallest ideal such that $\left\{x_1^{d_1}, \ldots, x_n^{d_n}\right\}$ is part of a minimal system of generators is $\left\langle x_1^{d_1}, \ldots, x_n^{d_n}\right\rangle$. The largest ideal such that $\left\{x_1^{d_1}, \ldots, x_n^{d_n}\right\}$ is part of a minimal set of generators is $\left\langle x_1^{d_1}, \ldots, x_n^{d_n}\right\rangle + \langle x_ix_j | i \neq j\rangle$.

Section 3.1

2. Compare x_1x_3 and x_2^2 in the two orderings.
3. Which is the smallest monomial?

Section 3.12

1. Show that $x_1^2, x_1x_2 - x_2^2, x_2^3 \in \langle x_1^2, x_1x_2 - x_2^2 \rangle$ and that all S-polynomials reduce to 0.

Section 4.3

1. Show that $x_1(1-x_1^2) \notin \mathfrak{a}$ and that $(x_1(1-x_1^2))^3 \in \mathfrak{a}$.
2. First show that every nilpotent element belongs to every prime ideal. Let a be an element which is not nilpotent. Show that there is a prime ideal which does not contain a. (Consider the set of ideals which does not contain a. This set is not empty, since $\langle 0 \rangle$ does not contain a. Take a maximal element in this set and show that it is a prime ideal.)
3. For an ideal $\mathfrak{a}$ consider the ring $\mathbb{C}[x_1, \dots, x_n]/\mathfrak{a}$ and use Exercise 2.

Chapter 5

1. If $x_{i_1}^{n_1} \cdots x_{i_k}^{n_k}, n_i > 0$ is a minimal generator for a monomial ideal $\mathfrak{a}$, show that $\mathfrak{a}$ is the intersection of $\mathfrak{a} + \langle x_{i_1}^{n_1} \rangle$ and $\mathfrak{a} + \langle x_{i_2}^{n_2} \cdots x_{i_k}^{n_k} \rangle$.
5. Show that (the image of) x is a nonzerodivisor in $k[x,y,z]/\mathfrak{p}^2$, but that no power of x belongs to $\mathfrak{p}^2$. To prove the last statement, one can e.g. calculate the Gröbner basis of $\mathfrak{p}^2$ in Lex with $z \succ y \succ x$.
6. An irreducible ideal is primary. Use Exercise 1.
8. Write the monomial ideal as an intersection of irreducible ideals. Show that the intersection of irreducible monomial ideals with the same radical is of the given form.

Section 6.3

1. The Gröbner basis in Lex with $x \succ y$ is $\{x^2 - 1, xy - x - y + 1, y^2 - 1\}$.
2. The Gröbner basis in Lex with $x \succ y \succ z$ is

$$\{x - yz, y^2 - z^2, yz^2 - y, z^3 - z\}.$$

3. The Gröbner basis in Lex with $x \succ y \succ z$ is $(x - 1, y, z - 1)$.
4. Determine the dimension of the vector space $k[a,b,c,d]/l(\mathfrak{a})$.
5. Calculate $\dim_k k[x_1, x_2, x_3]/l(\mathfrak{a})$.

Section 6.4

1b. Minimize $x^2 + y^2 + z^2$ when $xy + xz + yz - 1 = 0$. It is easiest first to determine λ.

Section 7.1.1

1. Calculate a Gröbner basis for $\mathfrak{a}$ in Lex with $x \succ y \succ z$.

Section 7.1.2

1. Show that $x, y, z \in \sqrt{\mathfrak{a}}$.

Section 7.5

3. Show that $x - 1, y, z + 1 \in \sqrt{\mathfrak{a}}$ and that $1 \notin \sqrt{\mathfrak{a}}$.
6. To show that R is a field, it suffices to show that $x^3 + x + 1$ is irreducible. If $x^3 + x + 1$ were reducible the equation would have a rational root.
9. Show that

$$k[t^3, t^4, t^5] = k[x, y, z]/ \left\langle y^2 - xz, yz - x^3, z^2 - x^2 y \right\rangle .$$

$k[t^3, t^4, t^5]$ is an integral domain.

Section 8.1

1. Show by induction on n that V is not the union of n proper subspaces. For the induction step suppose $V = \cup_1^n V_i$ and $\cup_1^{n-1} \not\subseteq V_n$ and $V_n \not\subseteq \cup_1^{n-1} V_i$. Take $x \in \cup_1^{n-1} V_i \setminus V_n$ and $y \in v_n \setminus \cup_1^{n-1} V_i$. Consider $x + y$. Then show that $\mathbb{Z}_2 \times \mathbb{Z}_2$ *is* the union of three proper subspaces.
2. x, y, and $x + y$ are the only homogeneous elements of degree 1. The set of zerodivisors is the union of $\langle x \rangle$, $\langle y \rangle$, and $\langle x + y \rangle$.

Section 8.3

3. The reduced Gröbner basis for the homogeneous ideal is

$$\left\{ x_1 x_2 - y^2, x_2 x_3 - x_1 y, x_3^2 - x_2 y - y^2, x_1^2 y - x_3 y^2, x_2^2 y - x_1 x_3 y + x_2 y^2 \right\} .$$

5. The reduced Gröbner basis for the homogeneous ideal in Deglex with $x \succ y \succ z \succ t$ is

$$\left\{ xy - zt, xz - yt, xt - yz, y^2 z - zt^2, y^2 t - z^2 t, yz^2 - yt^2, z^3 t - zt^3 \right\} .$$

Section 11.1

2a. Use the definitions of multiplication and derivation and calculate both sides.
2b. Use Exercise 2a and induction.
2c. $1 = G \cdot G^{-1}$. Use 2a.
2e. $G^{-n} = (G^{-1})^n$.

Section 11.2

2. Use Exercise 1 in this section and Exercise 2f in Section 11.1.

Section 11.3

1b. For each $i > n^N$ there are just as many such monomials as there are monomials $x_2^{m_2} \cdots x_n^{m_n}$ with $m_i < N$, i.e., $(n-1)^N$.
1c. There are $n(n-1)^N$ such monomials.
1d. Let c_j be the number of monomials $x_{k+1}^{m_{k+1}} \cdots x_n^{m_n}$ of degree j with $m_i < N$ for all i. The number of monomials $x_1^{m_1} \cdots x_n^{m_n}$ of degree i with $m_i \geq N$ for $1 \leq i \leq k$ and $\deg(x_{k+1}^{m_{k+1}} \cdots x_n^{m_n}) = j$, $m_j < N$ for $k+1 \leq i \leq n$ then equals c_j times the number of monomials $x_1^{m_1} \cdots x_k^{m_k}$ of degree $i - kN - j$, which is a polynomial $p_j(i)$ in i of degree $k-1$. The total number of monomials asked for is $\sum_{j=0}^{(n-k)^N} c_j p_j(i)$, which is a sum of polynomials of degree $k-1$, and thus a polynomial of degree $k-1$.
1g. This follows from Exercise 1e and Corollary 8.
2. Show that $\mathbb{C}[x_1, \ldots, x_n]/\mathfrak{a} \simeq \mathbb{C}[x_{k+1}, \ldots, x_n]$.
3. Use Exercise 2 in Section 11.2.
4. Use Lemma 5.
5. Use Lemma 5 several times.

Section 12.3

3. Show that $x_1 x_3^j x_4 \in l(\mathfrak{a})$ if and only if $q > j - 2$.

Chapter 16

Answers to Exercises

Section 1.2.3

1.

Table 16.1: Addition in $\mathbb{Z}_6$.

+	[0]	[1]	[2]	[3]	[4]	[5]
[0]	[0]	[1]	[2]	[3]	[4]	[5]
[1]	[1]	[2]	[3]	[4]	[5]	[0]
[2]	[2]	[3]	[4]	[5]	[0]	[1]
[3]	[3]	[4]	[5]	[0]	[1]	[2]
[4]	[4]	[5]	[0]	[1]	[2]	[3]
[5]	[5]	[0]	[1]	[2]	[3]	[4]

Table 16.2: Multiplication in $\mathbb{Z}_6$.

·	[0]	[1]	[2]	[3]	[4]	[5]
[0]	[0]	[0]	[0]	[0]	[0]	[0]
[1]	[0]	[1]	[2]	[3]	[4]	[5]
[2]	[0]	[2]	[4]	[0]	[2]	[4]
[3]	[0]	[3]	[0]	[3]	[0]	[3]
[4]	[0]	[4]	[2]	[0]	[4]	[2]
[5]	[0]	[5]	[4]	[3]	[2]	[1]

Table 16.3: Addition in $\mathbb{Z}_7$.

+	[0]	[1]	[2]	[3]	[4]	[5]	[6]
[0]	[0]	[1]	[2]	[3]	[4]	[5]	[6]
[1]	[1]	[2]	[3]	[4]	[5]	[6]	[0]
[2]	[2]	[3]	[4]	[5]	[6]	[0]	[1]
[3]	[3]	[4]	[5]	[6]	[0]	[1]	[2]
[4]	[4]	[5]	[6]	[0]	[1]	[2]	[3]
[5]	[5]	[6]	[0]	[1]	[2]	[3]	[4]
[6]	[6]	[0]	[1]	[2]	[3]	[4]	[5]

Table 16.4: Multiplication in $\mathbb{Z}_7$.

·	[0]	[1]	[2]	[3]	[4]	[5]	[6]
[0]	[0]	[0]	[0]	[0]	[0]	[0]	[0]
[1]	[0]	[1]	[2]	[3]	[4]	[5]	[6]
[2]	[0]	[2]	[4]	[6]	[1]	[3]	[5]
[3]	[0]	[3]	[6]	[2]	[5]	[1]	[4]
[4]	[0]	[4]	[1]	[5]	[2]	[6]	[3]
[5]	[0]	[5]	[3]	[1]	[6]	[4]	[2]
[6]	[0]	[6]	[5]	[4]	[3]	[2]	[1]

In $\mathbb{Z}_7$ we have $[1]^{-1} = [1], [2]^{-1} = [4], [3]^{-1} = [5], [4]^{-1} = [2], [5]^{-1} = [3]$, and $[6]^{-1} = [6]$.

Section 1.3.1

4. $\{[0]\}, \{[0],[2],[4]\}, \{[0],[3]\}, \mathbb{Z}_6$.

Section 1.3.3

5. $\gcd(x^4 + 3x^2 + 2, x^3 - 2x^2 + x - 2) = x^2 + 1$.
   $x^2 + 1 = (x^4 + 3x^2 + 2)/6 - (x^3 - 2x^2 + x - 2)(x + 2)/6$.
7. $\mathrm{lcm}(f, g) = (x - 1)^3(x - 2)(x - 3)(x^2 + 1)(x^2 + 2)$.

Section 1.4

11. $\langle 6 \rangle$.

Section 1.5

4. $\{a+bi \mid a,b \in \mathbb{Q}\}$.

Section 1.6

5. In $\mathbb{Z}_2[x]$: $x, x+1, x^2+x+1, x^3+x^2+1, x^3+x+1$.
In $\mathbb{Z}_3[x]$: $x, x+1, x+2, x^2+1, x^2+2x+2$.

Section 1.7

1. $\ker(\phi) = \langle x-c \rangle$.
2. $\ker(\phi) = \langle x_1 - c_1, \ldots, x_n - c_n \rangle$.
7. In the table we use a representative a for the class $a + \langle x^2, xy, y^2 \rangle$. We have omitted 0 and 1.

Table 16.5: Multiplication in $\mathbb{Z}_2[x,y]/\langle x^2, xy, y^2 \rangle$.

$\cdot$	x	y	$x+1$	$y+1$	$x+y$	$x+y+1$
x	0	0	x	x	0	x
y	0	0	y	y	0	y
$x+1$	x	y	1	$x+y+1$	$x+y$	$y+1$
$y+1$	x	y	$x+y+1$	1	$x+y$	$x+1$
$x+y$	0	0	$x+y$	$x+y$	0	$x+y$
$x+y+1$	x	y	$y+1$	$x+1$	$x+y$	1

9. In the table we use a representative a for the class $a + \langle x^2+1 \rangle$ and write the elements in $\mathbb{Z}_3$ as 0,1, and 2. We have omitted 0 and 1.

Table 16.6: Multiplication in $\mathbb{Z}_3[x]/\langle x^2+1 \rangle$.

$\cdot$	2	x	$x+1$	$x+2$	$2x$	$2x+1$	$2x+2$
2	1	$2x$	$2x+2$	$2x+1$	x	$x+2$	$x+1$
x	$2x$	2	$x+2$	$2x+2$	1	$x+1$	$2x+1$
$x+1$	$2x+2$	$x+2$	$2x$	1	$2x+1$	2	x
$x+2$	$2x+1$	$2x+2$	1	x	$x+1$	$2x$	2
$2x$	x	1	$2x+1$	$x+1$	2	$2x+2$	$x+2$
$2x+1$	$x+2$	$x+1$	2	$2x$	$2x+2$	x	1
$2x+2$	$x+1$	$2x+1$	x	2	$x+2$	1	$2x$

17. $x^4 - x^3 + 2x^2 - x + 1 = (x^2 - x + 1)(x^2 + 1)$.

Section 1.9

3. A basis is $\{1, \sqrt{2}\}$. The dimension is 2.
4. A basis is $\{1 + \mathfrak{a}, x + \mathfrak{a}, y + \mathfrak{a}\}$, where $\mathfrak{a} = \langle x^2, xy, y^2\rangle$ (and $1 = [1]$). The dimension is 3.
5. A basis is $\{1, x, y, x^2, xy, y^2, x^2y, xy^2, y^3, x^2y^2, xy^3, x^2y^2\}$. The dimension is 12.

Chapter 2

1. $\mathfrak{a} + \mathfrak{b} = \langle x_1^3, x_2^2, x_2x_4, x_3^2, x_3x_4\rangle$
   $\mathfrak{a}\mathfrak{b} = \langle x_1^3x_2^2, x_1^3x_3x_4, x_2^3x_4, x_2x_3x_4^2, x_2^2x_3^2, x_3^3x_4\rangle$
   $\mathfrak{a} \cap \mathfrak{b} = \langle x_1^3x_2^2, x_1^3x_3x_4, x_2^3x_4, x_2x_3x_4, x_2^2x_3^2x_3^2x_4\rangle$
   $\mathfrak{a} : \mathfrak{b} = \langle x_1^3, x_2x_4, x_3x_4, x_3^2\rangle$
   $\sqrt{\mathfrak{a}} = \langle x_1, x_2x_4, x_3\rangle$.
2. $\mathfrak{a} + \mathfrak{b} = \langle x_1^2, x_1x_2, x_2x_3\rangle$
   $\mathfrak{a}\mathfrak{b} = \langle x_1^3x_2, x_1^2x_2x_3, x_1^2x_2^3x_1x_2^3x_3\rangle$
   $\mathfrak{a} \cap \mathfrak{b} = \langle x_1^2x_2, x_1x_2^2\rangle$
   $\mathfrak{a} : \mathfrak{b} = \langle x_1^2, x_1x_2\rangle$
   $\sqrt{\mathfrak{a}} = \langle x_1\rangle$
3. $\langle x_1, x_2\rangle \cap \langle x_4\rangle$.
4. $\langle x_1^3, x_2^2, x_4^3\rangle \cap \langle x_1^3, x_2^3, x_3^3, x_4\rangle \cap \langle x_1^3, x_3^2, x_4\rangle$.

Section 3.1.1

2. $(1, 1, \ldots, 1), (1, 1, \ldots, 1, 1 - n), (1, 1, \ldots, 2 - n, 0), \ldots, (1, -1, 0, \ldots, 0)$.
3. $x_1^a x_2^b x_3^c \succ x_1^d x_2^e x_3^f$ if $a + b > d + e$ or if $a + b = d + e$ and $a > d$ or if $a = d, b = e$ and $c > f$.
4. $1 \prec x_1 \prec x_1^2 \prec x_2 \prec x_1x_2$.

Section 3.7

1. $\{x_1^2 + x_2, x_1x_2^2 + x_1, x_2^3 + x_2\}$.

Section 3.9

2. $\{f(x_1, x_2)(x_2, -x_1)\}$.

Section 3.10

1. $\{f_1(x_3, -x_2, 0) + f_2(x_3, 0, -x_1) + f_3(0, x_2, -x_1)\}$.

Section 3.13

1. In Lex: $\{x_1^2 - x_1x_3^2, x_2^2 - x_2x_3^2\}$.
   In Deglex: $\{x_1^2 - x_1x_3^2, x_2^2 - x_2x_3^2, x_1^2x_2 - x_1x_2^2\}$.
2. In Deglex: $\{x_1^2, x_1x_3 + x_2^2, x_1x_2^2, x_2^4\}$.
   In Degrevlex: $\{x_1^2, x_1x_3 + x_2^2\}$.

Section 4.1

1. $\langle y \rangle$.
2. $\langle y - x^2 \rangle$.
5. A finite set is irreducible if and only if it consists of at most one point.

Chapter 5

7. $\langle x, y^2 \rangle \cap \langle x^2, z \rangle \cap \langle y, z^2 \rangle \cap \langle x^2, y^2, z^2 \rangle$.

Section 6.3

1. $(x, y) = (1, 1)$ or $(-1, 1)$ or $(1, -1)$.
2. $(x, y, z) = (0, 0, 0)$ or $(1, 1, 1)$ or $(-1, -1, 1)$ or $(-1, 1, -1)$ or $(1, -1, -1)$.
3. $(x, y, z) = (1, 0, 1)$.
4. There are 10 solutions.
5. There are 7 solutions.
6. $(0, 0, 0), (-6/59, 9/118, -9/236), (0, 1/6, -1/12)$.

Section 6.4

1a. The maximum is $1/432$ for $x = 1/6, y = 1/3, z = 1/2$.
1b. The smallest distance is 1 for $x = y = z = \pm\sqrt{3}/3$.

Section 7.1.1

1. $2z^4 = (2z^2 - 2yz + y^2)(xz + z^2) - yz(x^2 + xz) + (xz + z^2 - yz)(xy + y^2)$.

Section 7.1.2

1. $\langle x, y, z \rangle$.

Section 7.1.3

1. $\sigma_1^2\sigma_2 - 2\sigma_2^2 - \sigma_1\sigma_3$.

Section 7.2

1. We use a k-basis for the factor ring, but 1 is omitted from the table. We denote an element $a + \mathfrak{a}$ just by a.

Table 16.7: Multiplication table using normal forms.

$\cdot$	x	y	z	xz	yz	z^2	z^3
x	z^2	$-xz - yz$	xz	$-z^3$	$-2z^3$	z^3	0
y	$-xz - yz$	z^2	yz	$-2z^3$	z^3	z^3	0
z	xz	yz	z^2	z^3	z^3	z^3	0
xz	z^3	$-2z^3$	z^3	0	0	0	0
yz	$-2z^3$	z^3	z^3	0	0	0	0
z^2	z^3	z^3	z^3	0	0	0	0
z^3	0	0	0	0	0	0	0

Section 7.3

1. $\mathfrak{a} = \langle y_1^5 - y_2y_3, y_1^3y_2 - y_3^2, y_1^2y_3 - y_2^2, y_1y_2^3 - y_3^3, y_1y_3^4 - y_2^5, y_2^8 - y_3^7\rangle = \langle y_1^5 - y_2y_3, y_1^3y_2 - y_3^2, y_1^2y_3 - y_2^2\rangle$.

Section 7.4.1

1. $\langle x^4 + x^2, x^2y + xy^2, xy^3 + xy, y^2 + y^4\rangle$.

Section 7.4.2

1. $\langle x^3 + x, y\rangle$.

Section 7.5

4. $\mathfrak{b} \cap k[x, y] = \langle x - y, x^2 - x\rangle$ and $\mathfrak{b} \cap k[x] = \langle x^2 - x\rangle$.
6. $y^3 - 3y^2 + 4y - 1 = 0$ has (the image of) $x + 1$ as a root.
7. $k[y_1, y_2]/\langle y_1^3 - 3y_1y_2^2 - y_2^4 - y_2^2\rangle$.
8. $k[y_1, y_2, y_3, y_4]/\langle y_1^2y_3 - y_2^3, y_1y_3^2 - y_2^2y_4, y_1y_4 - y_2y_3, y_2y_4^2 - y_3^3\rangle$.

Section 8.3

3. $\{x_1x_2 - 1, x_2x_3 - x_1, x_3^2 - x_2 - 1, x_1^2 - x_3, x_2^2 - x_1x_3 + x_2\}$
5. $\{xy - z, xz - y, yz - x, z_3 - z, y^2 - z^2\}$

Section 9.1

1. $\mathfrak{V}(\langle x_1^2 - x_2 y, x_1 x_2 - y^2, x_2^2 - x_1 y\rangle)$

Chapter 10

1. $m(\mathfrak{a}) = \langle x, y^3\rangle$.
3. The tangent cone is $\mathfrak{v}(\langle x^2\rangle)$.
4a. $\mathfrak{v}(\langle xy\rangle)$.
4b. $\mathfrak{v}(\langle x(x+y-1)\rangle)$.
4c. $\mathfrak{v}(\langle y\rangle)$.
5. $\mathfrak{v}(\langle x^2 + y^2\rangle)$.
6a. $\mathfrak{v}(\langle x, y^3\rangle)$.
6b. $\mathfrak{v}(\langle x+1, y+1\rangle)$.

Section 11.2

3a. $H_R(X) = 1 + 3X + 3X^2 + 3X^3 + \cdots = (1+2X)/(1-X)$, $h_R(i) = 3$.
3b. $H_R(X) = 1 + 3X + 3X^2 + X^3$, $h_R(i) = 0$.
3c. $H_R(X) = 1+3X+3X^2+4X^3+5X^4+\cdots = (1+X-2X^2+X^3)/(1-X)^2$, $h_R(i) = i+1$.

Section 12.3

1. $\{x_i x_j - x_j x_i, i < j\}$.
2. $\left\{x_1x_2 - x_2x_1, x_1x_3 - x_3x_1, x_2x_3 - x_3x_2, x_3x_2^j x_1, j = 1, 2, \ldots\right\}$.

Chapter 17

Bibliography

17.1 Books

W. Adams-P. Loustaunau, *An introduction to Gröbner bases*, Graduate studies in mathematics 3, AMS 1994.

T. Becker-V. Wiespfenning, *Gröbner bases. A computational approach to commutative algebra*, Springer 1991.

D. Cox-J. Little-D. O'Shea, *Ideals, varieties, and algorithms. An introduction to computational algebraic geometry and commutative algebra*, Springer 1992.

D. Eisenbud, *Commutative algebra with a view toward algebraic geometry*, Springer 1995.

R. Hartshorne, *Algebraic geometry*, Springer 1977.

I. Kaplansky, *An introduction to differential algebra*, Hermann, 1957.

E. R. Kolchin, *Differential algebra and algebraic groups*, Academic Press, 1973.

O. Zariski-P. Samuel, *Commutative algebra I,II*, Springer 1975/79.

17.2 Articles

S. Abhyankar-W. Li, *On the Jacobian conjecture: A new approach via Gröbner bases*, J. Pure Appl. Algebra **61** (1989), 211-222.

J. Backelin-R. Fröberg, *How we proved that there are exactly 924 cyclic 7-roots*, in S. M. Watt ed., Proc. ISSAC'91 pp. 103–111, ACM Press 1991.

D. Bayer-M. Stillman, *A theorem on refining division orders by the reverse lexicographic order*, Duke Math. J. **55** (1987), 321–328.

D. Bayer-M. Stillman, *Computation of Hilbert functions*, J. Symb. Comp. **14** (1992), 31–50.

G. Bergman, *The diamond lemma in ring theory*, Adv. Math. **29** (1978), 178–218.

A. Bigatti-M. Caboara-L. Robbiano, *On the computation of Hilbert-Poincaré series*, Proc. AAECC'2 (1991), 21–33.

A. Bigatti-P. Conti-L. Robbiano-C. Traverso, *A "divide and conquer" algorithm for Hilbert-Poincaré series, multiplicity, and dimension of monomial ideals*, Proc. AAECC (1993), Lect. Notes in Comp. Science **673** pp. 76–88, Springer 1993.

W. Böge-R. Gebauer-H. Kredel, *Some examples for solving systems of algebraic equations by calculating Gröbner bases*, J. Symb. Comp. **2** (1986), 83–98.

B. Buchberger, *Ein Algorithmus zum Auffinden der Basiselemente des Restklassenringes nach einem nulldimensionalen Polynomideal*, Univ. of Innsbruck Math. Inst., PhD thesis 1965.

B. Buchberger, *Gröbner bases: An algorithmic method in polynomial ideal theory*, in N. K. Bose ed., Multidimensional systems theory pp. 184–232, Reidel 1985.

B. Buchberger, *A criterion for detecting unnecessary reductions in the construction of Gröbner bases*, Lect. Notes in Comp. Science **72** pp. 3–21, Springer 1979.

G. Carrá Ferro, *Gröbner bases and differential algebra*, Proc. AAECC-6, Springer Lecture Notes in Computer Science **357** (1989), 152–163.

S. R. Czapor, *A heuristic strategy for lexicographic Gröbner bases?* in S. M. Watt ed., Proc. ISSAC'91, 39–48, ACM Press 1991.

S. R. Czapor, *Solving algebraic equations: Combining Buchberger's algorithm with multivariate factorization*, J. Symb. Comp. **7** (1989), 49–55.

J. C. Faugère-P. Gianni-D. Lazard-T. Mora, *Efficient computation of zero-dimensional Gröbner bases by change of ordering*, J. Symb. Comp. **16** (1993), 329–344.

R. Gebauer-H. M. Möller, *On an installation of Buchberger's algorithm*, J. Symb. Comp. **6** (1988), 275–286.

P. Gianni, *Properties of Gröbner bases under specializations*, in J. Davenport ed., Proc. Eurocal'87, Lect. Notes in Comp. Science, **358**, pp. 293–297, Springer 1987.

P. Gianni-B. Trager-G. Zacharias, *Gröbner bases and primary decomposition of polynomial ideals*, J. Symb. Comp. **6** (1988), 149–167.

A. Giovini-T. Mora-G. Niesi-L. Robbiano-C. Traverso, *"One sugar cube, please" or selection strategies on the Buchberger algorithm*, in S. M. Watt ed., Proc. ISSAC'91, 49–54, ACM Press 1991.

H.-G. Gräbe, *The tangent cone algorithm and homogenization*, J. Pure Appl. Algebra **97** (1994), 303–312.

H. Grassmann,-G.M. Greuel-B. Martin-W. Neumann-G. Pfister -W. Pohl-H. Schönemann-T. Siebert, *Standard bases, syzygies and their implementation in SINGULAR*, Beiträge zur angewandten Analysis und Informatik, pp. 69–96, Shaker Aachen 1994.

P. Gritzmann-B. Sturmfels, *Minkowski addition of polytopes: Computational complexity and applications to Gröbner bases*, SIAM J. Discrete Math **6** (1993), 246–269.

F. Hreinsdottir, *A case where choosing a product order makes the calculations of a Groebner basis much faster*, J. Symb. Comp. **18** (1994), 373–378.

D. Kapur, *Using Gröbner bases to reason about geometry problems*, J. Symb. Comp. **2** (1986), 399–408.

D. Kapur-K. Madlener, *A completion procedure for computing a canonical basis for a k-subalgebra*, in Computers and Mathematics, Cambridge 1989, pp. 1–11, Springer 1989.

D. Lazard, *Solving zero-dimensional algebraic systems*, J. Symb. Comp. **13** (1992), 117–131.

H. M. Möller-T. Mora, *New constructive methods in classical ideal theory*, J. Algebra **100** (1986), 138–178.

T. Mora, *An algorithm to compute the equations of tangent cones*, in J. Calmet ed., Proc. EUROCAM 82, Springer Lect. Notes in Comp. Science **144** (1982), 158–165.

F. Ollivier, *Standard bases of differential ideals*, Proc. AAECC-8, Springer Lecture Notes in Computer Science **508** (1990), 304–321.

L. Robbiano, *Term orderings on the polynomial ring*, Proc. Eurocal'85, Lect. Notes in Comp. Science **204**, pp. 513–517, Springer 1985.

L. Robbiano-M. Sweedler, *Subalgebra bases*, Proc. Salvador 1988, Lect. Notes Math. **1430**, 61–87, Springer 1990.

C. Traverso, *Gröbner trace algorithms*, in P. Gianni ed., Proc. ISSAC'88, Lect. Notes in Comp. Science **358** (1988), 125–138,

W. Trinks, *Über Buchbergers Verfahren, Systeme algebraischer Gleichungen zu lösen*, J. Number Theory **10** (1978), 475–488.

V. Weispfenning, *Constructing universal Gröbner bases*, in L. Huguet and A. Poli eds. Proc. AAECC 5, Springer Lect. Notes in Comp. Science **356** (1987), 336–347.

V. Weispfenning, *Comprehensive Gröbner bases*, J. Symb. Comp. **14** (1992), 1–29.

F. Winkler, *A p-adic approach to the computation of Gröbner bases*, J. Symb. Comp. **6** (1988), 287–304.

Index

PURE AND APPLIED MATHEMATICS
A Wiley-Interscience Series of Texts, Monographs, and Tracts

Founded by RICHARD COURANT
Editor Emeritus: PETER HILTON
Editors: MYRON B. ALLEN III, DAVID A. COX, HARRY HOCHSTADT, PETER LAX, JOHN TOLAND

ADÁMEK, HERRLICH, and STRECKER—Abstract and Concrete Categories
ADAMOWICZ and ZBIERSKI—Logic of Mathematics
AKIVIS and GOLDBERG—Conformal Differential Geometry and Its Generalizations
*ARTIN—Geometric Algebra
AZIZOV and IOKHVIDOV—Linear Operators in Spaces with an Indefinite Metric
BERMAN, NEUMANN, and STERN—Nonnegative Matrices in Dynamic Systems
BOYARINTSEV—Methods of Solving Singular Systems of Ordinary Differential Equations
*CARTER—Finite Groups of Lie Type
CHATELIN—Eigenvalues of Matrices
CLARK—Mathematical Bioeconomics: The Optimal Management of Renewable Resources, Second Editon
COX—Primes of the Form $x^2 + ny^2$: Fermat, Class Field Theory, and Complex Multiplication
*CURTIS and REINER—Representation Theory of Finite Groups and Associative Algebras
*CURTIS and REINER—Methods of Representation Theory: With Applications to Finite Groups and Orders, Volume I
CURTIS and REINER—Methods of Representation Theory: With Applications to Finite Groups and Orders, Volume II
*DUNFORD and SCHWARTZ—Linear Operators
Part 1—General Theory
Part 2—Spectral Theory, Self Adjoint Operators in Hilbert Space
Part 3—Spectral Operators
FOLLAND—Real Analysis: Modern Techniques and Their Applications
FRÖBERG—An Introduction to Gröbner Bases
FRÖLICHER and KRIEGL—Linear Spaces and Differentiation Theory
GARDINER—Teichmüller Theory and Quadratic Differentials
GREENE and KRATZ—Function Theory of One Complex Variable
*GRIFFITHS and HARRIS—Principles of Algebraic Geometry
GROVE—Groups and Characters
GUSTAFSSON, KREISS and OLIGER—Time Dependent Problems and Difference Methods
HANNA and ROWLAND—Fourier Series, Transforms, and Boundary Value Problems, Second Edition
*HENRICI—Applied and Computational Complex Analysis
Volume 1, Power Series—Integration—Conformal Mapping—Location of Zeros

Volume 2, Special Functions—Integral Transforms—Asymptotics—
Continued Fractions
Volume 3, Discrete Fourier Analysis, Cauchy Integrals, Construction
of Conformal Maps, Univalent Functions
*HILTON and WU—A Course in Modern Algebra
*HOCHSTADT—Integral Equations
JOST—Two-Dimensional Geometric Variational Procedures
*KOBAYASHI and NOMIZU—Foundations of Differential Geometry, Volume I
*KOBAYASHI and NOMIZU—Foundations of Differential Geometry, Volume II
LAX—Linear Algebra
LOGAN—An Introduction to Nonlinear Partial Differential Equations
LOUIS, MAAß and RIEDER—Wavelets: Theory and Applications
McCONNELL and ROBSON—Noncommutative Noetherian Rings
NAYFEH—Perturbation Methods
NAYFEH and MOOK—Nonlinear Oscillations
PANDEY—The Hilbert Transform of Schwartz Distributions and Applications
PETKOV—Geometry of Reflecting Rays and Inverse Spectral Problems
*PRENTER—Splines and Variational Methods
RAO—Measure Theory and Integration
RASSIAS and SIMSA—Finite Sums Decompositions in Mathematical Analysis
RENELT—Elliptic Systems and Quasiconformal Mappings
RIVLIN—Chebyshev Polynomials: From Approximation Theory to Algebra and Number
Theory, Second Edition
ROCKAFELLAR—Network Flows and Monotropic Optimization
ROITMAN—Introduction to Modern Set Theory
*RUDIN—Fourier Analysis on Groups
SENDOV—The Averaged Moduli of Smoothness: Applications in Numerical Methods
and Approximations
SENDOV and POPOV—The Averaged Moduli of Smoothness
*SIEGEL—Topics in Complex Function Theory
Volume 1—Elliptic Functions and Uniformization Theory
Volume 2—Automorphic Functions and Abelian Integrals
Volume 3—Abelian Functions and Modular Functions of Several Variables
STAKGOLD—Green's Functions and Boundary Value Problems
*STOKER—Differential Geometry
*STOKER—Nonlinear Vibrations in Mechanical and Electrical Systems
*STOKER—Water Waves: The Mathematical Theory with Applications
WESSELING—An Introduction to Multigrid Methods
WHITMAN—Linear and Nonlinear Waves
ZAUDERER—Partial Differential Equations of Applied Mathematics, Second Edition

*Now available in a lower priced paperback edition in the Wiley Classics Library.

Volume 2, Special Functions—Integral Transforms—Asymptotics—Continued Fractions
Volume 3, Discrete Fourier Analysis, Cauchy Integrals, Construction of Conformal Maps, Univalent Functions

*HILTON and WU—A Course in Modern Algebra
*HOCHSTADT—Integral Equations
JOST—Two-Dimensional Geometric Variational Procedures
*KOBAYASHI and NOMIZU—Foundations of Differential Geometry, Volume I
*KOBAYASHI and NOMIZU—Foundations of Differential Geometry, Volume II
LAX—Linear Algebra
LOGAN—An Introduction to Nonlinear Partial Differential Equations
LOUIS, MAAß and RIEDER—Wavelets: Theory and Applications
McCONNELL and ROBSON—Noncommutative Noetherian Rings
NAYFEH—Perturbation Methods
NAYFEH and MOOK—Nonlinear Oscillations
PANDEY—The Hilbert Transform of Schwartz Distributions and Applications
PETKOV—Geometry of Reflecting Rays and Inverse Spectral Problems
*PRENTER—Splines and Variational Methods
RAO—Measure Theory and Integration
RASSIAS and SIMSA—Finite Sums Decompositions in Mathematical Analysis
RENELT—Elliptic Systems and Quasiconformal Mappings
RIVLIN—Chebyshev Polynomials: From Approximation Theory to Algebra and Number Theory, Second Edition
ROCKAFELLAR—Network Flows and Monotropic Optimization
ROITMAN—Introduction to Modern Set Theory
*RUDIN—Fourier Analysis on Groups
SENDOV—The Averaged Moduli of Smoothness: Applications in Numerical Methods and Approximations
SENDOV and POPOV—The Averaged Moduli of Smoothness
*SIEGEL—Topics in Complex Function Theory
Volume 1—Elliptic Functions and Uniformization Theory
Volume 2—Automorphic Functions and Abelian Integrals
Volume 3—Abelian Functions and Modular Functions of Several Variables

STAKGOLD—Green's Functions and Boundary Value Problems
*STOKER—Differential Geometry
*STOKER—Nonlinear Vibrations in Mechanical and Electrical Systems
*STOKER—Water Waves: The Mathematical Theory with Applications
WESSELING—An Introduction to Multigrid Methods
WHITMAN—Linear and Nonlinear Waves
ZAUDERER—Partial Differential Equations of Applied Mathematics, Second Edition

*Now available in a lower priced paperback edition in the Wiley Classics Library.

But Steve also recognises the importance of the local community with many clients being based locally. To deliver his global "no job too big or small" philosophy he has ensured he employs local people.

Steve concludes, "Whether it's a low level print job or a full scale strategic plan, we devise a unique, ingenious solution that's right for our client and we have the pleasure of doing it right here in Wychavon."

Business support

If you're now inspired to start your own business or need some extra support, here are some useful contacts to help to get you on your way.

Wychavon District Council

We provide support for businesses, town centres, the rural economy, developments for land and business premises and attract inward investment. Advice is available on grants and funding opportunities. We also promote tourism, festivals and events, which also supports local businesses.

www.wychavon.gov.uk or 01386 565343

Worcestershire Business Central

Worcestershire Business Central is a single point of contact for all Worcestershire business support needs: start-ups, growing businesses, developing skills, finding property, networking or finance. With a dedicated web presence, helpline and enquiry system, Worcestershire Business Central brings all available business support into one place.

www.business-central.co.uk or 0300 123 1440

Herefordshire and Worcestershire Chamber of Commerce

The Chamber of Commerce is a business membership organisation supporting growth and development of businesses across the two counties.

www.hwchamber.co.uk or 0845 641 1641

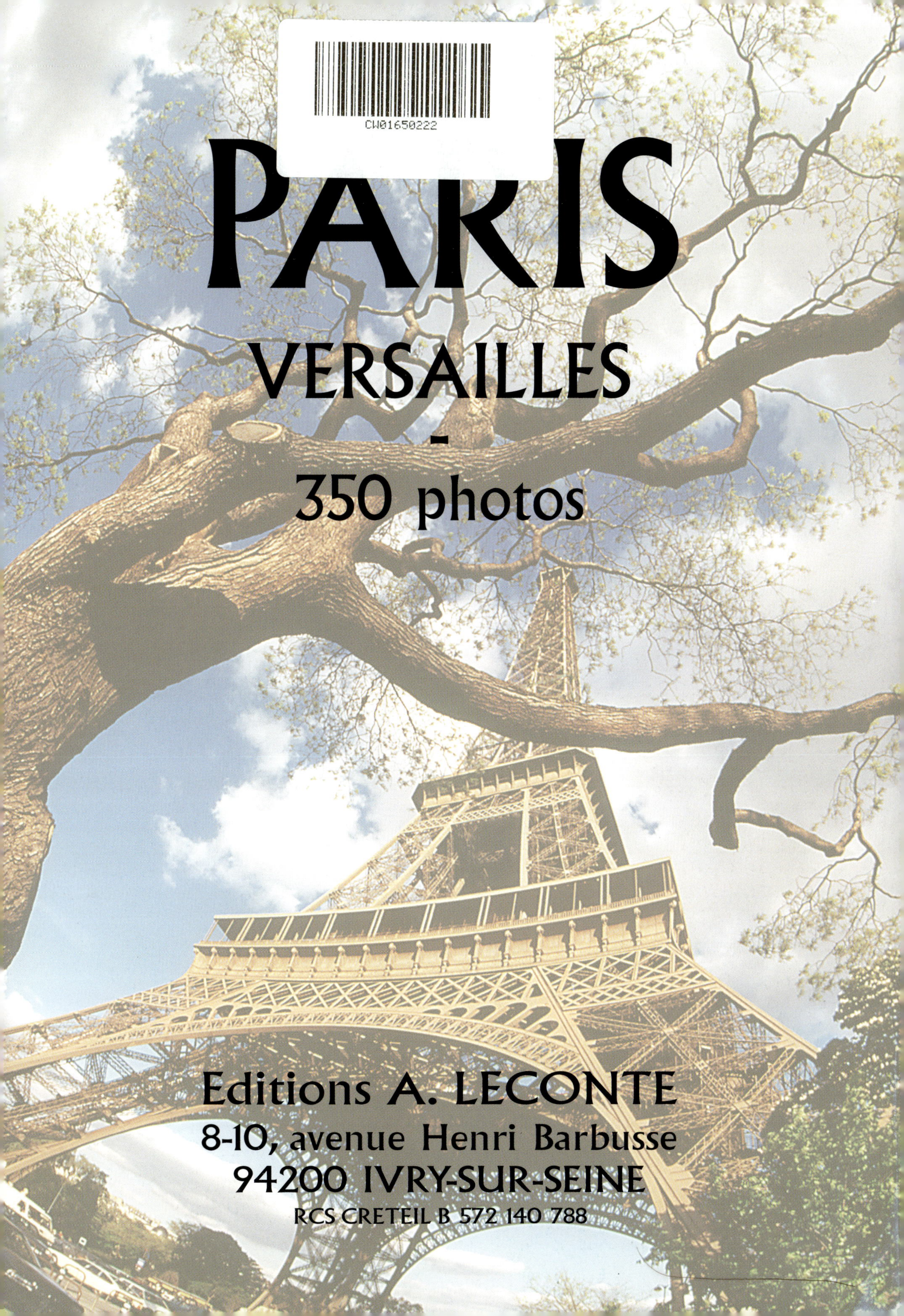

PARIS

VERSAILLES

-

350 photos

Editions A. LECONTE
8-10, avenue Henri Barbusse
94200 IVRY-SUR-SEINE
RCS CRETEIL B 572 140 788

TOURIST PARIS

Editions A. LECONTE

Registered design – Reproduction prohibited

The names of sites in red are illustrated in this book

A1 vers LILLE - BRUXELLES
Aéroports du Bourget et
Roissy-Charles de Gaulle
STADE DE FRANCE®
AUBERVILLIERS
PANTIN
PORTE DE CLIGNANCOURT
PORTE DES POISSONNIERS
PORTE DE LA CHAPELLE
PORTE D'AUBERVILLIERS
PORTE DE LA VILLETTE
CITE DES SCIENCES ET DE L'INDUSTRIE
LA GEODE
PARC DE LA VILLETTE
PORTE DE PANTIN
PORTE CHAUMONT
LE PRÉ - ST-GERVAIS
PORTE DU PRÉ ST-GERVAIS
PORTE BRUNET
LES LILAS
PORTE DES LILAS
BAGNOLET
PORTE DE BAGNOLET
vers LILLE - BRUXELLES Aéroport Roissy Ch. de Gaulle
18
19
20
9
10
2
3
4
11
5
12
13
BASILIQUE DU SACRE-CŒUR
MONTMARTRE
PLACE DU TERTRE
Pl. Saint Pierre
GARE DU NORD
GARE DE L'EST
ROTONDE DE LA VILLETTE (PLACE DE STALINGRAD)
CANAL SAINT MARTIN
PARC DES BUTTES CHAUMONT
PARC DE BELLEVILLE
PALAIS BRONGNIART
PLACE DE LA REPUBLIQUE
PALAIS ROYAL
QUARTIER DES HALLES
CENTRE G. POMPIDOU
ST-GERMAIN L'AUXERROIS
MUSEE DU LOUVRE
PLACE DU CHATELET
TOUR ST-JACQUES
ST-MERRI
PONT NEUF
CONCIERGERIE
ILE DE LA CITE
PALAIS DE JUSTICE
SAINTE CHAPELLE
HÔTEL DE VILLE
QUARTIER DU MARAIS
PLACE DES VOSGES
PLACE DE LA BASTILLE
CATHEDRALE NOTRE-DAME
FONTAINE SAINT-MICHEL
PL. ST-ANDRE DES-ARTS
MUSEE DE CLUNY
CIMETIERE DU PERE LACHAISE
PORTE DE MONTREUIL
MONTREUIL SS - BOIS
VINCENNES
PORTE DE VINCENNES
PORTE DE ST-MANDÉ
PORTE DE MONTEMPOIVRE
ST-MANDÉ
PORTE DORÉE
PARC ZOOLOGIQUE
BOIS DE VINCENNES
PORTE DE REUILLY
PORTE DE CHARENTON
PORTE DE BERCY
CHARENTON LE - PONT
DISNEYLAND® RESORT PARIS
PARC WALT DISNEY STUDIOS®
vers METZ - NANCY
PANTHEON
ARENES DE LUTECE
JARDIN DES PLANTES
QUARTIER MOUFFETARD
ST-MEDARD
GARE D'AUSTERLITZ
GARE DE LYON
PROMENADE PLANTEE
PALAIS OMNISPORT DE PARIS-BERCY
PARC DE BERCY
BIBLIOTHEQUE NATIONALE DE FRANCE
PORTE DE LA GARE
QUAI D'IVRY
PORTE DE VITRY
PORTE D'IVRY
IVRY - SUR - SEINE
PORTE DE CHOISY
PORTE D'ITALIE
PORTE DE GENTILLY
GENTILLY
A6b vers LYON - BORDEAUX (par A10) NANTES (par A11) Aéroport d'Orly
PARC MONTSOURIS
SEINE
PERIPHERIQUE
AUTOROUTE A4
Bd de Rochechouart
Bd de Belleville
Bd de Ménilmontant
Boulevard de Charonne
Rue du Faubourg Saint Antoine
Cours de Vincennes
Bd Saint Germain
Bd de Port Royal
Bd Saint Marcel
Boulevard Vincent Auriol
Place d'Italie
Place de la Nation
Place Gambetta
Rue de Rivoli
R. des Francs Bourgeois
Bd Beaumarchais
Boulevard Voltaire
Avenue de la République
Rue de Belleville
Rue de Ménilmontant
Bd Haussmann

FROM LUTETIA TO TODAY

Paris, City of Light, Paris, the most beautiful city in the world... What superlatives have not been used to describe our capital! We Parisians are also grateful to all the tourists, travellers and visitors who appreciate and admire our beautiful city. We wanted this book to reflect our own affection for Paris, hoping that through these pages the reader will come to share our feelings.

Originally a fishing village of the Parisii tribe on the largest island on the Seine, Lutetia, which rebelled against Caesar (52 BC), was attacked by Labienus who travelled down the Seine to the site of Saint-Germain l'Auxerrois, on the right bank. Camulogène the Gaul was camped on the left bank on the site of Saint-Germain-des-Prés. A stratagem gave victory to the Romans.

Under the first emperors, Lutetia was a pleasant provincial city. The Ile de la Cité devoted its temples to the gods and an imperial palace to power. The city grew up on the left bank, along the road to Orléans. Under Tiberius a new town covered the hill of Sainte-Geneviève. The centre of this district is now the Luxembourg Gardens. The arenas were on the other side of the hill, outside the walls.

From 258 the Barbarians laid waste to Lutetia. The inhabitants shut themselves into the hastily fortified Ile de la Cité. After a further invasion in 355, the Emperor Constans sent his brother-in-law Julian to be governor of northern Gaul. He settled in Lutetia, in the Cité, where his legionnaires proclaimed him an emperor in 360. One century later the Huns arrived at the walls of Lutetia. St Geneviève persuaded the inhabitants not to flee and saved the Cité.

The Merovingians built a capital, but the palace and wooden walls of these Franks vanished with their dynasty. The Carolingians moved their capitals to the east of the empire. Lutetia, now known as Paris, still in the Cité, in 885 barred the Seine to the Danes who wished to ravage the towns upstream. The defence of the Cité was commanded by Eudes, count of France. The Danes retreated and he received the title of king. He was the first Capetian, and this was the start of the fortune of Paris which was then to develop along both banks around the monasteries.

In 1183 Philippe Auguste had two great halls built; in 1186 the town was paved; in 1190 the city wall on the right bank was constructed and in 1209 that on the left bank. Building took place both inside and outside the walls and the King protected his rampart over the Seine with the huge dungeon of the Louvre...

Charles V (1364-1380) built a second enclosure on the right bank surrounding the Hôtel St-Paul, the king's new residence; the Célestins, Maison du Temple; Saint-Martin-des-Champs; Saint-Nicolas-des-Champs; Rue Montorgueil; Rue Montmartre. The wall connected the Seine and the Louvre, following a path corresponding to Rue d'Aboukir, Place des Victoires and the Palais Royal. This wall was fortified to the east by a fortress guarding the Vincennes road.

In the 16th and early 17th centuries, expansion of the city continued towards the west, so much so that Louis XIII extended the Paris of Charles V with a wall which encircled it from Porte Saint-Denis to the borders of the Tuileries.

Louis XIV, affected by the Frondes which humiliated him, installed his government at Versailles. Nonetheless he embellished his capital, transformed the Louvre, built the Hôtel des Invalides, Salpêtrière and the Observatoire, surrounding the town with a ring of triumphal gates and creating the squares of Victoires and Vendôme...

The Ile de la Cité, cradle of the capital: Notre Dame Cathedral represents zero kilometres, the hub of all the major routes radiating from Paris.

To Louis XV we owe the Concorde, the Ecole Militaire, the Hôtel de la Monnaie and the Palais de Justice, but it was in the reign of Louis XVI that Paris was completely transformed with the implementation of 3 measures: the width of new roads was set at 9.75 m, the construction of an enclosure in 1784, called 'Fermiers Généraux', which took the limits of Paris to the outer boulevards and the demolition of all the houses built on the bridges of Paris. Long views of the Seine appeared, much as we see it today.

The arenas of Lutetia (first century AD), whose shape and layout were designed to permit both theatrical performances and circus games in ancient times.

Napoleon I had a plan to make Paris "the most beautiful city possible". For him, architecture and town planning went hand in hand. He created not only bridges, embankments, market halls, abattoirs, the Arc de Triomphe de l'Etoile, the great highways and churches but also a water supply system, diverting the waters of the Ourcq to supply the city and constructing the Villette basin, aqueducts and numerous public fountains.

The outstanding development of the Restoration period was the formation of the aristocratic quarter of Faubourg Saint-Germain.

The last major changes, before 20th century Paris, were made under Napoleon III, with Baron Haussmann's considerable works, together with the anarchic industrialisation of the suburbs and faubourgs. Then there was the World Fair of 1889 and the appearance of an initially controversial, but later celebrated, monument, the Eiffel Tower... But Paris never ceases to change...

While it is true that "Paris will always be Paris", over the centuries the capital has also known how to dress itself up with some magnificent sites where kings and the men in high places found peace and quiet away from the noise and bustle of the big city. Thus we have Versailles, where the shadow of the Sun King still reigns...

... And nearby on a new and enchanting site, we have the **DISNEYLAND® RESORT PARIS** and **WALT DISNEY STUDIOS©** theme parks.

Enjoy your visit!

Paris past and present: in the Cour Napoléon, architect Ieoh Ming Pei's pyramid, with buildings constructed by Percier and Fontaine under the First Empire, then by Visconti and Lefuel under the Second Empire.

Notre Dame Cathedral, begun in 1163 on the orders of Bishop Maurice de Sully, was completed in 1330. The ravages of time and the Revolution made restoration necessary. This was carried out between 1841 and 1864 by Viollet-le-Duc, who in particular rebuilt the spire which reaches a height of 90m.

Notre Dame Cathedral, from the banks of the Seine (19th century engraving).

◀ In the Ile de la Cité: the Square Jean XXIII and Notre Dame Cathedral, the Police headquarters, the Commercial court, the Palais de Justice and the Sainte-Chapelle.

The Seine and its bridges (bottom to top, left to right): Pont de l'Archevêché, Pont Saint-Louis, Pont-au-Double, Pont d'Arcole, Petit-Pont, Pont Notre-Dame, Pont Saint-Michel, Pont-au-Change, Pont Neuf, Pont des Arts.

On the right bank: the Théâtre de la Ville and the Théâtre du Châtelet, the Louvre Palace and the Tuileries gardens.

"Charlemagne and his Liegemen" in the style of Formigé. The Emperor, on horseback, is led by Olivier and Roland. The latter's sword, Durandal, is an exact copy of the original, which is still kept in Madrid.

NOTRE DAME CATHEDRAL

The Parvis of Notre Dame graces the Cathedral whose towers rise 69 m above the ground. The façade is divided into four levels:

1 - the 3 portals, with left to right:

- The north portal, constructed first, in the second decade of the 13th century, in honour of the Virgin Mary.
- The central portal, known as the "Last Judgment", where Christ, having returned to judge the living and the dead, dominates the statues of St Michael and Satan.
- The south portal, known as St Anne's, placed under the sign of the Virgin presenting her child, framed by two thurifer angels behind whom can be seen, on the right, King Louis VII, and on the left, Maurice de Sully.

2 - the Kings' gallery, with a line of statues of the 28 kings of Judah and Israel;

3 - the rose window 9.60 m in diameter, framed by two arched recesses in which stand statutes of Adam on the left and Eve on the right ;

4 - a tracery gallery surmounted by a balustrade, a kingdom of chimeras, ghouls, demons and monsters in addition to the famous gargoyles, themselves dominated by two towers from which you can enjoy a magnificent view of the capital.

Under the square, the largest archaeological crypt in Europe narrates the history of Paris from recently conducted excavations.

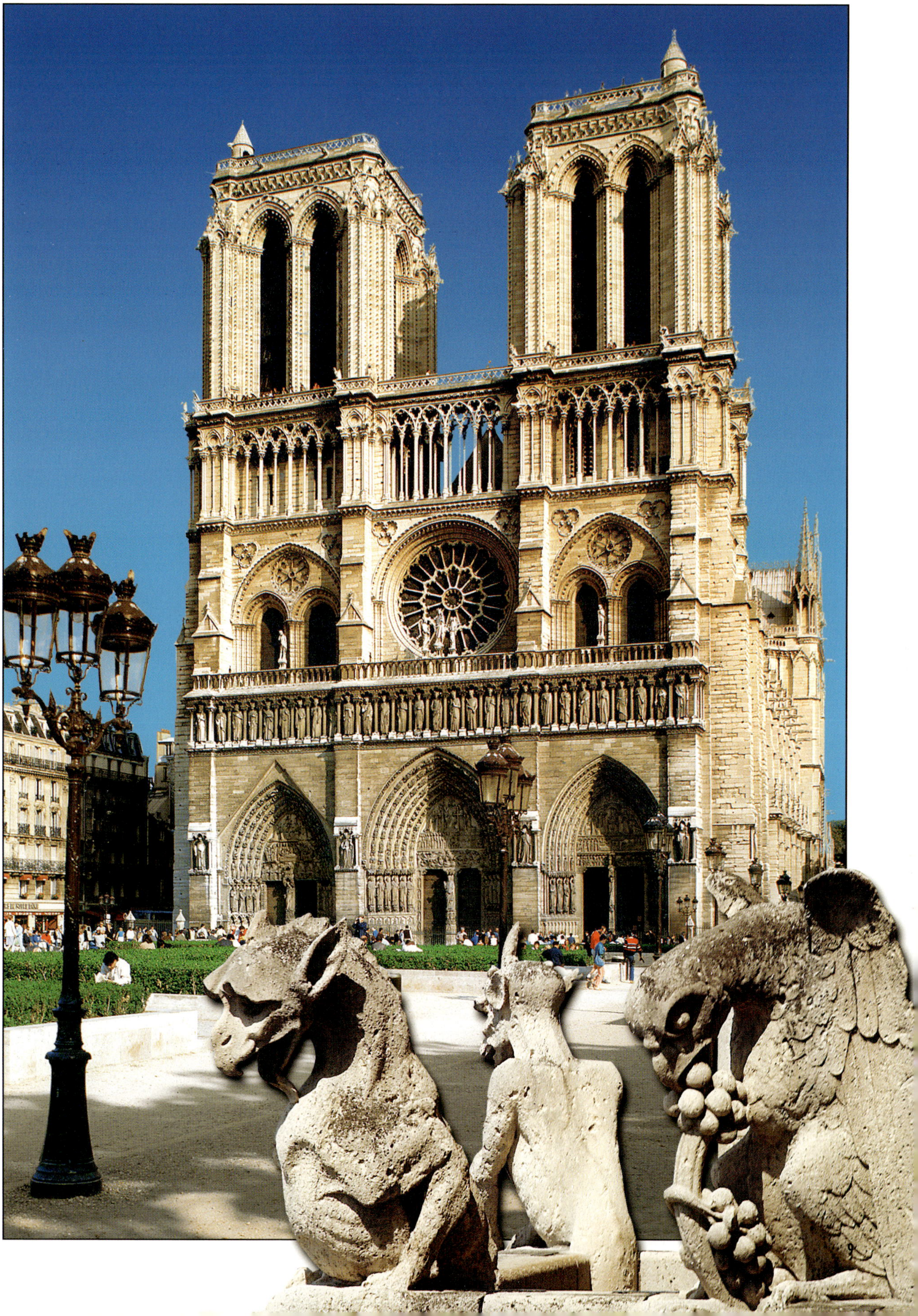

◀ The interior of the cathedral extending to a length of 127.50 m for a height of 32.50 m beneath the vaults and capable of holding -with the galleries- about 9000 of the faithful, according to the calculations of Viollet-le-Duc. In the centre of the transept, the podium is surmounted by a bronze altar consecrated in 1989, the work of Jean and Sébastien Touret.

The stained glass in the north rose window shows, in the centre, the Virgin holding the Holy Child to her bosom, surrounded by sixteen prophets. In the second circle are thirty-two kings and ancestors of Christ and, crowning the whole, thirty-two patriarchs and high priests of Israel. ▶

At the back of the Chancel, the high altar and the group of sculptures represents "the vow of Louis XIII": the king, right, and his son Louis XIV on the left, either side of the Pietà group by Nicolas Coustou (1723). A Cross of Glory has recently been added, symbolising the triumph of the Resurrection. ▼

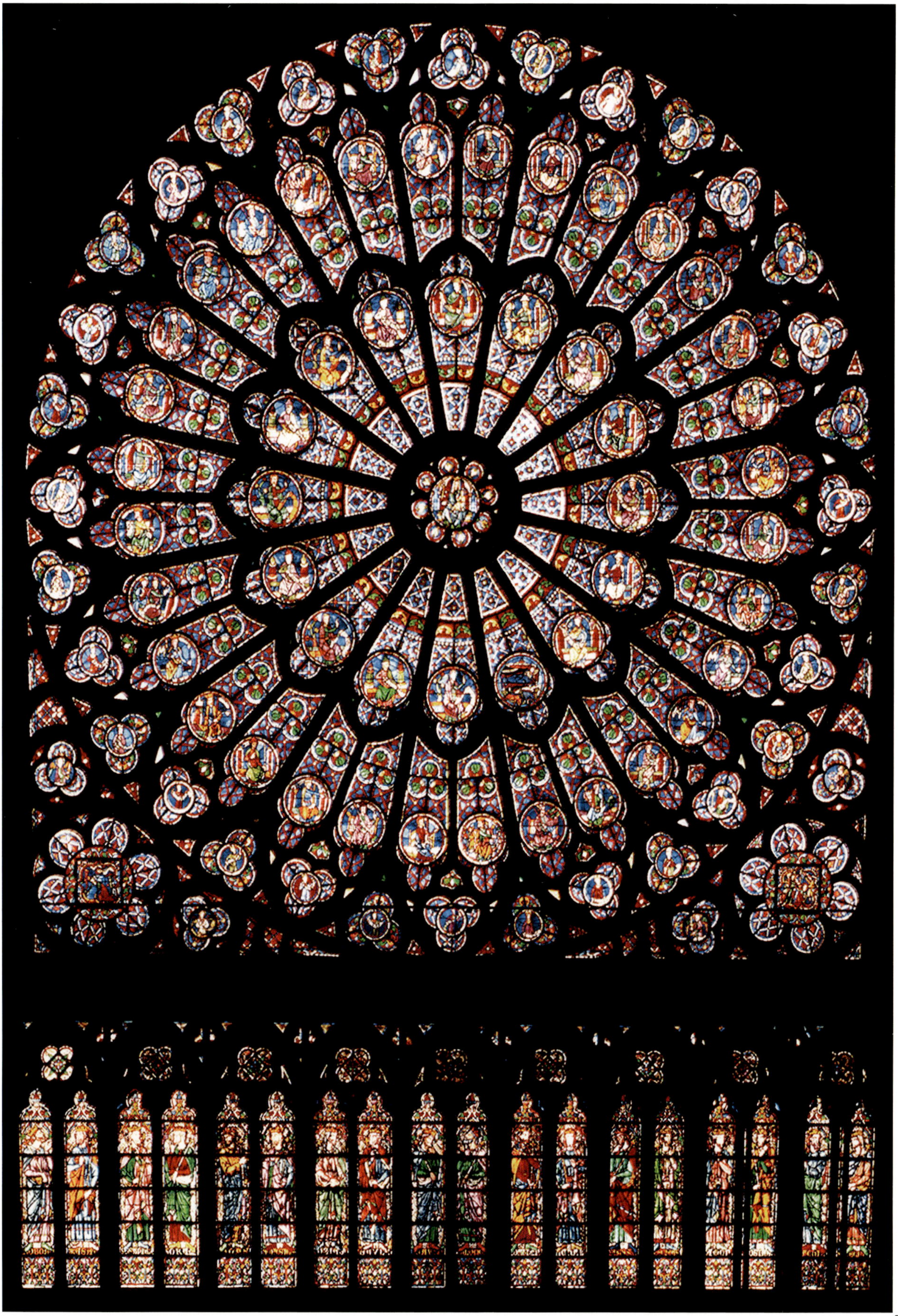

METROPOLITAIN

Place Louis Lépine: the flower market.

LAW COURTS

The entrance to the Palais de Justice, opening onto Cour de Mai [May Court] which owes its name to the fact that the clerks used to plant a tree here every year at that time. On this site were successively built the residence of the Roman governors of Lutetia, that of the Merovingian kings, then, under the name of Palais de la Cité, that of the kings of France, until 1358. Then the kings deserted it, following riots led by Etienne Marcel, who, forcing the gates of the future Charles V, then dauphin [crown prince], forced him to don a hood in the colours of Paris, and cut the throats of his advisors before his eyes. The place then became the palace of the Parliament, supreme court of the monarchy.

Today the site is divided into four parts: for the Police - the famous Quai d'Orfèvres, for Justice, for La Conciergerie, a place of incarceration from the end of the 14th century until 1914, and for the spiritual world of the Middle Ages, symbolised by la Sainte-Chapelle.

SAINTE-CHAPELLE

La Sainte-Chapelle, a gem of Gothic art, was constructed between 1246 and 1248, on the orders of St Louis, to house the holy relics ceded to the king of France by the Emperor of the East, Baudouin II - Christ's crown of thorns and a fragment of the cross. The building, attributed to Pierre de Montreuil, 36 m long, 17 m wide and 42.50 m high, contains two chapels one above the other. The upper chapel was reserved for the royal family and the great dignitaries of the court, and the lower chapel was for the servants.
The site is equally remarkable for its stained-glass windows which tell the story of the Old Testament, the Passion and the Resurrection of Christ, and the lives of the Virgin Mary, John the Baptist and John the Evangelist.

CONCIERGERIE

So called because this part of the Palace was placed under the authority of the "concierge", a high ranking royal official. As you travel downstream a square tower can be seen, known as the "Tour de l'Horloge" [clock tower] which houses the oldest clock in Paris (1370), two twin towers flanking the former entrance -Tour de César and Tour d'Argent, the latter containing the royal treasury - and the Bonbec tower, so called because prisoners were tortured there. Under the Revolution, 4174 prisoners were detained here, including Queen Marie-Antoinette from 2 August to 16 October 1793. Within, of particular interest are: the room known as the Chapelle des Girondins, the Cour des Femmes [Women's courtyard] and the courtyard known as Cour des Douze [Courtyard of the Twelve] where those condemned to death in 1793 and 1794 were assembled in groups of that number.

In the Ile de la Cité, the most famous and oldest bridge in Paris: the Pont Neuf.
The first stone was laid on 31 May 1578 by Henri III. Henri IV inaugurated it on 20 June 1603. Under Louis XIII it acquired a curious reputation:

"You, meeting place of charlatans,
Of rogues and 'passe-volans'
[soldiers in name only who attended parades
to justify excessive army accounts]
Pont Neuf, commonplace theatre
Of sellers of ointments and patches ;
Resting place for tooth pullers,
Secondhand clothes sellers, booksellers, wordsmiths,
Singers of new songs,
Pimps of young maidens
Cutpurses, the foul-mouthed,
Masters of all evil trades..."

(Les filouteries de Pont Neuf)
[The swindles of Pont Neuf]

Dominating the square du Vert-Galant, the equestrian statue of Henri IV, erected in 1818. It replaced its predecessor, destroyed on 14 August 1792, which had been put there in 1614, commissioned by Marie de Médicis.

An example of "Art nouveau" of the early 20th century, La Samaritaine was built between 1926 and 1928 to the plans of the architect Henri Sauvage, commissioned by Ernest Cognacq, nicknamed the "Napoléon du déballage" [the Napoleon of display]. Today it is still one of the capital's largest department stores.

The church of Saint-Germain l'Auxerrois, on the right, constructed in the 12th to 16th centuries and restored in the 19th. The belfry, in the centre, includes a 38-bell campanile, including the bell "Marie", which rang Saint-Bartholomew in the night of 23 to 24 August 1572. On the left is the town hall of the 1st arrondissement (1859).

TOUR SAINT JACQUES

Tour Saint-Jacques - height 52 m - is a remnant of the former church of Saint-Jacques-la-Boucherie (1508-1522) destroyed in 1797, during the Revolution. Its name was associated with the corporation of butchers, animal skinners and tanners, then later with Nicolas Flamel who was one of its main benefactors.

Nearby is Place du Châtelet, which takes its name from the old fortress called Grand Châtelet, a sinister prison which was razed to the ground shortly after the Revolution. The present fountain, known as "du Palmier" occupies the centre of this square. Begun in 1806, the column is decorated with palm leaves between which are inscribed the names of Napoleon Bonaparte's campaigns in Egypt and Italy. A winged statue of Victory, the work of Bosio, tops the column.

PALAIS ROYAL

It was Cardinal Richelieu who had the idea of this palace, in 1629. He entrusted the plans for it to his architect, Jacques Le Mercier, and the gardens to Degots. On his death, he left the house to King Louis XIII, which henceforth became the Palais Royal. Louis XIV inherited it, granted it to his brother Gaston d'Orléans, whose descendants retained it until 1848. In 1786 the Duke of Chartres had 60 pavilions added, the rents for which enabled him to pay his debts. The site became the theatre for various entertainments: "There was nothing there but some booksellers, poets, political activists and wordsmiths, some fashion retailers and lastly ladies of pleasure who came only in the evenings..." (Balzac - Lost Illusions).

Today the area is a haven of tranquillity. The buildings house the Council of State, the Constitutional Council, the Comédie Française and the Ministry of Culture.

The name Carrousel was given to this square in which Louis XIV gave a grand fête in 1662, in honour of Mademoiselle de La Vallière. Between 1806 and 1808, Percier and Fontaine built the present triumphal arch here, an imitation of that of Septimius Severus in Rome, to the glory of Napoleon I's victories in Italy.

LOUVRE

At first a royal château constructed by Philippe Auguste in 1204, transformed by Charles V from 1364 to 1380, it was completely rebuilt between the time of François I and the 19th century. The differing styles of four centuries have thus left their mark.

The new Louvre was built by Percier and Fontaine under the First Empire and by Visconti and Lefuel under the Second. It is here, in the cour Napoléon [Napoleon courtyard], that the Pyramid now stands, work of the architect Ieoh Ming Pei (Public section of the Grand Louvre).

Finally, the old Louvre, with its famous Cour Carrée [Square Courtyard], brings together some remarkable architectural features: Pierre Lescot's façade, Lemercier's 'pavillon de l'Horloge' [the Clock Pavilion] and the colonnade on the east façade by Claude Perrault.

This mecca of the arts, today houses one of the largest museum collections in the world.

The Pyramid at the Louvre (Etablissement public du Grand Louvre – architect: Ieoh Ming Pei).

Interior of the museum reception area (Etablissement public du Grand Louvre – architect: Ieoh Ming Pei).

HELLENIC GREEK ART
(late 3rd century – 1st century BC):
"Venus de Milo".

Jean-Antoine WATTEAU (1684–1721):
"Gilles".

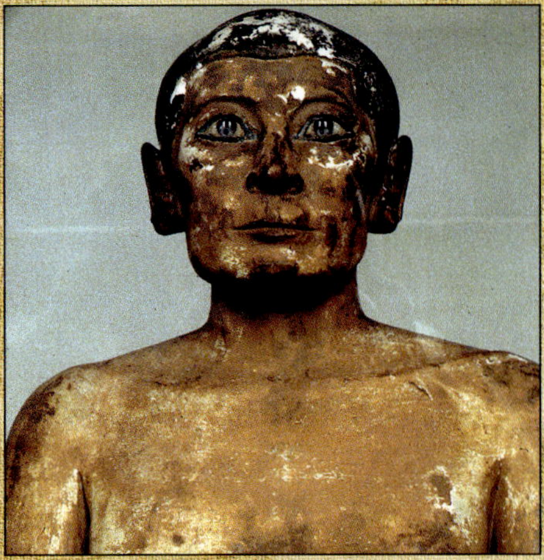

EGYPTIAN ART
(c. 2500 BC – 5th dynasty):
"The Seated Scribe".

Giovanni Battista TIEPOLO (1696–1770): "Jesus Curing the Paralytic".

Jean FOUQUET (1415/1420–1478/1481): "Charles VII".

LEONARDO da Vinci (1452–1519): "Mona Lisa, la Gioconda".

Jean Honoré FRAGONARD (1732–1806): "The Bolt".

ROMAN ANTIQUITIES:
"Statue of Emperor Augustus".

Sandro FILIPEPI, known as BOTTICELLI (1445–1510): "The Virgin and Child with Five Angels".

Francisco de GOYA Y LUCIENTES (1746–1828):
"The Countess del Carpio, Marquesa de la Solana".

Georges de LA TOUR (1593–1652): "The Cheat with the Ace of Diamonds".

Frans HALS (1580/1585–1666): "The Bohemian Girl".

Elisabeth VIGEE, Madame VIGEE-LEBRUN (1755–1842): "The Artist and her Daughter, Jeanne-Lucie".

Jean-Baptiste Camille COROT (1796–1875):
"Bridge at Mantes".

FONTAINEBLEAU SCHOOL (late 15th century):
"Gabrielle d'Estrée and the Duchess of Villars".

EGYPTIAN ANTIQUITIES (c. 1290–1224 BC):
"Seated Statue of Ramses II".

Tiziano VECELLIO, known as TITIAN (1488/1490–1576):
"The Virgin and Child, St Stephen, St Ambroise, St Maurice".

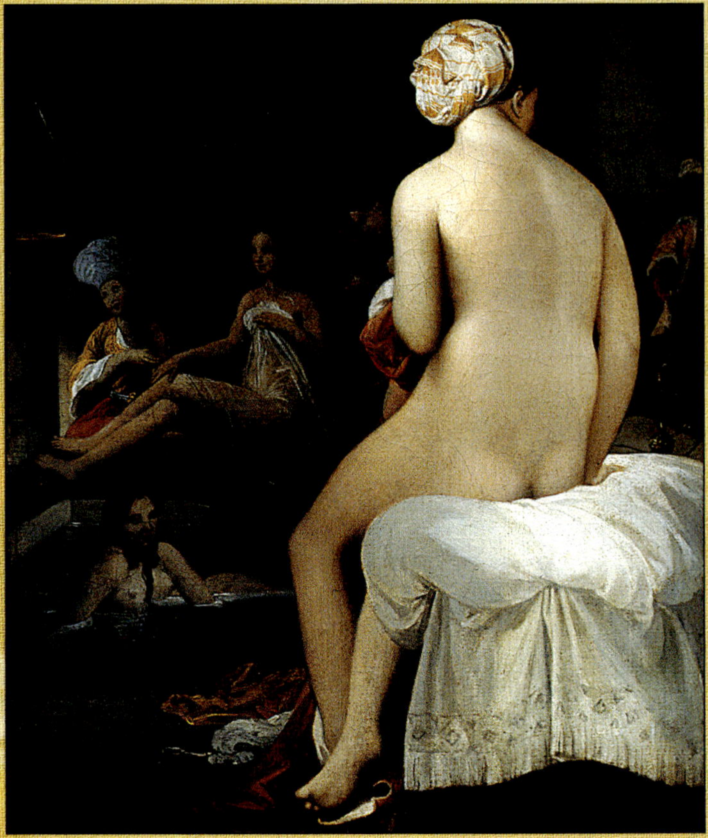

Jean Auguste Dominique INGRES (1780–1867): "The Young Bather, Harem Interior".

Eugène DELACROIX (1798–1863): "Liberty Guiding the People".

LE NAIN (...Antoine †1648, Louis †1648, Mathieu †1677): "Pilgrims at Emmaus".

Nicolas POUSSIN (1594–1665): "The Shepherds of Arcadia".

Jean CLOUET (1485 ?–1540/1541): "Portrait of Francis I".

Louis DAVID (1748–1825): "The Oath of the Horatii".

Antoine, Baron GROS (1771–1835): "Bonaparte on the Arcole Bridge".

Antonio PISANO, known as PISANELLO (1395 ?– c. 1455): "Portrait of a Princess of the House of Este".

Claude GELLEE, known as Claude LORRAIN or LE LORRAIN (1600–1682): "Seaport at Sunset".

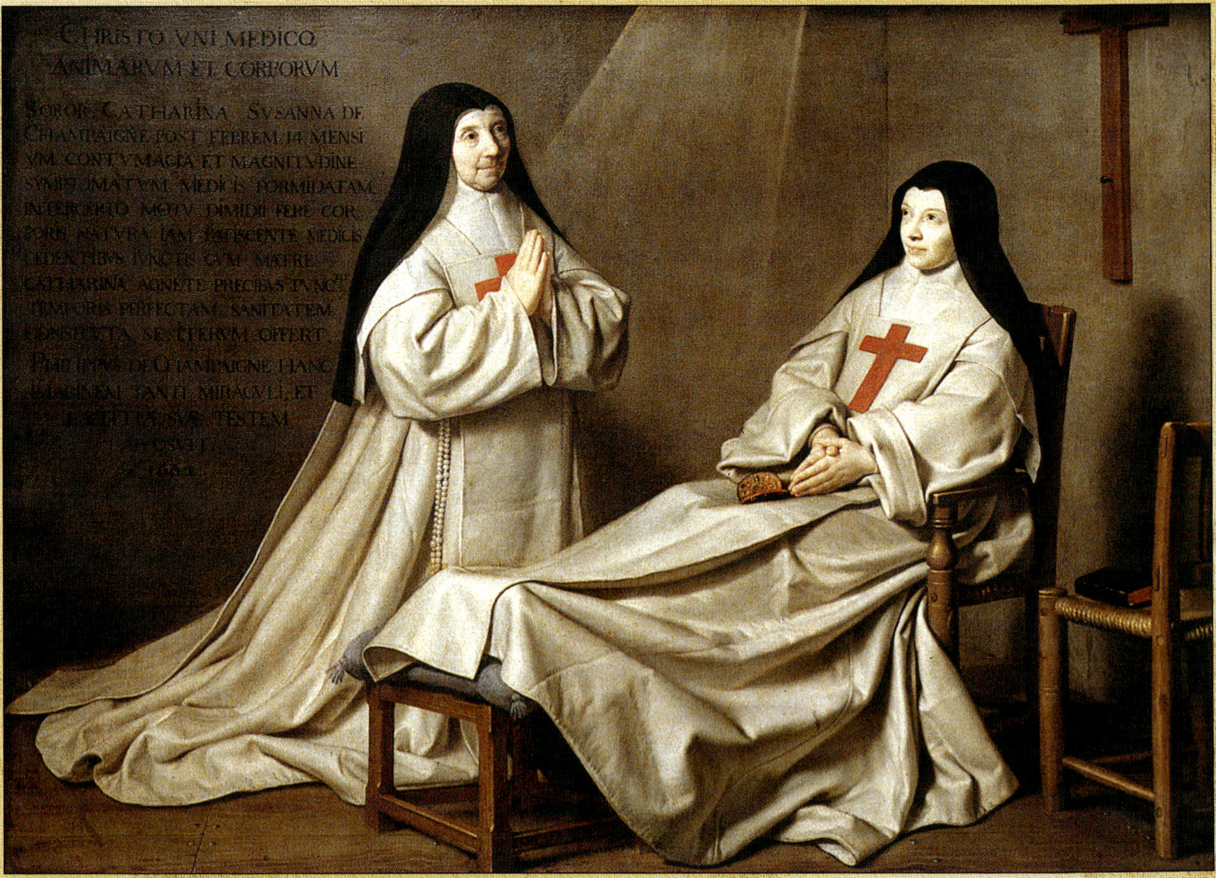

Philippe de CHAMPAIGNE (1602–1674): "The Ex-Voto of 1662".

Jean Siméon CHARDIN (1699–1779): "The Ray".

François BOUCHER (1703–1770): "Diana Leaving her Bath".

The Pyramid at the Louvre (architect: Ieoh Ming Pei – Etablissement public du Grand Louvre).

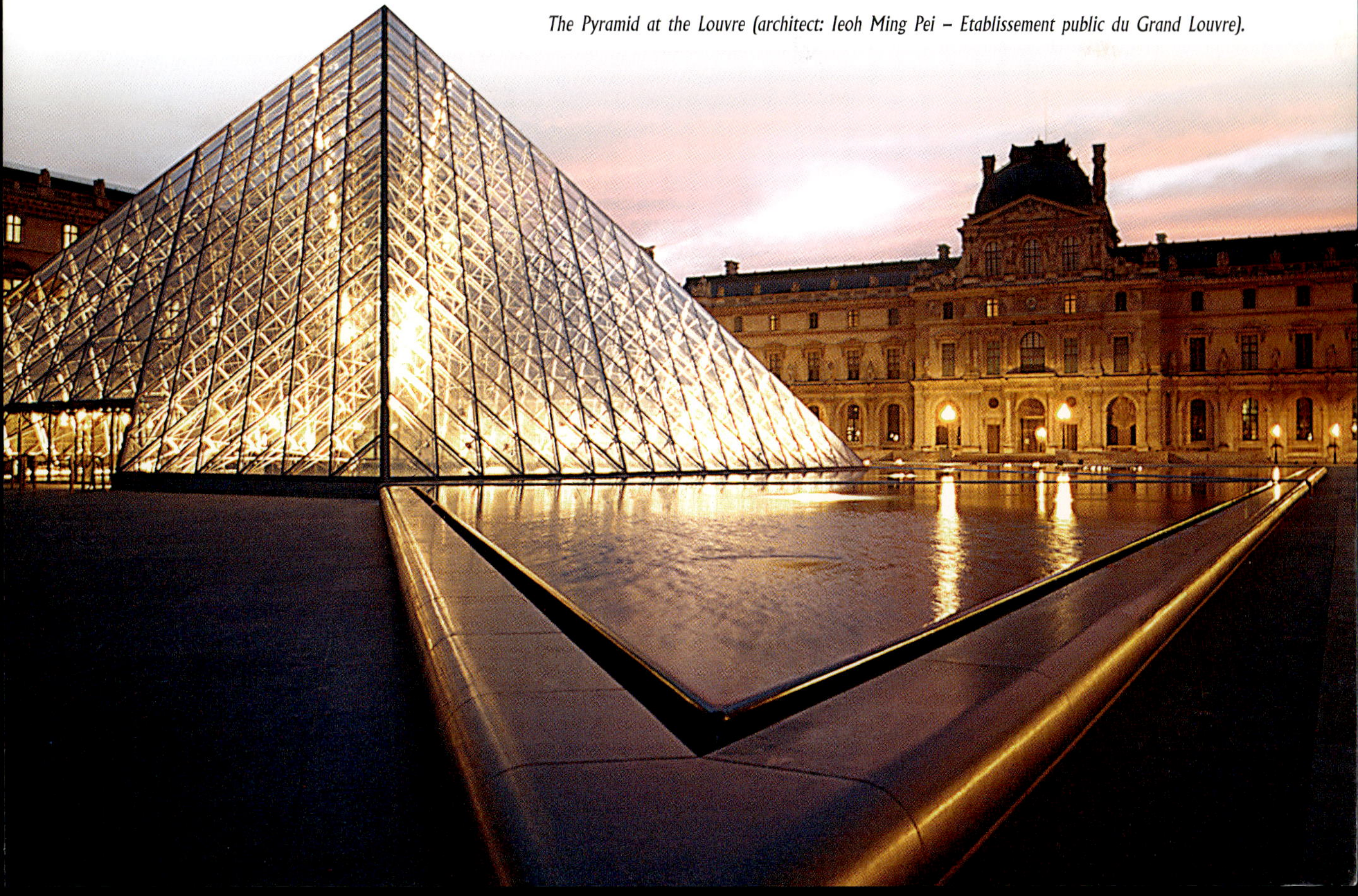

TUILERIES

LE NÔTRE

The Gardens of the Tuileries, one of the largest public parks in Paris, with its 28 hectares and 2800 trees, retains to this day the appearance given to it by Le Nôtre, the great gardener to King Louis XIV. Besides its lawns and shrubberies, the gardens contain museums - l'Orangerie and Jeu de Paume - and statues - works by Coustou, Coysevox, Carpeaux, Barrois, Barrias, Cain, Maillol and others - making it a kind of open air antechamber to the Louvre museum. Running parallel to the gardens is the rue de Rivoli which was constructed on a commission from Napoleon Bonaparte in 1804. Beneath its arcades are many souvenir shops, cafés and tea rooms for the delight of passers-by and tourists. In the centre, between the Louvre and the Tuileries is Place des Pyramides, where a statue of Joan of Arc by Frémiet is a reminder that the heroine was wounded on this site during an attack to deliver Paris in 1429.

PLACE DE LA CONCORDE

The largest square in Paris - 84,000 m^2 - was laid out between 1754 and 1763 to the plans of the architect Gabriel as the site for the equestrian statue of Louis XV. The statue was removed in 1790, and Place Louis XV renamed Place de la Révolution. On 21 January 1793, a scaffold was erected here for the execution of King Louis XVI, and in two years 1119 of the condemned were guillotined here. In 1795 the site took the name Place de la Concorde, became Place Louis XV again in 1814, then Place Louis XVI in 1826, before reverting to its present name in 1830. In 1836 the engineer Jean-Baptiste Apollinaire Lebas erected an obelisk from the temple of Luxor here, given to King Louis-Philippe in 1831 by the Viceroy of Egypt, Mohamet Ali. This pink syenite monolith, 23 m high and weighing 230 tonnes, stands in the centre of the square between two fountains. Facing the Seine, two houses designed by Gabriel can be admired from either side of the Rue Royale: Hôtel de Crillon on the left and the Ministry of Naval Affairs on the right. Framing the entrance to the Champs Elysées are Guillaume Coustou's Marly horses facing the winged horses of Mercy, Mercury and Fame by Antoine Coysevox, which stand by the gates of the Tuileries Gardens.

The fountains of Place de la Concorde are laid out to designs by Jacques-Ignace Hittorff, to whom we also owe the eighteen plinthed columns between the balustrades as well as the eight statues he was responsible for ordering and which symbolise the cities of Lille, Strasbourg, Lyons, Marseilles, Bordeaux, Nantes, Brest and Rouen, in the four corners of the square.

Bottom left,
the Rond–Point on the Champs Elysées,
designed by Le Nôtre in 1667.

Bottom right,
the Statue of Clémenceau, "The Father of Victory",
by François–Victor Cogné (1932).

Petit Palais (arch. Charles Girault), a museum housing the City of Paris art collections.

Grand Palais (arch. Deglane, Louvet and Thomas).

Between the Rond-Point and the Place de la Concorde, the Avenue Champs Elysées becomes a place to stroll, as laid down by decree in 1667. Its development was entrusted to Le Nôtre. The green area was enhanced in the 19th century by theatres - known as Panoramas at that time - restaurants, and, at the end of the century, two palaces, the Petit Palais and the Grand Palais, built for the World Fair of 1900 and today museums and art galleries. In our times the Marigny and Renaud-Barrault theatres, l'Espace Pierre-Cardin and the Ledoyen, Pavillon de l'Elysée, Pavillon Gabriel and Laurent restaurants offer hospitality to theatre-goers and enlightened gourmets.

Promenade des Champs Elysées.

Théâtre Marigny, designed by Charles Garnier in 1883.

ARC DE TRIOMPHE

Dominating "the most beautiful avenue in the world", the Arc de Triomphe de l'Etoile stands at the centre of a perspective which starts at the Louvre Palace, passes through the Arc de Triomphe du Carrousel, the Tuileries Gardens and Place de la Concorde to finish at the Grande Arche, in the La Défense district. This work, some 50 m high and 40 m wide, was constructed by a decree of Napoleon I of 18 February 1806. The architect Jean-François Chalgrin was entrusted with its completion, which was accomplished in 1836. The monument, dedicated to the glory of the imperial armies, is decorated with some famous high reliefs on the four pillars: "Departure of the volunteers of 1792" or "la Marseillaise" by François Rude, "the Triumph of 1810" by Jean-Pierre Cortot, and "Peace" and "Resistance" by Antoine Etex. The stone is also engraved with the names of the greatest victories of the Republic and the Empire together with the names of 558 generals. The ashes of Napoleon I passed beneath its vaulted arch in 1840, as did the mortal remains of Victor Hugo in 1885 and victorious soldiers on 14 July 1919. In memory of the dead of the Great War, the body of the Unknown Soldier was laid to rest here on 11 November 1920 on whose Tomb burns a flame which is rekindled each evening.

Conceived by Colbert following a decree by Louis XIV, the Tuileries walk took the name Champs Elysées [Elysian Fields] in 1694. Three centuries later major works gave the famous avenue all the majesty the minister of the Sun King wished for. A triumphal way, a place for strolling, or an obligatory route for Heads of State on official visits, the Champs Elysées is also the prestigious shop-window of Paris. The 628 buildings lining the service roads include cinemas, luxury shops, cafés, restaurants, banks, motor manufacturers and airlines, vying with each other in imagination to give the avenue its glamour and enhance the capital's reputation.

The Arc de Triomphe de l'Etoile from Avenue Foch.

La Grande Arche (architect: Johan Otto von Spreckelsen); right, the CNIT (architects: Camelot, de Mailly and Zehrfuss).

LA DEFENSE

In the 18th century King Louis XV wished to link the Louvre to the palace at Versailles and thus extend the royal route carved out by Le Nôtre. By the end of his reign, after passing through Neuilly and crossing the Seine, the road reached the knoll of Chantecoq where a great intersection was laid out. Later a statue of Napoleon in its centre was taken down in 1870 and replaced in 1883 by a bronze group by Louis-Ernest Barrias, "la Défense de Paris", commemorating Parisian resistance to the Prussian siege during the winter of 1870-1871. The statue gave its name to the site which has been developed in the communes of Puteaux and Courbevoie. Various office and residential property developments, conceived from 1950 onwards, have resulted in the townscape we see today: tower blocks, the highest of which reach some hundred metres, the CNIT building (1958), the Quatre Temps shopping centre (1981), and the Grande Arche (1989). This monument in the shape of a hollow 110 m cube, weighing 300,000 tonnes, is the work of the Danish architect Johan Otto von Spreckelsen. On the esplanade, in addition to lines of trees, the works of some sixty artists can be seen, forming an open air museum of modern art. Amongst them are Alexandre Calder's "Stabile" and Juan Miró's "Personnages". Today the area has 30,000 residents and 100,000 people come here to work each day.

Between the pillars of the Grande Arche, the "Sail" by the engineer Peter Rice.

BOIS DE BOULOGNE

Château de Longchamp.

Longchamp racecourse.

The lower lake.

Monument des Fusillés.

Château de Bagatelle.

Châlet des Iles.

PARC DE MONCEAU

Once the property of the seigneury of Clichy, in the 18th century the Mousseaux site was acquired in lots by a few lords and financiers of the period. Here they constructed buildings known as "follies". Amongst them, "Folie de Chartres", the property of Louis-Philippe of Orleans who, having bought the land in 1769, had Carmontelle construct a garden of illusions in 1778, scattered with "fabriques" ['creations'] of which some still survive today: the "Naumachie", the "Egyptian pyramid " and others. On 22 October 1797 Garnerin made the first parachute drop from a balloon floating 1000 m above the ground. The city of Paris became owners of the site in 1870 and Napoleon III later asked Alphand to remodel the 8.5 h grounds giving it the appearance it has today. The entrances, from the avenues Van Dick and Velasquez, are protected by magnificent gates, the work of Davioud. On the Boulevard Malesherbes, Avenue Velasquez side, the Cernuschi museum houses a collection of Chinese art.

QUARTIER DE L'EUROPE

House of Monsieur Loo. Fernand Block built this residence between 1926 and 1928 for a Chinese antiquarian, Ching Tsai Loo, who turned it into a gallery devoted to Asian art.

St-Augustin church: beneath its stone mantle, Victor Baltard concealed a metal structure supporting the interior of the church, built between 1860 and 1871.

Gare Saint-Lazare, terminus of the first railway line opened in 1837, later removed to its present site in 1841. The buildings, constructed by Alfred Armand and fitted out by Juste Lisch for the 1889 World Fair, are a terminus for trains travelling to and from Normandy.

Cercle National des Armées, built in 1927 by Charles Lemaresquier on the site of the former Pépinière barracks.

Chapelle Expiatoire, built between 1815 and 1826 by the architect Fontaine, commissioned by Louis XVIII, occupies the site of the former cemetery of la Madeleine de la Ville l'Evêque, in which around 3000 victims of the Revolution were buried. Among them, Louis XVI and Marie-Antoinette, whose remains were transferred to the Saint-Denis basilica in 1815.

MADELEINE

Dedicated in 1492 by King Charles VIII to St Mary-Magdalen "who, from time immemorial, has been especially honoured in the borough", over the years a succession of buildings was to see the light of day, culminating in the present church, whose construction was decided by Napoleon I in 1806. The building, whose plans were commissioned by Pierre-Alexandre Vignon, were to be consecrated to the glory of the Great Army and the architecture inspired by Greek temples. Louis XVIII continued the works in accordance with the original plans, but the temple, completed in 1842, was used for worship. Supported by 52 Corinthian columns, the building's triangular façade presents the "Last Judgment" by the sculptor Philippe-Henri Lemaire. Intersecting with Rue Royale, made famous by a certain well-known restaurant, the street of Faubourg Saint-Honoré is the home of numerous boutiques selling haute couture and internationally famous luxury goods.

Madeleine gallery (1845).

PLACE VENDOME

From an idea of the minister, Louvois, the architect Jules Hardouin-Mansart designed the square completed in 1720 to be surrounded by buildings with identical façades. At the centre of Place des Conquêtes, later renamed place Louis-le-Grand, there stood until 1792 an equestrian statue of Louis XIV, the work of Girardon. It was replaced in 1806 by a column inspired by the Trajan Column in Rome on the orders of Napoleon I. The bronze plaques illustrating the campaign of 1805 were made from melting down 1200 cannons captured at the battle of Austerlitz. A statue of the Emperor crowned the edifice which was knocked down on the orders of Gustave Courbet during the Commune in 1871. It was re-erected at the painter's expense in 1873.

OPERA

Following the attack on Napoleon III perpetrated by Orsini on 14 January 1858, the Emperor decided to construct a new opera house for Paris. Charles Garnier, winner of a competition which attracted 171 entries, began work in 1861. It was officially opened on 5 January 1875 under MacMahon's presidency. A real architectural masterpiece of the Second Empire, the Opéra is sumptuously decorative. Raised above a wide terrace with ten steps, seven fully-arched arcades flanked by sculpted groups, including "la Danse" by Jean-Baptiste Carpeaux, open into the lobbies. On the first floor, a loggia with eight double Corinthian pillars in Ravières stone opens into the foyer, echoing the small columns that frame the window openings surmounted with gilded bronze busts of composers. An attic storey dominates the loggia, edged with 53 antique masks in gilded bronze linked by a garland and sculpted by Jean-Baptiste Klagmann. Above, two groups in bronze by Charles-Alphonse Gumery, symbolising Harmony on the right and Poetry on the left. Finally, the dome in copper verdigris heightened with gold and surmounted by a composition by Aimé Millet "Apollo between Dance and Music raising the golden lyre above his head with both hands".

Main staircase.

Orchestra pit and main curtain.

Fountain beneath the small rotunda.

General view of the auditorium from the stage.

The great chandelier and the ceiling of the auditorium painted by Marc Chagall.

Inside the building the same profusion of materials, decoration and colour is to be found. This desire to achieve grandeur is successfully confirmed in the magnificent grand staircase which is divided in two in front of the orchestra entrance. The main foyer is richly decorated with columns, statues, lamps and allegorical paintings by Paul Baudry. The auditorium, despite the size of the building with a floor area of 110,000 m^2, has an audience capacity of only 2158. From the ceiling, since 1964 covered by a painting by Marc Chagall, hangs a massive lamp designed by Garnier himself. The fore-stage curtain, appearing to be a red velvet drape decorated with gold braid, is in fact a 'trompe l'oeil' painting by Auguste Alfred Rubé and Philippe Chaperon. Behind it, the Italianate stage is approximately 1200 m^2 permitting up to 450 performers on stage at any one time; the dancers' greenroom, communicating with it, is decorated with paintings by Boulanger, and the back wall is fitted with mirrors. Legend has it that an underground river, la Grange Batelière, flows beneath the Opéra.

The main foyer.

The dancers' greenroom.

The Opéra, the national academy of music and dance [Académie de Musique et de Danse], is a place of wonder and marvels, full of the passions expressed here - for art and life - in a kind of temple, where the world finds itself recreated through singing and dancing. Composers and choreographers, instrumentalists, distinguished singers and star dancers have performed here, at the height of their talent, sometimes achieving the sublime and sometimes falling to the depths of despair. Here the most spectacular plots of a Romanesque world have been hatched or unravelled, on stage, in the auditorium and the boxes, adorned with the finest worldly glories, in the corridors of this fabled palace.

In the Place de l'Opéra, at the corner of Boulevard des Capucines, the terrace of the Café de la Paix is reputed to be the largest in Paris. It has always, since the Second Empire, been an international meeting place for the literary, artistic and political world. ▼

ST-EUSTACHE CHURCH

The construction of Saint-Eustache church on the site of a chapel dedicated to St Agnes began in 1532. The building, consecrated in 1637, has the highest nave in Paris. The plan of the interior is similar to that of Notre-Dame. Remarkable acoustics permit the organ - 85 stops, 7000 pipes - to give of its best. A special feature of the church is that it contains the greatest number of famous sepulchres. Still to be seen today is Colbert's tomb, sculpted by Coysevox and Tuby, to Le Brun's designs. The former parish of the corporations of Les Halles is today dominated by Forum des Halles, a commercial, cultural and leisure complex completed in 1986.

The Fountain of the Innocents, constructed by the sculptor Jean Goujon between 1547 and 1549, commissioned, it is thought, by Philippe Auguste. It stands at the corner of Rue Saint–Denis and Rue Berger (formerly Rue aux Fers).

FORUM DES HALLES

The Forum des Halles, a vast quadrilateral in the shape of an inverted pyramid is the work of architects Claude Vasconi and Georges Penchréac'h. Spread over four levels are numerous shops, as well a large multiplex cinema.

CENTRE POMPIDOU

In December 1969 President Georges Pompidou decided to establish a museum devoted to modern art which he wanted to be accessible to the greatest possible number of people. The plans of architects Renzo Piano and Richard Rogers, assisted by G. Franchini, were selected in 1971 and have resulted in an original construction where the interior space is totally usable and adaptable, since the functional elements -"the pipes"- have all been banished to the exterior. The building consists of a 15,000 tonne metal framework, housing the collections of the modern art museum, a library, various performance and conference rooms, an industrial design centre and the I.R.C.A.M. [Institut de Recherche et Coordination Acoustique/Musique - an institute for research and development in contemporary music]. An obligatory stop for package tourists, 8 million people every year visit the Centre National d'Art et de Culture Georges Pompidou.

St-Merri church, rebuilt between 1515 and 1612, has a remarkable interior and houses the oldest bell in Paris (1331).

▼

HOTEL DE VILLE

The present Renaissance-style building, was reconstructed by Ballu and Deperthes on the model of the former building, devastated by the flames of the Paris Commune in 1871. It had been designed by "Il Boccoro" for François I, and succeeded the former "Maison aux Pilliers" [Pillared House] acquired in 1357 by Etienne Marcel to house the Municipal Assembly.

MARAIS

In the heart of the Marais, a favoured district in which numerous lords took up residence from the 15th century, is the Place des Vosges, consisting of 36 brick and freestone houses arranged around a quadrangle. Building began in 1605 on the orders of Henri IV and it was officially opened in 1612 under the name Place Royale. It took the name Place des Vosges in 1800, to honour the département of that name whose citizens were the first to pay their taxes. The house inhabited by Victor Hugo in the last century can be seen at No. 6, now converted into a museum.

Notre-Dame des Blancs Manteaux church, rebuilt in 1685 on plans by Dom Antoine de Machy, takes its name from the white cloaks worn by the mendicant order of the Serfs of the Virgin, whose monastery stood on this site. The interior contains a remarkable German pulpit (1749) made of wood inlaid with ivory and pewter marquetry.

The old houses on rue François Miron are reminiscent of the Middle Ages. Presumed to be 15th century, they were restored in 1967.

Standing behind the elm where justice was dispensed, is the classical façade (1616–1621) of the Saint-Gervais-Saint-Protais church, begun in 1494 in Flamboyant Gothic style. The organ (17th–18th century) was for many generations in the charge of the Couperin family.

Hôtel de Sens: Tristan de Salazar, Bishop of Sens, had this mansion built between 1475 and 1519. It is an example of Parisian medieval architecture, in spite of significant restorations. Since 1961, it has housed the Forney Library, specialising in decorative arts, crafts and industrial techniques (collections of posters, wallpapers, periodicals...).

Hôtel de Sully: its most illustrious owner, Maximilien de Béthune, Duc de Sully, former minister of Henri IV, bought it in 1634 and left it his name. It was originally built for the financial controller Mesme Gallet, by Jean Androuet du Cerceau, around 1625, in the spirit of the Renaissance. Abandoned in the 19th century, it was later bought by the State, which restored it in an exemplary fashion. It is now the headquarters of the Centre des Monuments Nationaux.

Hôtel de Lamoignon, built in the 16th century by Henri II for his legitimate daughter, Diane of France, Duchess of Angoulême, is named after the first president of the Parliament of Paris.
The Bibliothèque Historique de la Ville de Paris [City of Paris History Library], heir to the legacy of Antoine Moriau in 1759, has been housed here since 1968.

Hôtel de Soubise: François de Rohan, Prince of Soubise, acquired the Hôtel de Guise in 1700 and entrusted its transformation to Pierre-Alexis Delamain. This magnificent residence, whose apartments are open to visitors, has housed the Musée de l'Histoire de France since 1867.

Hôtel Hérouet: the building, badly damaged by bombing in August 1944 and since rebuilt, only faintly resembles the mansion built in the early 16th century for a king's counsellor to Parliament.

The locks museum (Musée Bricard): restored and converted by the Société Bricard in 1960, this mansion was built for his own use by Libéral Bruant, the architect of the Hôtel des Invalides, in 1685.

Hôtel Carnavalet: built in 1548 for Jacques de Ligneris, the mansion was taken over in 1578 by the widow of François de Kernevenoy, renamed Carnavalet by the people of Paris. Claude Boislève bought it in 1654 and commissioned François Mansart to embellish it. From 1677 to 1696, the Marquise de Sévigné was its illustrious tenant. Owned by the City of Paris since 1866, the capital tells its own story here through the Musée de l'Histoire de Paris.

Hôtel de Clisson: of the mansion built around 1375 by the constable Olivier de Clisson only this door remains, a unique example of 14th century civil architecture in Paris. Bought in 1553 by the Guise family, it kept that name until 1688, on the death of the last descendent, Marie.

Hôtel Aubert de Fontenay – known as Hôtel Salé: built in 1656 by Jean Boullier de Bourges for Pierre Aubert de Fontenay, the name probably derives from his job as salt-tax collector. The building now houses the Musée Picasso, which opened in 1985.

BERCY - BASTILLE

Place de la Bastille, view from the Arsenal marina.

Parc de Bercy.

The July column, topped by the "Spirit of Liberty", by Dumont, this work by Alavoine rises to a height of over 50 m. Louis-Philippe, when unveiling it in 1840, wished to commemorate the victims of the "Trois Glorieuses" (the three-day Revolution of 27-28-29 July 1830), to which were later added those of 1848, all of them buried in the foundations of the monument.

Palais Omnisport [Sports Centre] at Paris-Bercy, the facilities inside are designed for both sporting events and variety shows (architects Andrault, Guvan and Parat).

Wishing to improve water distribution in Paris and the movement of boats between the lower and upper Seine, a law was passed on 29 Floreal year X (1802) specifying the layout of the canals.

The Saint-Martin canal was started twenty years later, under the supervision of Charles-Eduoard de Villiers and the engineers Tarbé and Brémentier, and officially opened on 4 November 1825. Beginning at the Villette basin, it flows in the open for 2.431 km with a width of 27 m and draught of 1.90 m; from La Douane footbridge it runs underground for 2.123 km before again finding itself in the open in the Port de l'Arsenal - now a yacht marina - before rejoining the Seine. Five bridges, five footbridges, three swing bridges and nine locks add to the charm of its towpaths which are very popular with walkers.

Not far away, in the centre of the square of the same name, is the statue of the Republic, the work of the brothers Léopold and Charles Morice, unveiled on 14 July 1883.

BUTTES CHAUMONT

On the site of some disused gypsum quarries which had been worked since the 13th century, in 1863 Baron Haussmann decided to create a park, the plans for which he entrusted to the engineer Alphand. This pleasure garden, backing on to the "Mont Chauve" mound - today Buttes Chaumont - has a lake at its centre, in the middle of which is a promontory surmounted by a Greek monopteral temple, the work of Davioud. Two bridges link it to the shore, one of brick 22 m above the water and the other a suspension bridge 65 m long. A grotto was created at the entrance to one of the old quarries, decorated with mock stalactites and a waterfall 30 m high. A café-restaurant and some snack-bars have been built in the park in which many species of trees planted by Barillet-Deschamps in 1866 still stand.

La Villette rotunda, the remains of 54 tax offices built by Claude-Nicolas Ledoux in the late 18th century.

© CSI / A. LEGRAIN

THE CITE DES SCIENCES ET DE L'INDUSTRIE

The Cité des sciences et de l'industrie, Paris's museum of science and industry, is an immense parallelepiped standing to the north of the Ourq canal in La Villette park, and at 100,000 m^2 Europe's biggest institution popularising science and technology. Constructed from former abattoir buildings, the main structure of la Cité des sciences was designed by the architect Adrien Fainsilber, together with the glittering Geode sphere facing it. A genuine resource for

© CSI / A. LEGRAIN

© CSI / A. LEGRAIN

© CSI / M. LAMOUREUX

© CSI / A. LEGRAIN

learning about the great scientific and technological advances of our day, when the Cité des sciences was officially opened in 1986 it inaugurated a new concept for science museums. Through its many activities it transports the visitor to a world combining knowledge and the joys of discovery. This world is made up of many different places - the permanent and temporary exhibitions at the Explora museum and its planetarium; the Cité des enfants [Children's City] for 3 to 12 year-olds; Cyber-base, a space offering introductory and advanced courses in web surfing as well as information and communication technologies; the media library, open to everyone, offers books, magazines, educational software, CD Roms...; the Cité des métiers, the Cité de la santé. In the park, there are the Geode and its spectacular shows, the Cinaxe and its simulation cabin and the undersea Argonaute.

SACRE-COEUR BASILICA

In the 17th century, Sister Marguerite-Marie, a humble nun, was chosen by Our Lord to ask France to erect a monument "to the glory of his Divine Heart". There was a wait of two centuries before two devout Christians, Alexandre-Félix Legentil and his brother-in-law Hubert Rohault de Fleury, vowed, in December 1870, to build a church dedicated to the Sacred Heart, in answer to the wish expressed by Christ.

With a large following of the faithful, the approval of Monseignor Guilbert, Archbishop of Paris and the blessing of Pope Pius IX, l'Oeuvre du Voeu National [Works of the National Vow] was adopted on 25 July 1873 by the National Assembly. The cost of construction, which rose to more than forty million francs, was to be paid by the faithful through an initial subscription allowing each subscriber to buy one stone, including some which, depending on how much had been paid, gave the right for it to be inscribed with the donor's initials. The plans of Paul Abadie were selected and the first stone was laid on 18 June 1875 in the presence of Marshal MacMahon, President of the Republic. The basilica was to be officially opened on 28 June 1889, but was not consecrated until 16 October 1919. Meanwhile, Honoré Daumet, Charles Laisné, Henri Rauline, Lucien Magne and Louis-Jean Hulot supervised the realisation of the initial plans, interpreting and sometimes radically altering them. The building, 85 m long, 35 m wide and 83.33 m high, was constructed in Château-Landon stone, which on contact with rainwater secretes a white substance, lime, and gives the building its unsullied appearance. Inspired by the Roman-Byzantine style, it comprises a vast central rotunda, around which a short nave, two transepts and a chancel are laid out to form the arms of a cross. On the façade, constructed in the image of the Romanesque churches of the Charentes, and above the pediment, Gustave Michel's statue of the Sacred Heart can be seen. On either side of the porch are the equestrian statues of St Joan of Arc and St Louis, cast by Hippolyte Lefebvre.

Interior of the Sacré–Coeur basilica: top left, the statue of Sacré–Coeur d'Argent, by Eugène Benet; the solid silver statue of the Holy Virgin holding her son, made by Paul Brunet in 1876 (top right); below, the great organ of Aristide Cavaillé–Coll, constructed in 1898: 78 stops – 4 keyboards of which three are enclosed.

Opposite: the mosaic laid to the drawings of Marcel Magne and Luc–Olivier Merson. Covering an area of 475 m², it represents the Sacred Heart of Jesus glorified by the Catholic Church and France.

JESV GALLIA POENITENS ET

PLACE DU TERTRE

From the 14th century, the Place du Tertre was owned by Montmartre Abbey which had its whipping post there. In the 18th century it became the heart of the Montmartre commune, which in 1790 was the site of the first town hall, at No. 3. This house was the residence of the first mayor, Félix Desportes de Blinval. The commune was then annexed to Paris in 1860, by a decree issued by Baron Haussmann. Nevertheless, the site preserved its village and bohemian appearance and attracted large numbers of artists including Corot, Delacroix, Géricault, Degas, Renoir, Cézanne, Manet, Van Gogh and Toulouse-Lautrec amongst others, who depicted the atmosphere of Montmartre in their work. In our day, the square is still popular with painters, portrait artists and caricaturists, who, with the many cafés, restaurants and cabarets, create that special Montmartre atmosphere: no-one knows Paris if he has not climbed "la butte".

A general view of the Sacré-Coeur basilica; on the left the funicular railway which climbs the 45 m incline and links Place Saint-Pierre and Rue Saint-Eleuthère (cabins designed by R.Talon - station architect: F. Deslaugiers). ▼

MONTMARTRE

Amongst the cabarets which make the name of Montmartre is the well-known "Lapin Agile" at No. 22 Rue des Saules at the corner of Rue Saint-Vincent. In 1860 it was a notorious dance hall called "Au rendez-vous des voleurs" [Thieves' meeting place] then "Cabaret des assassins" [Murderers' Cabaret] at the period when Salz became its owner. In 1886, a former cancan dancer, Adèle Decerf, acquired it and renamed the place "Ma campagne" [My Country]. Aristide Bruant bought the property in 1903 and entrusted its management to Frédéric Gérard - "Le père Frédé". Meanwhile the famous rabbit sign painted by the cartoonist André Gill had given the establishment the name by which we know it today.

MOULIN ROUGE

At the foot of the mount, "between Pigalle and Blanche", Boulevard de Clichy shamelessly flaunts its unique establishments, popular with night-time revellers. Amongst the boutiques, cafés, cinemas and cabarets, every night the Moulin Rouge marks the parade by turning its illuminated sails. Opened on 6 October 1889, the Moulin Rouge dance hall rapidly became very successful. Special attractions were also offered to its audience, the most famous of which was the "quadrille naturaliste", danced by colourful figures such as La Goulue, Jane Avril and Valentin-le-Désossé, who were to be immortalised by the pencil of Henri de Toulouse-Lautrec. After several refurbishments at the beginning of the century, the establishment came to specialise in revues: that of 1927 "Ça c'est Paris" [That's Paris], led by Mistinguett, was a triumph. In 1937, the dance hall was converted into a night club, ultra-modern for its time. Edith Piaf and Yves Montand appeared here in July 1944. The establishment was refurbished again in 1951 by Henri Mahé, and again in 1959 with the creation of the "dinner-show", finally in 1962 the dancers turned mermaids performed for the first time in a giant aquarium. Throughout its history, the Moulin Rouge dance hall has attracted glittering audiences, notably members of the British royal family, as well as leading artists from each period. Today it is still a venue popular with lovers of music-hall.

The French Cancan, the proper original version, was danced for the first time in 1889 at the Moulin Rouge. This dance, based on the quadrille, had a success that knew no borders and it was on the other side of the Channel that Englishman Charles Morton gave it the name by which it is still known today. Well over a century old, this wild dance still reveals to the audience the legs of the prettiest girls of Paris… and elsewhere!

Notre–Dame Cathedral, viewed from the Pont de l'Archevêché.

Saint–Julien Le Pauvre church.

Rue du Chat qui Pèche.

JARDIN DES PLANTES

The Jardin des Plantes, originally the "royal garden of medicinal herbs", was founded in 1626 by Héroard and Guy de la Brosse, doctors to King Louis XIII. Buffon, appointed steward in 1739, laid it out and developed the interest in the natural sciences by surrounding himself with a great many academics, such as Jussieu, Daubenton and others. By decree of the Convention, the garden later became the national natural history museum in 1793. The gardens are planted with numerous plant species - including the famous Cedar of Lebanon smuggled back by Jussieu in 1734; they also contain a menagerie and exhibition galleries: geology, mineralogy, botany, anatomy and comparative palaeontology and zoology.

Coming down from Place de la Contrescarpe, Rue Mouffetard and its lively, picturesque market on the forecourt of Saint-Médard church.

PANTHEON

Dedicated to Sainte Geneviève, the royal church designed by Jacques-Germain Soufflot, commissioned by Louis XV, was completed in 1790. One year later, the Revolution made it the last resting place of "grand hommes" [great men], the Panthéon. Napoleon yielded it to the church in 1806, but in 1830, Louis-Philippe converted the building back into a Pantheon, which became the Temple d'Humanité in 1848 and the National Basilica in 1853, in the name of St Geneviève. The Third Republic again converted it back into a Pantheon for the funeral of Victor Hugo on 1 June 1885. The building is constructed in the shape a Greek cross 110 m long by 84.50 m wide and 83 m high. The monumental portico supports the triangular pediment carved by Pierre-Jean David d'Angers and bears the inscription in gold lettering: "Aux grand hommes la Patrie reconnaissante" [To great men, their grateful country]. Inside, the dome is decorated with a fresco by Antoine-Jean Gros depicting the ascension of St Geneviève. It was here that Foucault conducted his pendulum experiments in 1851. In the necropolis lie the remains of some sixty distinguished people, among them Sadi Carnot, Emile Zola and Jean Jaurès.

LUXEMBOURG PALACE AND GARDENS

It was François of Luxembourg, Prince of Tingry, who gave his name to the building - the present Petit Luxembourg - by acquiring it in 1570. Marie de Médicis bought it in 1612, together with the surrounding land, and had a palace to suit her built by Salomon de Brosse, while conserving the original residence. The property then passed through the hands of: Gaston d'Orléans, Mademoiselle de Montpensier, the Duchess of Guise, Madame de Maintenon, Philippe d'Orléans, the Duchess of Berry and the Count of Provence, who all lived there. Bonaparte granted the building to the Senate in 1800, and it became the Chambre des Pairs [equivalent of the House of Lords] under the Restoration. From the Third Republic, the Luxembourg again housed the Senate. The French classical-style garden was commissioned by Marie de Médicis. It is adorned with fountains and some fifty statues, those of the terrace representing the queens of France and illustrious ladies.

HOTEL DE CLUNY

In the heart of the Latin Quarter, Cluny is evidence of the origins of the capital: the Gallo-Roman baths, constructed in the late 2nd and early 3rd centuries, to which the Hôtel de Cluny is attached, a rare example of civic architecture of the late Middle Ages. Today the entire building is a national museum, established by an Act passed on 24 July 1843. The baths proper occupy a quadrangle some 100 m long by 65 m wide, and are built to the traditional Roman bathing sequence: a room for exercising before bathing communicating with the central frigidarium - this room, 14.50 m in height, is remarkable both for its exceptional state of preservation and for the presence of consoles with the motif of the prow of a merchant ship, a symbol attributed to the corporation of Nautes - this in turn opens on to the tepidarium (for warm individual baths) and the caldarium for hot baths. The Hôtel de Cluny, constructed by Jacques d'Amboise, elder brother of the Louis XII's cardinal and minister and Abbot of Cluny from 1485 to 1510, today houses a collection of medieval art.

SAINT-MICHEL

Under the Second Empire, Baron Haussmann, wishing to orient the Boulevard Saint-Michel into the perspective of the view from Sainte-Chapelle, had the north-south road diverted from its axis. As the view was then obstructed by the gable wall of the buildings at the corner of Rue Danton, he had the idea of covering this with a decorative fountain. The monument, the work of the sculptor Gabriel Davioud, was unveiled on 5 August 1860. Measuring 26 m high and 15 m wide, the fountain is formed from four large bowls superimposed in steps, flanked by two crouching winged dragons. In the central recess, surrounded by four Corinthian columns of Languedoc red marble, stands the bronze statue of "St Michael defeating the dragon" by Francisque-Joseph Duret. Above, the polychromatic marble attic is ornamented with four statues symbolising "Prudence", "Strength", "Justice" and "Temperance". At the apex, the pediment is broken by an escutcheon embellished with a helmet and a sceptre, flanked by Auguste Debay's allegorical "Power" and "Moderation".

On 15 Prairial in the year XIII, by a decision of Minister Champagny, the site of the former church of Saint-André-des-Arts became a square of the same name. The church itself, built in the 13th century and commissioned by the Abbot of Saint-Germain, was remarkable both architecturally and for its sculptures. Having become the property of the state, the church was sold on 4 Fructidor in the year V (21 August 1797) and demolished a short time later.

SAINT-GERMAIN DES PRES

According to legend, King Childebert I, on the advice of Germain, Bishop of Paris, in 542 had a basilica constructed to house the treasures he had brought back from Saragossa: St Vincent's tunic, a gold Toledo cross and many precious objects. The Merovingian kings were buried here, as was St Germain in 576. After being pillaged by the Normans in 845, the basilica, which had been endowed with a monastery, was demolished in the 9th century. On its foundations Abbot Morard built a Romanesque church, of which the steeple porch constructed between 990 and 1014, still retains a considerable proportion of its original features. Inside, are to be found the tombstones of Descartes, Montfaucon and Mabillon.

Built in 1699, Rue de Furstenberg takes its name from the Cardinal, Abbot of Saint-Germain-des-Prés.

SAINT-SULPICE

The first church dedicated to St Sulpice was built in about 1211 on the site of the present church. When it became too small, Anne of Austria decided in 1646 to have a larger building constructed to the plans of Christophe Gamart, which was to be enlarged again by Le Vau in 1649. The building was not officially opened until 1745. The façade, which we owe to J.N. Servandoni, is flanked by two towers, the work of Chalgrin, one 73 m high and the other 68 m, which remains unfinished. Claude Chappe installed two telegraphs here, which were removed in 1850. The interior has a white marble obelisk and a copper-plated sundial permitting calculation of the dates of the March equinox and Easter Sunday.

Pont des Arts

L'Institut de France and Pont des Arts.

MUSEE D'ORSAY

The Orleans railway company, wishing to take its lines further into Paris, acquired the site occupied by the Orsay Palace, destroyed during the Paris Commune in 1871. The station and its hotel, built by Victor Laloux, were officially opened on 14 July 1900. A ministerial committee in 1977 converted the buildings into a museum dedicated to the art of the second half of the 19th century. Opened in 1986, the Musée d'Orsay - exterior designed by Renaud Bardon, Pierre Colboc and Jean-Paul Philippon, and interior by Gae Aulenti - houses collections of paintings, sculpture, architecture and town planning, from 1848 to 1914.

Auguste RENOIR (1841–1919): "The Reading".

Edouard MANET (1832–1883): "On the Beach".

Vincent VAN GOGH (1853–1890): "Portrait of Doctor Gachet".

Gustave CAILLEBOTTE (1848–1894): "Rooftops in the snow".

Camille PISSARRO (1830–1903): "The Village Steeple".

Alfred SISLEY (1839–1899): "The Canal du Loing".

Henri de TOULOUSE-LAUTREC (1864–1901): "Woman Combing Her Hair".

Henri de TOULOUSE-LAUTREC (1864–1901): "Jane Avril Dancing".

Paul CEZANNE (1839–1906): "The Card Players".

Edgar DEGAS (1834–1917): "Women Ironing".

Edouard MANET (1832–1883): "Woman with Fans Nina de Callias".

Edgar DEGAS (1834–1917): "Dancing Class".

Jean–Baptiste Camille COROT (1796–1875): "Ville d'Avray, the Cabassud Houses".

Vincent VAN GOGH (1853–1890): "Church at Auvers–sur–Oise".

Camille PISSARRO (1830–1903): "Woman Hanging the Washing".

PALAIS BOURBON

Louise-Françoise de Bourbon, legitimate daughter of Louis XIV and Madame de Montespan, had the palace which bears her name constructed between 1722 and 1728. Louis XV was its owner in 1756, the Prince de Condé in 1764. In the meantime, the building underwent a number of alterations, until 1791 when the Bourbon Palace became the property of the nation. The Council of the Five Hundred held its first session there on 21 January 1798, as the legislative body of the Empire subsequently did. Since 1814 the Palace, now the Assemblée Nationale [French National Assembly] has housed the Chambre des Députés [Chamber of Deputies - the lower house]. The main façade, the work of Poyet, looks over Place de la Concorde, and is reflected at the end of Rue Royale in the church of the Madeleine. The tympanum, sculpted in 1842 by J.P. Cortot, represents France surrounded by Liberty and Order.

Constructed between 1897 and 1900, Pont Alexandre III, the work of the engineer Jean Résal, was to symbolise the Franco-Russian friendship instigated by Tsar Alexander III and President Sadi Carnot. The first stone was laid on 7 October 1896 in the presence of Tsar Nicholas II, son of Alexander III and the President of the Republic, Félix Faure. Richly decorated, the 40 m wide bridge supports on the keystone of the vault of the arch, two embossed copper fronted motifs by the sculptor Georges Récipon: the arms of Paris surrounded by Seine nymphs upstream, the arms of Russia and the figure of la Néva surrounded by his nymphs downstream. At each end, two square pylons are surmounted by gilded bronze equestrian groups symbolising Pegasus held by Fame. The work, lit by two rows of fourteen three-lamp candelabras, is the finishing touch to the view of the Hôtel des Invalides.

EIFFEL

INVALIDES

Louis XIV, moved by seeing his glorious soldiers condemned until then to beg for their bread and to stretch out their mutilated bodies on Pont Neuf, decided in 1670 to found the Hôtel des Invalides. This was built between 1671 to 1676 to the plans of the architect Libéral Bruant, and the work was supervised by Louvois, Minister of War. The main façade, 196 m long, extends from both sides of a huge portal, embellished with an equestrian statue of Louis XIV by Guillaume Coustou. Today the buildings house the Musée de l'Armée, the Musée de l'Artillerie, the Musée des Plans-Reliefs [the Army, Artillery and Relief Map museums] and the military authorities of Paris. The soldiers' church, dedicated to St Louis, was deemed too small by the Sun King and was enlarged by Jules Hardouin-Mansart to whom we owe the current church of the Dôme, completed in 1708. In its crypt rest the ashes of

Napoleon I, brought back from the island of St Helena by the Prince de Joinville in 1840. The sarcophagus of red porphyry from Finland, made by Visconti, contains six coffins of which the last contains the remains of the Emperor. In front is his statue in Coronation dress, beneath a slab is the body of his son, the King of Rome, brought back to France in 1940 on the orders of Adolf Hitler.

"I want my ashes to lie
on the banks of the Seine,
amidst the French people
I have so loved".

Statue of Napoleon I, by Seurre, formerly standing on top of the column in Place Vendôme.

The bronze door.

Eglise du Dôme: the high altar.

The Emperor's tomb.

Exterior of the Eglise du Dôme.

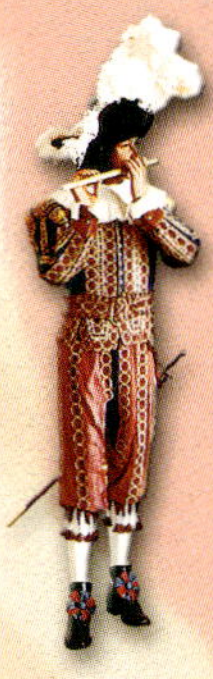

Former monarchy

First Republic

First Empire

Restoration

July Monarchy

Edouard Detaille Collection

MUSEE DE L'ARMEE

The Musée des Plans-Reliefs has on display a hundred or so models (1: 600) of fortresses and fortified villages of France.

Relief map of Mont Saint-Michel made in 1701.

Relief map of the palace of the kings of Mallorca in Perpignan made in 1701.

©M56

MONTPARNASSE TOWER

When the former Montparnasse station no longer met the needs of SNCF, it was knocked down and the Tour Montparnasse and its shopping centre built on the cleared site. The tower was built between 1969 and 1974. All of its 58 floors, including 54 floors of offices measuring nearly 2000 m^2 each represents a superstructure of 112,000 m^2, with a flat roof creating a 59th floor, open to the public. Architects: Beaudouin, Cassan, de Marien and Saubot. Height: 210 m - Weight: 120,000 tonnes - Lifts: 25 – Reinforced-concrete core: 70m deep foundations. Tour Montparnasse, one of the landmark buildings of the 20th century, stands at the centre of the capital. The top affords an extraordinary panorama on 2 floors open to the public: the 56th floor, roofed and air-conditioned, reached by the fastest lift in Europe (195 m in 38 seconds); the 59th floor, the highest and biggest roof terrace in Paris, the only point from which it is possible to view the capital through 360° and distances of up to 40 km.

ECOLE MILITAIRE
(former royal officers' training school)

Founded by edict of King Louis XV in 1751, it was intended to train 500 cadets of modest means. It was built between 1751 and 1773 by the architect Jacques-Ange Gabriel who revised his plans several times, aiming to outdo Les Invalides. After having been the summit of French horsemanship, from 1878 the building housed the military academy. In front of it, looking towards the Champ de Mars, stands the equestrian statue of Marshal Joffre, hero of the Great War.

Bust of Gustave Eiffel, by Bourdelle. Born in 1832, and a product of the Ecole Centrale, Gustave Eiffel rapidly earned a reputation as a great expert in metal construction. We owe to him notably, Pest Station in Hungary, the Maria–Pia viaduct in Porto, the Gabarit viaduct and the structural framework of the Statue of Liberty in New York. At the end of his life he conducted some in–depth experiments into meteorology and aerodynamics in his small apartment at the top of the tower.

EIFFEL TOWER

In 1884, under the Third Republic, the French government decided to hold a World Fair in 1889, for which it wished to have constructed a monument unlike anything ever seen before. Several projects arrived on the desk of the minister, Lockroy, including plans by Bourdais, architect of the Trocadéro, who proposed the construction of a "Sun Tower" one thousand feet high. Maurice Koechlin and Emile Nouguier, colleagues of Eiffel, with whom the architect Stephen Sauvestre was to be associated, had the idea of a metal tower some 300 m high. Interested in the project, Gustave Eiffel took it over and won the competition held by the government in 1886. Work began on 26 January 1887, despite the protests of many prominent people, including Guy de Maupassant, Charles Gounod, Victorien Sardou and others. The site employed hundreds of workers and it necessitated 700 drawings to make the 18,000 parts of the tower. These were assembled using 2,500,000 rivets and work progressed at an average 15 m per month. It was officially opened on 30 March 1889. The result was spectacularly light: despite weighing 10,000 tonnes, the tower only exerts the same pressure on the ground as a man seated in a chair (4 kg/cm^2)! Nevertheless, in order to reach the first floor at 57 m, 360 steps must be climbed, then 380 to reach the second at 155 m, and another 1062 to reach the top at a height of 300 m. Fortunately lifts were installed whose machinery is still functioning today. Built as part of the exhibition, the tower was to have been demolished at the end of the concession obtained by Eiffel, in 1909. It was saved by radiotelegraphy and its military applications, among others: the interception of coded messages during the Great War, which most notably led to the arrest of the spy Mata-Hari. Continuing its career on the airwaves, the tower was improved in 1957 with television equipment which took its height to 320.75 m.

105

Love goes like this running water
Love goes
Slow as life
And violent as Hope

Days pass and weeks pass
Neither time past
Nor love comes back
beneath Mirabeau bridge [where] flows the Seine

Come the night sound the hour
Days go by [yet] I remain

(Extract from a poem by Guillaume Apollinaire,
Mirabeau Bridge)

At the southern tip of Allée des Cygnes, the statue of "Liberty Lighting the World", a bronze miniature of Bartholdi's original.

Palais de Chaillot (architects: Boileau, Carlu and Azéma) built for the 1937 World Fair.

VERSAILLES

Motivated equally by his concern to distance himself from the bourgeois and the people of Paris, whose inclination to insurrection and riot had punctuated his childhood, and to keep the great nobles, indeed all the nobility, under his control, Louis XIV transformed Louis XIII's hunting lodge so that he could move his government there permanently, creating a magnificent royal walled town, a temple to absolute power, where ritualised etiquette ruled the life of the court of the world's most powerful monarch.

Entry to the château is via Place des Armes, enclosed by entrance gates surmounted with the royal coat of arms. After the forecourt with its large round paving stones, you enter the Cour des Ministres [Ministers' Courtyard], passing the equestrian statue of Louis XIV. You then reach the Cour de Marbre [Marble Courtyard], enclosed by Louis XIII's small château where the Sun King lived, in the apartments behind the three high windows of the balcony. The architects Le Vau, François d'Orbay and Jules Hardouin-Mansart extended the wings of the original building, giving it a new splendour.

A TOUTES LES GLOIRES DE LA FRANCE

From the gardens created by the landscape gardener Le Nôtre, the magnificence of the building can be fully appreciated. Before it, the Tapis Vert [Green Carpet] unfurls, with the waters of the Apollo Basin adding interest.

Portrait of Louis XIV, by Rigaud.

Grand Trianon, constructed in 1687 by Jules Hardouin-Mansart.

Galerie des Batailles [Gallery of Battles].

Galerie des Glaces [Hall of Mirrors].

The Queen's hamlet in the gardens of the Petit Trianon.

Parterres du Midi [southern section of the open terraces].

Below the Parterres du Midi are huge steps leading down to another masterpiece by Jules Hardouin-Mansart, the Orangerie. Taking advantage of its aspect, Le Nôtre planted a group of unusual trees on the steep slope. And the architect here constructed a glasshouse of absolutely classical simplicity and power.

The King's apartments. ▶

◀ *Temple de l'Amour [Temple of Love]*

Bassin de Latone. [Latona Basin] ▼

KING'S KITCHEN GARDEN

At the initiative of the Sun King, who wished to have excellent fruit and vegetables served at his table, Jean-Baptiste de la Quintinie created a garden between 1678 and 1683 to match the appetite of his sovereign. Laid out near the Pièce d'eau des Suisses, on the site of the "l'Etang puant" [stinking pond], it was designed to be "in a convenient location for the walks and the satisfaction of the King". Occupying an area of twenty-five arpents (about nine hectares), its layout has changed very little to this day: the Grand Carré, composed of sixteen vegetable patches or "carrés" arranged around a central pool, surrounded by twenty-nine orchards of fruit trees. A terrace overlooks the whole kitchen garden from where Louis XIV would have had a grandstand view of the crops and the endless dance of the forty or so gardeners employed at the time. To regale the royal palace, La Quintinie carried out trials with plants to improve the look of the produce and achieve early cropping. Adjusting exposure to the sun, using manure from the nearby stables, effectively treating any plant diseases that appeared, using covers and cloches to protect fragile plants, he obtained spectacular results, envied by the courts of Europe. On La Quintinie's death in 1688, several families of gardeners succeeded one another, and kept the tradition of using the Potager du Roi as a place of experimentation. The Le Normand family developed the cultivation of asparagus, introduced the first coffee bushes and built a Dutch greenhouse (1732) enabling Louis XV, three years later, to taste the first pineapple. The Potager du Roi became an experimental garden of the school of Versailles in 1798, under the iron rule of botanist Antoine Richard, appointed by the French National Convention. Between 1801 and 1815 a "National Nursery" was set up to select the best fruit varieties. More greenhouses were put up, enabling a banana tree to produce its first fruits in 1840.

Used by the National Agronomy Institute (1848) for applied horticulture, the kitchen garden became the National Horticulture School in 1874, with the addition, at the initiative of Auguste Hardy, of 1900 vegetable species and a rose garden with 800 varieties of roses. The school then gained an international reputation at the same time as moving into landscape art. This discipline became increasingly important and a decree of 1976 founded the Ecole Nationale Supérieure du Paysage (ENSP) concentrating on landscape gardening, while the Ecole Nationale Supérieure d'Horticulture (ENSH) left Versailles for Angers.

Alongside the Potager, the Parc Balbi (currently being restored), designed by the architect Chalgrin (1785) on the Anglo-Chinese model, allowed the Comte de Provence to enjoy the country air, sheltered by a pavilion which was demolished following the Revolution. The site, like the Potager du Roi, has been classified a historic monument since 1926.

A
LA QUINTINYE

STADE DE FRANCE®

In 1988, France ran as a candidate to host the last Football World Cup of the 20th century in 1998. To qualify it had to equip itself with a large stadium seating at least 60,000 people. The Saint-Denis site was finally chosen for this new Parisian temple to sport.

The Stade de France® is today the biggest modular stadium in the world and the biggest open-air theatre in France. Designed by four French architects, Macary, Zubléna, Regembal and Costantini, it has 80,000 seats under cover. It is a multipurpose events venue, which hosts the biggest sporting events (6 Nations Tournament, the matches of the France football and rugby teams...), mega-concerts and varied shows.

Put up in record time (31 months), the Stade de France® has a number of remarkable architectural features including a suspended elliptical roof of an area equivalent to the Place de la Concorde and a mobile grandstand seating 25,000 which can be moved back 15 m to uncover the full athletics track.

Open to the general public since 1998, this exceptional building is open every day to visitors wishing to discover its secrets.

SUD
Stade de France®

DISNEYLAND® RESORT PARIS

Located in Marne-la-Vallée, 32 km east of Paris, DISNEYLAND® RESORT PARIS has become a must for tourists visiting the capital. Young and old will find that this magical world created by Walt Disney in an enchanted setting with rides and attractions, has something for everyone. A range of refreshment stalls and restaurants ensure visitor comfort, and for those lucky enough to be staying longer, hotels offer themed accommodation with staff in costume. An array of shops ensures that everyone can take home a souvenir of their journey to the kingdom of the imagination.

Covering an area of 57 hectares, the park is divided into 5 "lands" each with rides, attractions, food outlets and shops on a particular theme:

- **"MAIN STREET, U.S.A."**, creates the atmosphere of small-town America at the turn of the century.

- **"FRONTIERLAND"**, a real boom town of the West during the gold rush era.

- **"ADVENTURELAND"**, Africa, the Caribbean, the jungle - the meeting place for all explorers.

- **"FANTASYLAND"**, the world of make-believe and fairy tales which you enter through ***le Château de la Belle au Bois Dormant.***

- **"DISCOVERYLAND"**, the future world in a setting which Jules Verne might have created.

Located between the theme park and the hotels, near the RER/TGV station, Disney® Village recreates a typically American atmosphere with bars, shops, discos and restaurants; Dinner and Show at ***"La Légende de Buffalo Bill"***.

Le Château de La Belle au Bois Dormant *in Fantasyland.*

Indiana Jones™ et le Temple du Péril... à l'envers! *in Adventureland.*

Captain Hook's Galley *in Adventureland.*

Le Château de La Belle au Bois Dormant *in Fantasyland.*

Big Thunder Mountain *in Frontierland.*

Orbitron, Space Mountain *in Discoveryland.*

PARC WALT DISNEY STUDIOS©

"Behind the scenes" at the cinema

Opened in 2002, this new theme park adjoins the DISNEYLAND® RESORT PARIS park and offers visitors a full day of family fun. You will also find shops and restaurants on a

© Disney

FRONTLOT Disney Studio 1.

tour offering plenty of activities.

The site is divided into four "production zones" which reveals what goes on behind the camera at each one: sets, rigging, animation, special effects, costumes...

© Disney

-**"FRONT LOT"**: a Hollywood set on a boulevard of dreams at Disney Studio 1.

-**"ANIMATION COURTYARD"**: the art of Disney-style animation or inspired by the manipulation technique originating in Japan with ***Animagique***® or based on animated drawing as in ***Les Tapis Volants***.

-**"PRODUCTION COURTYARD"**: ***CinéMagique*** transports the spectator right inside a film, ***Television Production Tour*** reveals what goes on behind the scenes of Disney Chanel France, while ***Studio Tram Tour***® illustrates in a very realistic way the art of set design and special effects, notably at ***Catastrophe Canyon***®.

-**"BACKLOT"**: special effects with ***Armageddon***, the speed and music of ***Rock'n Roller Coaster avec Aerosmith***, the blockbuster stunts of ***Moteurs... Action! Spectacle de Cascades.***

Between two attractions, improvising actors - the Streetmospheres – interact with the visitor and let him play the actor in turn in humorous sketches or "real mock filming", thus fulfilling Walt Disney's vow for: "each family to find themselves plunged in a succession of extraordinary adventures and live shows".

© Disney

BACKLOT: Moteurs... Action! Spectacle de Cascades.

© Disney

PRODUCTION COURTYARD:
Studio Tram Tour® Catastrophe Canyon®.

BACKLOT: *Moteurs... Action! Spectacle de Cascades.*

© Disney

© Disney

ANIMATION COURTYARD:
Les Tapis Volants.

© Disney

PRODUCTION COURTYARD: *CinéMagique.*

© Disney

BACKLOT: *Moteurs... Action! Spectacle de Cascades.*

CONTENTS

Photos: Editions GUY, and courtesy of
ANTRANIK : p.40 (a), p.41 (a), p.65 (c) – ©BAL DU MOULIN ROUGE: p.78–79 – BOUTIER Fabienne: p.61 (a) – CAMERON Robert W.: p.100 (c) CHARDONNET Gérard: p.14, p.15 (d), p.40 (c), p.44–45, p.99 (tomb) – ©CHARLES Jean-Marc: p.54 (d), p.60 (a), p.122 (a), p.123 (a) – CORBIN Marcel: p.74 (b) DAVIGNON Agnès: p.42 (pigeons) – ©DISNEY: p.124 à 127 – Rights reserved: p.7 (a) – DUBOS Laurent: p.22 (b), p.58 (a), p.96 (c) – ©ENSP / LEROUX: p.120–121 FAGGIANO J.: p.53 (d) – GODARD Franck: p.85 (b, c), p.98 (a), p.101 (c) – JAUSSERAND Pascale: p.81 (c) – JORDI Xavier: p.12–13, p.21 (b), p.46 (a), p.47 (a), p.56 (c and background), p.58 (b), p.71 (b), p.106 (a), p.107 (b), p.108 – LAMONTAGNE Daniel: p.26, p.89 (a), p.112 (a, b), p.116 (a), p.119 (a) – LAMOUREUX Michel: p.69 (c) ©CSI LE GAC Hervé: p.39 (b) – LEGRAIN Arnaud: p.68, p.69 (a, b) ©CSI – MANCEAU Marie-Hélène: p.49 (b) – M L: p.5 (b), p.16, p.17 (c), p.21 (a), p.22 (a), p.23 (c), p.24, p.27 (b, c, d), p.38 (b, c), p.40 (b), p.42, p.43, p.47 (b), p.50 (a, b, c, d, e, g), p.51 (d, e), p.52 (a, b, c, d, f, g), p.53 (b, c), p.54 (b), p.55, p.59 (a), p.61 (b), p.62 (d, e), p.63 (b, d, e, f, g), p.64 (a, b, c, d, e), p.65 (a, d, e), p.66 (c), p.80 (b, c), p.81 (b), p.86, p.96 (b), p.97 (a, c), p.100 (a, b), p.103 (a) – ©M. 56: p.100 (b) NICOLAS Olivier: p.19 (a), p.38 (a, d), p.62 (a) – PEACHE: p.65 (b) – PETIT Jean-Pierre: p.46 (b) – ROBINSON Marcus: p.122–123 – SABAT Jean-Michel: p.76 (b) SIMON Pierre: p.67 (c), p.76 (c) – TARGAT Gilles: p.54 (c) – VIARD Paul: p.8 (b), p.9 (b), p.15 (b), p.59 (c), p.60 (b), p.102, p.105 – WERNER Jan: p.4, p.5 (a, c), p.8 (a), p.18 (a), p.20 (a), p.25 (a), p.37 (b), p.38 (e), p.39 (a), p.47 (c), p.74 (a), p.76 (d), p.96 (a), p.107 (a), p.115 (b) – ZENOBEL Jean-Pierre: p.28, p.30 à 36, p.37 (a), p.90 à 95
Front cover: Editions GUY, WERNER Jan, Pyramid at the Louvre (arch. Ieoh Ming Pei – Etablissement public du Grand Louvre), JORDI Xavier.
Back cover: Editions GUY, LAMONTAGNE Daniel, ©CSI/Arnaud LEGRAIN – Architect Adrien FAINSILBER, ZENOBEL Jean-Pierre, ©STADE DE FRANCE® – Architects: Macary, Zublena et Regembal, Costantini, ADAGP, Paris 2005 – Photos ©Jean-Marc CHARLES, Marcus ROBINSON.
Text: Marc LECONTE – Translations: ILS Paris Service de Langues